ADVANCED MATERIALS-BASED FLUIDS FOR THERMAL SYSTEMS

Advanced Materials-Based
FLUIDS FOR THERMAL SYSTEMS

Emerging Technologies and Materials in Thermal Engineering

Edited by

HAFIZ MUHAMMAD ALI

Mechanical Engineering Department, King Fahd University of Petroleum and Minerals, Dhahran, Saudi Arabia

Interdisciplinary Research Center for Sustainable Energy Systems (IRC-SES), King Fahd University of Petroleum and Minerals, Dhahran, Saudi Arabia

ELSEVIER

Elsevier
Radarweg 29, PO Box 211, 1000 AE Amsterdam, Netherlands
125 London Wall, London EC2Y 5AS, United Kingdom
50 Hampshire Street, 5th Floor, Cambridge, MA 02139, United States

Copyright © 2024 Elsevier Inc. All rights are reserved, including those for text and data mining, AI training, and similar technologies.

No part of this publication may be reproduced or transmitted in any form or by any means, electronic or mechanical, including photocopying, recording, or any information storage and retrieval system, without permission in writing from the publisher. Details on how to seek permission, further information about the Publisher's permissions policies and our arrangements with organizations such as the Copyright Clearance Center and the Copyright Licensing Agency, can be found at our website: www.elsevier.com/permissions.

This book and the individual contributions contained in it are protected under copyright by the Publisher (other than as may be noted herein).

Notices
Knowledge and best practice in this field are constantly changing. As new research and experience broaden our understanding, changes in research methods, professional practices, or medical treatment may become necessary.

Practitioners and researchers must always rely on their own experience and knowledge in evaluating and using any information, methods, compounds, or experiments described herein. In using such information or methods they should be mindful of their own safety and the safety of others, including parties for whom they have a professional responsibility.

To the fullest extent of the law, neither the Publisher nor the authors, contributors, or editors, assume any liability for any injury and/or damage to persons or property as a matter of products liability, negligence or otherwise, or from any use or operation of any methods, products, instructions, or ideas contained in the material herein.

ISBN: 978-0-443-21576-6

For information on all Elsevier publications visit our website at https://www.elsevier.com/books-and-journals

Publisher: Megan Ball
Acquisitions Editor: Fran Kennedy-Ellis
Editorial Project Manager: Joshua Mearns
Production Project Manager: Sruthi Satheesh
Cover Designer: Mark Rogers

Typeset by TNQ Technologies

Contents

Contributors   *ix*

1. Introduction to advanced fluids   1

Saeed Esfandeh

   1.   What is nanofluid?   1
   2.   A brief history of nanofluids   2
   3.   Nanofluid preparation methods   2
   4.   Nanofluid stability   3
   5.   Classification of nanofluids   7
   6.   Nanofluids thermophysical properties   8
   7.   Nanofluid's open challenges   9
   References   10

2. Impact of nanoparticle aggregation and melting heat transfer phenomena on magnetically triggered nanofluid flow: Artificial intelligence—based Levenberg—Marquardt approach   13

R.J. Punith Gowda, Ioannis E. Sarris, R. Naveen Kumar and
B.C. Prasannakumara

   1.   Introduction   14
   2.   Mathematical formulation   16
   3.   Results and discussions   20
   4.   Conclusions   30
   Nomenclature   32
   References   32

3. Applications of nanofluids in refrigeration and air-conditioning   35

Jahar Sarkar

   1.   Introduction   35
   2.   Nanofluids as secondary fluid   38
   3.   Nanorefrigerants   46
   4.   Nanolubricants   51
   5.   Nanoabsorbents   53
   6.   Miscellaneous applications   54
   7.   Challenges and future scope   55
   Nomenclature   56
   References   57

4.	**Heat transfer enhancement with ferrofluids**	**61**

Zouhaier Mehrez

1.	Introduction	61
2.	Preparation of ferrofluid	63
3.	Thermophysical properties of ferrofluids	77
4.	Mathematical formulation of FHD	83
5.	Heat transfer enhancement using ferrofluids	86
6.	Conclusion	95
	Nomenclature	97
	References	98

5.	**Nanofluids—Magnetic field interaction for heat transfer enhancement**	**101**

Brahim Fersadou, Walid Nessab and Henda Kahalerras

1.	Introduction	101
2.	Nanofluids	101
3.	Magnetic nanofluids	105
4.	Ferrohydrodynamics	106
5.	Magnetohydrodynamics	108
6.	Applications	109
7.	Conclusion	129
	Nomenclature	129
	References	131

6.	**Impact of Ohmic heating and nonlinear radiation on Darcy—Forchheimer magnetohydrodynamics flow of water-based nanotubes of carbon due to nonuniform heat source**	**135**

Khilap Singh, Padam Singh and Manoj Kumar

1.	Introduction	135
2.	Mathematical modelling of carbon nanotubes flow and heat transfer	140
3.	Numerical methods of solution	146
4.	Result and discussion	149
5.	Conclusions	160
	Nomenclatures	161
	References	162

Contents

vii

7. Thermos-physical properties and heat transfer characteristic of copper oxide—based ethylene glycol/water as a coolant for car radiator 169

Alhassan Salami Tijani and Muhammad Yus Azreen Bin Mohd Yusoff

1. Introduction 169
2. Modelling and simulation 172
3. Theoretical background 172
4. Discussion of findings 176
5. Conclusion and recommendation 184
Abbreviation 184
References 185

8. Discussion on the stability of nanofluids for optimal thermal applications 187

Taoufik Brahim and Abdelmajid Jemni

1. Nanofluids discussion 187
2. Mechanisms to increase the stability of nanofluids 196
3. Characterization of nanofluid stability 200
4. Conclusion 204
Nomenclature 205
References 205

9. Entropy optimization of magnetic nanofluid flow over a wedge under the influence of magnetophoresis 209

Kalidas Das and Md Tausif Sk

1. Literature review 209
2. Mathematical formations 213
3. Entropy generation 216
4. Numerical solution methodology 216
5. Result and discussion 216
6. Conclusions 228
Nomenclature 229
References 230

10. Nonaxisymmetric homann stagnation-point flow of nanofluid toward a flat surface in the presence of nanoparticle diameter and solid—liquid interfacial layer 233

Kalidas Das, Shib Sankar Giri and Nilangshu Acharya

1. Introduction 233
2. Mathematical formulation 236
3. Numerical experiment 240

4.	Results and discussion	241
5.	Conclusions	251
Nomenclature		252
References		252

11. On the hydrothermal performance of radiative Ag—MgO—water hybrid nanofluid over a slippery revolving disk in the presence of highly oscillating magnetic field

255

Nilankush Acharya and Kalidas Das

1.	Introduction	255
2.	Mathematical formulation	260
3.	Numerical method and code validation	269
4.	Results and discussion	270
5.	Conclusion	284
Nomenclature		284
References		286

12. Application of nanofluids and future directions

289

Saeed Esfandeh

1.	Application of nanofluids in energy and electricity sector	289
2.	Industrial application of nanoparticles in lubrication improvement	294
3.	Applications of nanotechnology in solar water heaters	297
4.	Applications of nanotechnology in solar water desalination	299
5.	Application of nanotechnology in oil and gas wells	301
6.	Future directions	306
Nomenclature		307
References		307

Index *311*

Contributors

Nilankush Acharya
NCP Umasashi High School, Kolkata, West Bengal, India

Nilangshu Acharya
Department of Mathematics, P.R. Thakur Govt. College, Ganti, West Bengal, India

Muhammad Yus Azreen Bin Mohd Yusoff
School of Mechanical Engineering, College of Engineering, Universiti Teknologi MARA (UiTM), Shah Alam, Selangor Darul Ehsan, Malaysia

Taoufik Brahim
University of Sousse, Higher Institute of Applied Sciences and Technology of Sousse (ISSAT-Sousse-Tunisia), Sousse, Tunisia

Kalidas Das
Department of Mathematics, Krishnagar Government College, Krishnanagar, West Bengal, India

Saeed Esfandeh
Department of Mechanical Engineering, Jundi-Shapur University of Technology, Dezful, Iran

Brahim Fersadou
Faculty of Mechanical and Process Engineering, Houari Boumediene University of Sciences and Technology (USTHB), Algiers, Algeria

Shib Sankar Giri
Department of Mathematics, Bidhannagar College, Kolkata, West Bengal, India

Abdelmajid Jemni
University of Monastir, National Engineering School of Monastir, Laboratory Studies of Thermal and Energy Systems- LESTE, Monastir, Tunisia

Henda Kahalerras
Faculty of Mechanical and Process Engineering, Houari Boumediene University of Sciences and Technology (USTHB), Algiers, Algeria

Manoj Kumar
Department of Mathematics, Statistics and Computer Science, G. B. Pant University of Agriculture and Technology, Pantnagar, Uttarakhand, India

Zouhaier Mehrez
Faculty of Sciences of Tunis, Laboratory of Energy, Heat and Mass Transfer (LETTM), Department of Physics, El Manar University, El Manar, Tunisia; Gabes Preparatory Engineering Institute, Gabes, Tunisia

R. Naveen Kumar
Department of Mathematics, Dayananda Sagar College of Engineering, Bangalore, Karnataka, India

Walid Nessab
Faculty of Mechanical and Process Engineering, Houari Boumediene University of Sciences and Technology (USTHB), Algiers, Algeria

B.C. Prasannakumara
Department of Mathematics, Davangere University, Shivagangotri, Davangere, Karnataka India

R.J. Punith Gowda
Department of Mathematics, Bapuji Institute of Engineering & Technology, Davanagere, Karnataka, India

Jahar Sarkar
Department of Mechanical Engineering, Indian Institute of Technology (BHU) Varanasi, UP, India

Ioannis E. Sarris
Department of Mechanical Engineering, University of West Attica, Athens, Greece

Khilap Singh
Department of Mathematics, H. N. B. Government Post Graduate College, Khatima, Uttarakhand, India

Padam Singh
Department of Mathematics, Galgotias College of Engineering and Technology, Greater Noida, Uttar Pradesh, India

Md Tausif Sk
Department of Mathematics, A. B. N. Seal College, Cooch Behar, West Bengal, India

Alhassan Salami Tijani
School of Mechanical Engineering, College of Engineering, Universiti Teknologi MARA (UiTM), Shah Alam, Selangor Darul Ehsan, Malaysia

CHAPTER ONE

Introduction to advanced fluids

Saeed Esfandeh
Department of Mechanical Engineering, Jundi-Shapur University of Technology, Dezful, Iran

Highlights
- A brief about history, classification, thermophysical properties and preparation methods of nanofluids
- Attraction and repulsion forces on nanoparticles and their effect on nanofluid stability
- Main open challenges against nanofluids development

1. What is nanofluid?

Nanofluid can be considered as advanced heat transfer fluid that is manufactured by dispersing the nano-sized particles (1—100 nm) in the host or base fluids in the form of a colloidal solution [1]. To improve the thermophysical properties of working fluids, most often utilized nanoparticles have high thermophysical properties to solve the problem. Next to improving the heat transfer properties that its result is energy saving, reducing the volume and dimension of heat transfer equipment could be another accomplishment of nanofluids. In one sentence, the main mission of nanoparticles in thermal engineering is improving the thermal conductivity, thermal diffusivity, viscosity, and convective heat transfer coefficients as main thermophysical and thermal properties of conventional working fluids.

Although the above-mentioned general characteristics are positive aspects of nanofluids, there are challenges in the way of nanofluid applications in thermal systems. The first limitation is the rheological behavior of nanofluids that it could be negative aspect of nanofluid characteristics because nonsuitable rheological behavior or nonoptimized viscosity behavior of nanofluids may cause energy loss. Also nanofluid stabilization can be another big challenge in front of nanofluid application development.

2. A brief history of nanofluids

Nanofluid science as one of the categories of nanotechnology has not a long-term history and is a young research field. Taking a look on history shows researchers that intended to increase the thermal conductivity of conventional fluids took the first step toward nanofluid invention.

The first revolutionary idea in the way of nanofluid invention stated by Maxwell [2] in 1873. He proposed dispersing solid particles in conventional working fluids to enhance and improve thermal characteristics of fluids. The story started from adding solid particles in microsize and millimeter size to the fluids. Although Maxwell step was a brilliant step toward nanofluid invention, the result was not promising. In fact, adding solid particles improved the thermal characteristics, but there were other problems like erosion in pipes, high pressure drop, clogging, and sedimentation. Maxwell didn't answer to this barrier on that time.

The next scientist who continued the progress path was Choi and his colleague Eastman [3] who proposed a new idea to solve the above-mentioned problems. They introduced nanoscale metallic particles and also nanotubes as alternative for Maxwell's microsize and millimeter size solid particles. Choi and his colleague did many experiments on various fluids and nanoparticles combinations, and as expected, the result was great. Although this step was the most important step in nanofluid science, there were many ambiguities in full recognition of nanofluids' behavior. So the process of nanofluid utilization in practical scale was stopped. After Choi and Eastman findings, many researchers have worked and are working on solving the barriers in front of nanofluid development.

3. Nanofluid preparation methods

Although there are various innovative methods for nanofluid preparation, there are only two main methods for nanofluid preparation that are named by two-step and single-step method. In the following, the process of two-step and single-step nanofluid preparation will be discussed. Nanofluids have two main parts that are the base fluid and the nanoparticles. As stated above, two general and most common methods can be defined as nanofluid preparation methods. The first method will be done in two steps in which nanoparticles will be synthesized in the first step and they will be dispersed in the base fluid as the second step. On the other hand, the second method will be done in one step in which the synthesis and dispersion of nanofluids in the base fluid will be done simultaneously.

Two-step method is more common than the single step in nanofluid production. Each of both methods has advantages and disadvantages. In two-step method, nanoparticles will be produced in powder form then the synthesized particles will be dispersed in the host fluid by applying magnetic stirring and ultrasonic waving to the suspension. These stabilizing mixing methods are essential in two-step method because generally the prepared nanofluid with this method has serious problems in stability. Low cost of preparation is one of the main advantages of two-step method, and on the other hand, the nanoparticles accumulating during powder preparation steps like drying and storage, creating and keeping stability of prepared nanofluid during time, and poor control on nanoparticles size and shape can be disadvantages of two-step nanofluid preparation.

But the condition is different for single-step method. In this method, the nanoparticle synthesis and dispersion will be done simultaneously in the fluid. In fact, synthesis and dispersion of nanoparticle will be done in a single step. In single-step method, there is no need to storing, drying, and dispersing in a separated step after synthesis that will result in lower possibility of nanoparticles' accumulation. Also controlling the shape and size of nanoparticle is possible compared to two-step method. Another positive feature of this method is its natural stability and no need for stabilization process. On the other hand, its high cost and impossibility of large quantity production are the main barriers and disadvantages against single-step method.

4. Nanofluid stability

As mentioned in previous sections, one of the main barriers against nanofluid development in practical applications is the probability of agglomeration and sedimentation of nanoparticles in the base fluid (Fig. 1.1). In other word, having a stable nanofluid is the first step of achievement in nanofluid preparation. Based on search results and colloidal theory, a critical radius or critical diameter can be defined for nanoparticles in which the applied forces to the particles cancel each other. As more detail in critical diameter or below critical diameter of nanoparticles, the Brownian forces counterbalance the gravity forces, so the sedimentation will not occur. Although the story is not easy enough, because nanoparticles have to be in an optimize diameter. On the other hand, although smaller size of nanoparticles makes them more applicable for various practical utilizations,

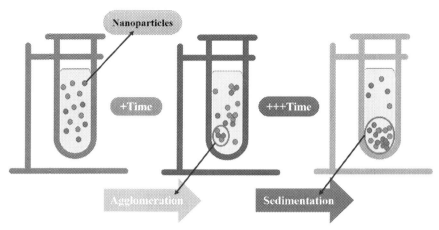

Figure 1.1 Nanoparticles agglomeration and sedimentation over time.

the smaller size will result in higher surface area and the higher surface area is equal to increasing the agglomeration happening chance [4,5]. So the size of particles is a serious challenge in the stability of prepared nanofluid. Nanoparticles type and concentration, base fluid viscosity, ultrasound, and stirring time are among other effective parameters on nanofluid stability. Adding surfactants, PH regulation, applying ultrasonic waves, and also magnetic stirring to prepared nanofluid are the other methods that can control and regulate the applied forces to the nanofluid.

Instability of a nanofluid can affect and weaken its improved thermal properties, and it would be the final and big failure for prepared nanofluid. In a simple expression, the main reason of instability of nanofluids is because of the imbalance of attractive and repulsive forces between nanoparticles. A stronger attractive force between nanoparticles compared to the repulsive forces will result in nanoparticles aggregation and agglomeration. Any mechanism that enhances the repulsive forces over attractive forces would be the key way of salvation for nanofluid against instability.

Two groups of forces play an important role in particles' attraction and repulsion. The first group of forces are short-range forces like van der Waals attraction and surface forces that play a role in every interaction [6], and the second group of forces are long-range repulsive forces. Forces like van der Waals and surface forces attract the nanoparticles to each other and will be categorized as the first group. Although the Brownian motion plays an important role in creating the attraction forces because it moves the nanoparticles and thus provides a condition for the attraction of the nanoparticles towards each other. Fig. 1.2 shows the schematic of Brownian motion.

To prevent agglomeration, aggregation, and closing particles to each other, the second group of forces act as neutralizing force against the first group and keeps the nanoparticles apart. There are two main mechanisms that can enhance the repulsive forces that are: (1) electrostatic stabilization and (2) steric stabilization. There are various theories about the type of long-range repulsive forces. The DLVO theory (named after Boris **D**erjaguin, Lev **L**andau, Evert **V**erway, and Theodoor **O**verbeek as developers of the theory) considers and introduces the electrostatic force as the only repulsive force, but the steric repulsion and solvent forces may play important role as other repulsive forces [7,8]. The DLVO theory focuses on the balance between van der Waals attraction and electrostatic repulsions in a liquid medium.

Interaction between the electrical double layers that have surrounding nanoparticles in a solvent is the source of electrostatic repulsion. An electrical double layer forms around each nanoparticle surface when a charged surface of nanoparticles surrounds by a liquid (Fig. 1.3). The Stern layer that is the first layer of electrical double layer appears around each charged nanoparticle because of chemical interaction reasons. Also the second layer of electrical double layer that has a weak connection with the nanoparticle surface compared to the first layer is formed by attracted ions to the surface of nanoparticles with the help of Coulomb forces. Based on chemical rules, nanoparticles attract the fluid's ions that having opposite charge to that of the nanoparticles' surface charge. But this attraction power weakens with increasing the distance from nanoparticle surface, so the concentration of

Figure 1.2 Schematic of Brownian motion.

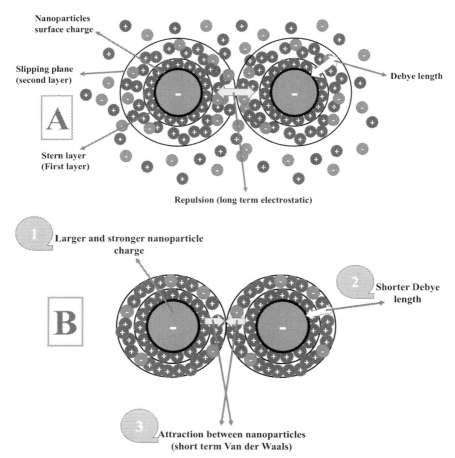

Figure 1.3 (A and B) Ion distribution in the proximity of a negatively charged particle, (A) interaction between particles with long Debye length, and (B) interaction between particles with short Debye length.

counterions around the nanoparticle reduces by increasing the distance from the charged nanoparticle surface and the reduction will be continued until reaching an ion concentration equilibrium in solvent bulk. The above-mentioned two layers around each nanoparticle in solvent create repulsive force between nanoparticles, so the nanoparticles can't come closer than twice of double layer length because of electrical double layer overlap [9]. The other name of electrical double layer length is Debye length. As the result to have more stable nanofluid, the stronger and larger nanoparticle surface charge and also longer Debye length are determinative parameters (Fig. 1.3) [10,11]. Although it should be considered that the larger

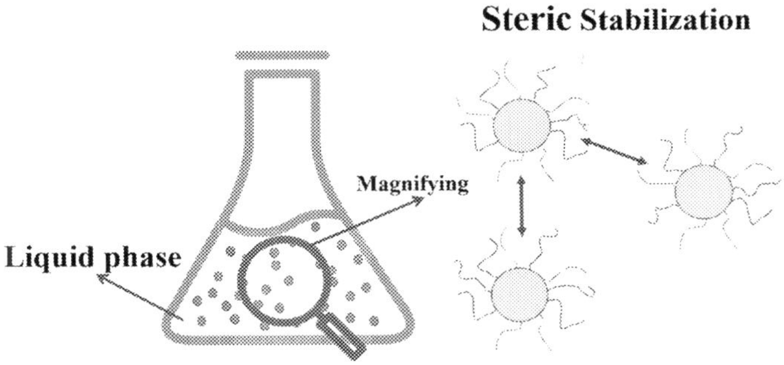

Steric Stabilization

Figure 1.4 Sterically stabilized nanoparticle.

nanoparticle charge will result in electrical double-layer compression, a balance is needed between amounts of nanoparticle surface charge and Debye length to reach the minimum nanoparticles aggregation and minimum instability (Fig. 1.3) [12]. Besides the electrostatic repulsion, the steric repulsive force is another repulsion mechanism. In steric stabilization, the macromolecules' (polymers, surfactant, etc.) attachment on nanoparticle surfaces keep them away from each other (Fig. 1.4) [13].

5. Classification of nanofluids

Nanofluids can be classified in various groups based on the type of applied nanoparticle and also number of nanoparticle types dispersed in the base fluid. The nanofluids are classified into four groups that are: (1) metal oxide-based, (2) metal-based, and (3) carbon-based and hybrid metal-based. Metal oxide and carbon-based nanofluids are more common than the others because of high stability, acceptable thermal properties, and low cost in synthesize process for metal oxide-based nanofluids and because of brilliant thermal properties for carbon-based nanofluids. Low cost of metal oxide-based nanofluids compared to carbon-based one makes them a more suitable choice for industrial and commercial applications.

Next to all above-mentioned nanofluids, the widespread tendency of researchers and scientists about hybrid nanofluids should not be neglected. Hybrid nanofluids are produced by mixing various nanoparticles in a base fluid. The most common type is mixing two types of nanoparticles, but there are few studies with mixing three types of nanoparticles in one base fluid. The main goal and idea of hybrid nanofluid production was simultaneous using of chemical and physical advantages of two, three, or more separate nanomaterials. This is because of no single nanoparticle has all positive physical and chemical features of an ideal nanoparticles. So mixing and combination of two or more nanoparticle known as hybrid nanofluid could be a good method to produce the more effective and more applicable nanofluid. Based on many search results [14], hybrid nanofluids have considerable thermophysical and thermal advantages compared to individual nanofluids.

6. Nanofluids thermophysical properties

It is obvious that by adding nanoparticles to conventional base fluids, thermophysical properties of the produced nanofluids will change. The main thermophysical properties are thermal conductivity, viscosity, and specific heat. Also other thermal and thermophysical properties are density, convective heat transfer, and thermal diffusivity [15] (Fig. 1.5).

Thermal conductivity as the most important of property for thermal performance improving of nanofluids is the ability level of prepared nanofluid in heat transition. Many experimental and theoretical-based studies have been conducted with the aim of thermal conductivity measurement of nanofluid, but still there are many unknowns and doubts about thermal behavior of nanofluid, because thermal conductivity of nanofluids depends on many parameters like volume fraction, size, shape, thermal conductivity, Brownian motion of nanoparticles, and type of nanoparticle combination, especially in hybrid nanofluids. Reviewing published reports and articles about nanofluid thermal conductivity shows that many of researchers have tried to propose analytical-theoretical-based formulas to predict thermal

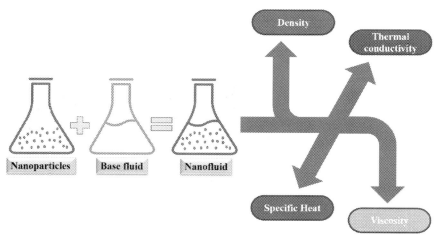

Figure 1.5 Nanofluids thermophysical properties.

conductivity [16–31], and this is the work that Maxwell did in 1881 [2] as one of the pioneers. Although he proposed a mathematical formula for some of spherical particles dispersed in the base fluid without considering the interaction between dispersed particles, it was an important step in the way of nanofluid science improvement.

Viscosity is another main thermophysical property of nanofluids that has an indirect effect on thermal performance of nanofluids. Although the viscosity doesn't have effect on thermal conductivity, its amount is very important to control and optimize needing pumping power in thermal cycles that contain fluids. By adding nanoparticles, the viscosity of base fluid will increase except in exception cases [32], although it should be noticed that this increment is a negative and unwanted phenomenon that need to be controlled. Same as thermal conductivity, researchers have tried to propose analytical-theoretical formulas for prediction viscosity behavior of nanofluids, the work that was started by Einstein [33] to determine the dynamic viscosity of nanofluids.

As the third important thermophysical property of nanofluid, the specific heat of base fluid experiences some changes by adding nanoparticles. Its amount has a significant effect on heat transfer rate of nanofluids because based on its description, the heat capacity is the heat quantity that is needed to enhance the temperature of one gram of nanofluid by one degree centigrade [34]. Based on some studies [35], adding nanoparticles to the base fluid lowers the specific heat capacity that is equivalent with enhancement of heat transfer rate. There are various theoretical-analytical formulas for specific heat capacity calculation, but the first simple formula was proposed by Pak and Cho in their research study [36].

7. Nanofluid's open challenges

Although the nanofluid science has a high potential in the development process of other sciences like heat transfer, there are too many open challenges against the rapid progress of this science. Some of the challenges and barriers against the development of nanofluids in real life are mentioned below:

1. Despite lots of published research articles and reports about the study on thermophysical properties of nanofluids, in some cases, the results of studies don't confirm each other and are inconsistent. Experimental errors, uncertainties, and lack of uniform standard for experimental tests are the main reasons of inconsistent experimental results, but to product a

commercial and industrial nanofluid, we need precise information about its thermophysical properties. Also high cost of nanofluid experimentations is another issue that limits the progress speed of this science.

2. Variety in candidate nanoparticles, especially for hybrid nanofluids with two or more nanoparticles, has caused confusion in the selection of best nanoparticle(s) combination as base fluid additives.

3. Two-step method for nanofluid preparation as the more commercial method has a limited control over size and shape of nanoparticles. On the other hand, the shape and size of nanoparticles affect the thermal properties of the produced nanofluids. So nanoparticle shape and size control is another big challenge in the development of commercial nanofluids. Although there is no necessary clarity about the mechanism of nanoparticle shape and size effect on thermal properties, and most of the stated results in scientific reports are speculative.

4. Producing a high stable nanofluid over long time and also over various working conditions like at high/low temperature is another big challenge and one of the big barriers against nanofluids to be more applicable in real life. As it is clear, agglomeration, aggregation, and then sedimentation of nanoparticles in the base fluid weaken the thermal properties of nanofluid and almost there is no scientific report that guarantee the nanofluid behavior in real world and in working condition.

5. Increase in pressure drop and more energy is required for nanofluid pumping as working fluid in heat transfer cycles is another undeniable problem. Certainly, by adding nanoparticles to the base fluid, the density of the working fluid will increase, so it needs spending more energy for pumping. Therefore, in addition to paying attention to the improvement of thermal properties, other features such as the optimal density of the produced nanofluid need to be considered. This is an issue that has received less attention in majority of published research papers. Many of conducted researches in the field of nanofluid only have focused on improving the thermal properties.

References

[1] S.U.S. Choi, Enhancing thermal conductivity of fluids with nanoparticles, in: D.A. Siginer, H.P. Wang (Eds.), Developments and applications of non-newtonian flows, ASME, 1995, pp. 99–105. FED-231/MD-66.

[2] J.C. Maxwell, A treatise on electricity and magnetism, Dover Publications, 1891.

[3] S.U. Choi, J.A. Eastman, Enhancing thermal conductivity of fluids with nanoparticles(No. ANL/MSD/CP-84938; CONF-951135-29), Argonne National Lab.(ANL), 1995.

[4] S. Witharana, I. Palabiyik, Z. Musina, Y. Ding, Stability of glycol nanofluids—the theory and experiment, Powder Technol. 239 (2013) 72–77.

[5] S. Jailani, G.V. Franks, T.W. Healy, The potential of nanoparticle suspensions: effect of electrolyte concentration, particle size and volume fraction, J. Am. Ceram. Soc. 91 (2008) 1141–1147.

[6] S.C. Endres, L.C. Ciacchi, L. Mädler, A review of contact force models between nanoparticles in agglomerates, aggregates, and films, J. Aerosol Sci. 153 (2021) 105719.

[7] S. Song, C. Peng, Thickness of solvation layers on nano-scale silica dispersed in water and ethanol, J. Dispers. Sci. Technol. 26 (2005) 197–201.

[8] S. Jafari Daghlian Sofla, L.A. James, Y. Zhang, Insight into the stability of hydrophilic silica nanoparticles in seawater for Enhanced oil recovery implications, Fuel 216 (2018) 559–571.

[9] T. Tadros, Electrostatic repulsion and colloid stability, in: Encyclopedia of colloid and interface science, Springer, 2013, p. 363.

[10] R.A. French, A.R. Jacobson, B. Kim, S.L. Isley, L. Penn, P.C. Baveye, Influence of ionic strength, pH, and cation valence on aggregation kinetics of titanium dioxide nanoparticles, Environ. Sci. Technol. 43 (2009) 1354–1359.

[11] N. Bukar, S.S. Zhao, D.M. Charbonneau, J.N. Pelletier, J.F. Masson, Influence of the Debye length on the interaction of a small molecule-modified Au nanoparticle with a surface-bound bioreceptor, Chem. Commun. 50 (2014) 4947–4950.

[12] A.M. Smith, A.A. Lee, S. Perkin, The electrostatic screening length in concentrated electrolytes increases with concentration, J. Phys. Chem. Lett. 7 (2016) 2157–2163.

[13] H. Zhu, C. Zhang, Y. Tang, J. Wang, B. Ren, Y. Yin, Preparation and thermal conductivity of suspensions of graphite nanoparticles, Carbon 45 (1) (2007) 226–228.

[14] B. Sun, Y. Zhang, D. Yang, H. Li, Experimental study on heat transfer characteristics of hybrid nanofluid impinging jets, Appl. Therm. Eng. 151 (2019) 556–566.

[15] T.P. Otanicar, P.E. Phelan, R.S. Prasher, et al., Nanofluid-based direct absorption solar collector, J. Renew. Sustain. Energy 2 (2010).

[16] I.H. Rizvi, A. Jain, S.K. Ghosh, P.S. Mukherjee, Mathematical modelling of thermal conductivity for nanofluid considering interfacial nano-layer, Heat Mass Transf. 49 (2013) 595–600.

[17] M.H. Esfe, S. Esfandeh, S. Saedodin, H. Rostamian, Experimental evaluation, sensitivity analyzation and ANN modeling of thermal conductivity of ZnO-MWCNT/EG-water hybrid nanofluid for engineering applications, Appl. Therm. Eng. 125 (2017) 673–685.

[18] M. Hemmat Esfe, S. Esfandeh, M. Rejvani, Modeling of thermal conductivity of MWCNT-SiO 2 (30: 70%)/EG hybrid nanofluid, sensitivity analyzing and cost performance for industrial applications: an experimental based study, J. Therm. Anal. Calorim. 131 (2018) 1437–1447.

[19] M.H. Esfe, S. Esfandeh, M.K. Amiri, M. Afrand, A novel applicable experimental study on the thermal behavior of SWCNTs (60%)-MgO (40%)/EG hybrid nanofluid by focusing on the thermal conductivity, Powder Technol. 342 (2019) 998–1007.

[20] M.H. Esfe, D. Toghraie, S. Esfandeh, S. Alidoust, Measurement of thermal conductivity of triple hybrid water based nanofluid containing MWCNT (10%)-Al2O3 (60%)-ZnO (30%) nanoparticles, Colloids Surf. A Physicochem. Eng. Asp. 647 (2022) 129083.

[21] M.H. Esfe, S. Esfandeh, D. Toghraie, Investigation of different training function efficiency in modeling thermal conductivity of TiO_2/Water nanofluid using artificial neural network, Colloids Surf. A Physicochem. Eng. Asp. 653 (2022) 129811.

[22] M.H. Esfe, S. Alidoust, S. Esfandeh, D. Toghraie, H. Hatami, M.H. Kamyab, E.M. Ardeshiri, Theoretical-Experimental study of factors affecting the thermal

conductivity of SWCNT-CuO (25: 75)/water nanofluid and challenging comparison with CuO nanofluids/water, Arab. J. Chem. 16 (5) (2023) 104689.

[23] S. Singh, S. Kumar, S.K. Ghosh, Development of a unique multi-layer perceptron neural architecture and mathematical model for predicting thermal conductivity of distilled water based nanofluids using experimental data, Colloids Surf. A Physicochem. Eng. Asp. 627 (2021) 127184.

[24] M. Ramezanizadeh, M. Alhuyi Nazari, Modeling thermal conductivity of Ag/water nanofluid by applying a mathematical correlation and artificial neural network, Int. J. Low Carbon Technol. 14 (4) (2019) 468—474.

[25] M. Molana, R. Ghasemiasl, T. Armaghani, A different look at the effect of temperature on the nanofluids thermal conductivity: focus on the experimental-based models, J. Therm. Anal. Calorim. 147 (2021) 1—25.

[26] E. Abu-Nada, Effects of variable viscosity and thermal conductivity of CuO-water nanofluid on heat transfer enhancement in natural convection: mathematical model and simulation, J. Heat Transf. 132 (2010) 052401.

[27] D.S. Saini, S.P.S. Matharu, Developing a mathematical model and an optimal artificial neural network to predict the thermal conductivity of zirconium oxide nanolubricant by comparing experimental and numerical data, Int. J. Interact. Des. Manuf. 17 (2022) 1—18.

[28] B. Dandoutiya, A. Kumar, Comparison of mathematical models to estimate the thermal conductivity of titanium oxide-water based nanofluid: a review, Therm. Sci. 224 (2021) 579—591.

[29] S. Uribe, N. Zouli, M.E. Cordero, M. Al-Dahhan, Development and validation of a mathematical model to predict the thermal behaviour of nanofluids, Heat Mass Transf. 57 (2021) 93—110.

[30] I. Mugica, S. Poncet, A critical review of the most popular mathematical models for nanofluid thermal conductivity, J. Nanoparticle Res. 22 (5) (2020) 113.

[31] A. Komeilibirjandi, A.H. Raffiee, A. Maleki, M. Alhuyi Nazari, M. Safdari Shadloo, Thermal conductivity prediction of nanofluids containing CuO nanoparticles by using correlation and artificial neural network, J. Therm. Anal. Calorim. 139 (2020) 2679—2689.

[32] M.H. Esfe, A.A.A. Arani, S. Esfandeh, Improving engine oil lubrication in light-duty vehicles by using of dispersing MWCNT and ZnO nanoparticles in 5W50 as viscosity index improvers (VII), Appl. Therm. Eng. 143 (2018) 493—506.

[33] A. Einstein, Investigations on the theory of the Brownian movement, Dover Publications, 1956.

[34] M. Gupta, V. Singh, R. Kumar, et al., A review on thermophysical properties of nanofluids and heat transfer applications, Renew. Sustain. Energ Rev. 74 (2015) 638—670.

[35] M. Saeedinia, M.A. Akhavan-Behabadi, P. Razi, Thermal and rheological characteristics of CuO-Base oil nanofluid flow inside a circular tube, Int. Commun. Heat Mass 39 (2012) 152—159.

[36] B.C. Pak, Y.I. Cho, Hydrodynamic and heat transfer study of dispersed fluids with submicron metallic oxide particles, Exp. Heat Transf. 11 (1998) 151—170.

CHAPTER TWO

Impact of nanoparticle aggregation and melting heat transfer phenomena on magnetically triggered nanofluid flow: Artificial intelligence—based Levenberg—Marquardt approach

R.J. Punith Gowda[1], Ioannis E. Sarris[2], R. Naveen Kumar[3] and B.C. Prasannakumara[4]

[1]Department of Mathematics, Bapuji Institute of Engineering & Technology, Davanagere, Karnataka, India
[2]Department of Mechanical Engineering, University of West Attica, Athens, Greece
[3]Department of Mathematics, Dayananda Sagar College of Engineering, Bangalore, Karnataka, India
[4]Department of Mathematics, Davangere University, Shivagangotri, Davangere, Karnataka India

Highlights

- The impact of nanoparticle aggregation on nanoliquid flow across a stretching disk is presented.
- The Krieger—Dougherty and Maxwell—Bruggeman models of nanoparticle aggregation are utilized for precise calculation of the viscosity and thermal conductivity of the particles.
- Study inspects the impact of uniform horizontal magnetic field and melting heat transfer phenomena.
- The heat transport is analyzed by utilizing the intelligent computing paradigm via Scheme of Artificial Levenberg Marquardt back propagated neural networks.
- The external validation criteria are employed to address the modeling overfitting.

Advanced Materials-Based Fluids for Thermal Systems
ISBN: 978-0-443-21576-6
https://doi.org/10.1016/B978-0-443-21576-6.00005-4

1. Introduction

Nanofluids (NFs) which contain suspended nanoparticles (NPs) can significantly alter the base fluid's thermal transport characteristics, potentially enhancing thermal conductivity. The base liquid is commonly a conductive liquid, while the NPs that constitute NFs are formed of nonmetals (graphite, carbon nanotubes), oxides (Al_2O_3), carbides (SiC), metals (Al, Cu), and nitrates (AlN, SiN). Suspension of NPs can have an extensive impact on the thermal characteristics of the base fluid. For their use in applications such as electronic chip cooling, power generation, solar collectors, heat storage systems, cancer-infected tissue treatment, laser-assisted drug delivery, microprocessors, hybrid machines, and sophisticated nuclear systems, nanoliquids have attracted enormous interest. Recently, Sharma et al. [1] swotted the magnetic NFs flow with variable properties of fluid on a gyrating disk. Gowda et al. [2] inspected the flow of NF with magnetic dipole on a sheet. Gul et al. [3] analyzed the stream of a hybrid NF inside the gap among the gyrating disk surface and the cone. Tassaddiq et al. [4] swotted the mass and heat transference in the stream of a hybrid NF. Using two vertical plates, Hajizadeh et al. [5] swotted the convective stream of NFs considering damped thermal flux.

Many physical phenomena can alter how NPs are suspended in a base liquid. These physical parameters have a considerable impact on the behavior of NPs in suspension and, consequently, the characteristics of NFs. Research demonstrates that NP aggregation has a major impact on rheological and thermal properties. The van der Waals force and surface charge cause NPs to bind together. Due to this property, distribution of heat is significantly more in NPs aggregation than in NPs. Wang et al. [6] scrutinized the NF stream with radiation and NP aggregation between the gap of a cone and disk. Rana et al. [7] analyzed the NPs aggregation features on the flow of NF. Mahmood and Khan [8] explored the time-dependent stream of hydrogen oxide-based NFs with stagnation point and NP aggregation. Yu et al. [9] swotted the aggregation of NP on the micropolar NF flow with thermophoretic particle deposition through a stretching sheet. Chen et al. [10] conferred the NP aggregation feature on the NFs stream with thermal radiation.

The process of melting is a fundamental phenomenon in which a substance changes its state from a solid to a liquid form due to the absorption of heat energy. This phase change process is a key area of study in fluid flow over surfaces because it plays a dynamic role in many natural and industrial processes. Uses of the melting phenomenon include geothermal energy recovery, heat exchangers, silicon wafer fabrication, permafrost melting, thermocouples, heat engines, and hot extrusion. Singh et al. [11] inspected the melting and chemical reaction upshot on the stream of micropolar fluid over a stretchable surface. Muhammad et al. [12] swotted the radiative and melting heat transference in NFs over an elastic plate. Singh et al. [13] educed the melting upshot on the stream of magnetic NF through a permeable stretching cylinder. Hayat et al. [14] examined the melting impact on an entropy-optimized dissipative stream with an effective Prandtl number. Jawad et al. [15] conferred the melting effect on the hybrid NF motion through stretching surface with stagnation point.

When a fluid flows through a uniform magnetic field (UMF), it encounters a force called as the Lorentz force. Both the direction in which the magnetic field is oriented and the way in which the velocity of the fluid is moving are perpendicular to the Lorentz force. As a direct consequence of this, the flow of the liquid is redirected, and the path that the fluid particles take is changed. There are many different fields of research and technology that make use of fluid flow with a UMF, such as the study of plasma physics and magnetohydrodynamics, as well as the construction of magnetic flow meters that are used to measure the flow rate of conductive fluids. The major reason for applying a UMF in the oblique direction on a gyrating disk is to suppress the mean field's magnitude to stabilize it. But the decreased heat transport rate from the disk surface causes the disk to overheat. Several researchers have investigated how a UMF affects the flow over an infinite disk that is revolving. Turkyilmazoglu [16] studied the flow and heat subject to a horizontal UMF over a gyrating disk. Pal and Kumar [17] scrutinized the convective flow with a UMF. Prasannakumara and Gowda [18] investigated the flow of radiative fluid with a thermophoretic particle deposition and UMF. Tagawa [19] explored the natural convective stream with the impact of a UMF in a long vertical rectangular enclosure. Wang et al. [20] examined the impact of NPs aggregation and UMF in the NF flow with melting.

The heat transport problem over a rotating disk continues to receive a lot of attention from researchers because it has so many real-world applications in aeronautical science and other engineering fields. The electronic components cooling and aerospace systems are just some of the many practical applications that may be accomplished by using NFs in the flow that occurs over spinning disks. However, due to the intricate structure of NF flow across a spinning disk, comprehensive knowledge of the underlying physics is required. In addition, more study is required to fully exploit the potential of these flows in practical applications. Naganthran et al. [21] analyzed the liquid stream subject to variable liquid features on a gyrating disk. Kumar et al. [22] educed the dusty NF stream with a heat source/sink due to stretching permeable gyratory disk. Zhou et al. [23] inspected the stream of Maxwell NF via stretching permeable gyratory disk. Between two shrinking disks, Bhandari [24] deliberated the heat transference in unsteady flow of the ferrofluid. Rauf et al. [25] scrutinized the hybrid ferrofluid stream with magnetic field due to a nonlinearly stretchable gyrating disk.

According to the best of author's knowledge, the effects of a UMF and the melting effect on the flow of NF past a stretching disk with aggregation of NPs have not yet been explored in any of the publications that were cited earlier so far. It is commonly known that there is a wide range of strategies that may be used to produce a number of apt solutions to problems of this kind. There has never been any kind of numerical answer to the flow that has been stated. The most essential objective of this study is to conduct numerical simulations of the flow that was discussed before.

2. Mathematical formulation

The steady incompressible MHD laminar boundary layer flow of TiO_2-ethylene glycol-based NF on an infinitely gyrating disk with an externally applied UMF is the focus of our attention in this particular study. The disk is positioned at $z = 0$ and rotates at a fixed angular velocity Ω. A disk is subjected to a UMF of strength B. Due to this, the Lorentz force acting over the TiO_2-ethylene glycol-based NF flow due to the UMF can be expressed as $J \times B = \sigma(V \times B) \times B$. (see Turkyilmazoglu [16]). Our working hypothesis is that the temperature of the melting surface is less than the ambient temperature $(T_m < T_\infty)$. The governing expressions are listed as follows:

$$\nabla . V = 0, \tag{2.1}$$

$$(V.\nabla)V = \nu_{nf}\nabla^2 V + \frac{1}{\rho_{nf}}J \times B, \tag{2.2}$$

$$(V.\nabla)T = \alpha_{nf}\nabla^2 T, \tag{2.3}$$

The boundary conditions are:

$$u = C_0 r = U_w(r), v = \Omega r, k_{nf}\frac{\partial T}{\partial z} = \rho_{nf}(C_s(T_m - T_0) + \lambda)w, T = T_m \text{ at } z = 0, \\ u \rightarrow 0, v \rightarrow 0, w \rightarrow 0, T \rightarrow T_\infty \text{ as } z \rightarrow \infty.$$

(2.4)

2.1 Thermophysical properties for aggregation approach

The values of thermophysical features for the considered base fluid and NP are tabulated in Table 2.1. Based on the results of experiments, it is widely believed that NFs have a high thermal conductivity. In addition, the random motion of NPs or the NPs aggregation that generates percolation act could be exploited to advance the thermal feature of the material. The Brownian randomness diminutions in comparison to the aggregation process, which increases the aggregates mass. On the other hand, the percolation nature of aggregates might lead to an upsurge in the thermal conductivity of the aggregates. As a result of this, the thermophysical features of NF for NP aggregation will need to be redefined and are stated as follows (see Ellahi et al. [27], Acharya et al. [28] and Benos et al. [29]):

$$\mu_{nf} = \mu_f \left(1 - \frac{\phi_{agg}}{\phi_{max}}\right)^{-2.5*\phi_{max}}$$

(2.5)

$$\rho_{nf} = \left(1 - \phi_{agg}\right)\rho_f + (\phi\rho)_{agg}$$

(2.6)

$$\left(\rho C_p\right)_{nf} = \left(1 - \phi_{agg}\right)\left(\rho C_p\right)_f + \phi_{agg}\left(\rho C_p\right)_{agg}$$

(2.7)

$$k_{nf} = k_f \left(\frac{k_{agg} + 2k_f + 2\phi_{agg}\left(k_{agg} - k_f\right)}{k_{agg} + 2k_f - \phi_{agg}\left(k_{agg} - k_f\right)}\right)$$

(2.8)

Table 2.1 Thermophysical properties for base liquid and NP considered in the study are as follows (see Mackolil and Mahanthesh [26]).

	Base fluid	Nanoparticle
C_p	2415	686.2
μ	0.0157	—
k	0.252	8.9538
ρ	1114	4250

The correlations for the viscosity and thermal conductivity have been determined, respectively, via the modified Krieger and Dougherty model and the modified Maxwell model with (see Ellahi et al. [27], Acharya et al. [28] and Benos et al. [29]):

$$\phi_{max} = 0.605, D = 1.8, \frac{R_{agg}}{R_p} = 3.34$$

$$\phi_{agg} = \frac{\phi}{\phi_{int}}, \phi_{int} = \left(\frac{R_{agg}}{R_p}\right)^{D-3} \tag{2.9}$$

$$\rho_{agg} = (1 - \phi_{int})\rho_f + \phi_{int}\rho_s \tag{2.10}$$

$$\left(\rho C_p\right)_{agg} = (1 - \phi_{int})\left(\rho C_p\right)_f + \phi_{int}\left(\rho C_p\right)_s \tag{2.11}$$

$$k_{agg} = \frac{k_f}{4}\left([3\phi_{int} - 1]\frac{k_s}{k_f} + [3(1 - \phi_{int}) - 1] + \left[\left[(3\phi_{int} - 1)\frac{k_s}{k_f} + (3(1 - \phi_{int}) - 1)\right]^2 + \frac{8k_s}{k_f}\right]^{\frac{1}{2}}\right) \tag{2.12}$$

Similarity variables for the current study are as follows:

$$u = r\Omega F(\eta), v = r\Omega G(\eta), w = \sqrt{v_f\Omega}H(\eta), \eta = \sqrt{\frac{\Omega}{v_f}}z,$$
$$\theta(\eta) = \frac{T - T_m}{T_\infty - T_m}$$

The following set of nondimensionless nonlinear differential equations are obtained:

$$H' + 2F = 0, \tag{2.13}$$

$$\left(\frac{\left(1 - \frac{\phi_{agg}}{\phi_{max}}\right)^{-2.5*\phi_{max}}}{\left(1 - \phi_{agg}\right) + \phi_{agg}\frac{\rho_{agg}}{\rho_f}}\right)F'' + G^2 - F^2 - HF'$$

$$-\frac{\sigma_{nf}}{\sigma_f}\left(\frac{1}{\left(1 - \phi_{agg}\right) + \phi_{agg}\frac{\rho_{agg}}{\rho_f}}\right)M \sin \alpha[F \sin \alpha - G \cos \alpha] = 0, \tag{2.14}$$

$$\left(\frac{\left(1 - \frac{\phi_{agg}}{\phi_{\max}}\right)^{-2.5*\phi_{\max}}}{\left(1 - \phi_{agg}\right) + \phi_{agg}\frac{\rho_{agg}}{\rho_f}} \right) G'' - HG' - 2FG$$

$$+ \frac{\sigma_{nf}}{\sigma_f} \left(\frac{1}{\left(1 - \phi_{agg}\right) + \phi_{agg}\frac{\rho_{agg}}{\rho_f}} \right) M[F \sin \alpha - G \cos \alpha]\cos \alpha = 0, \tag{2.15}$$

$$\left(\frac{\left(1 - \frac{\phi_{agg}}{\phi_{\max}}\right)^{-2.5*\phi_{\max}}}{\left(1 - \phi_{agg}\right) + \phi_{agg}\frac{\rho_{agg}}{\rho_f}} \right) H'' - \left(\frac{1}{\left(1 - \phi_{agg}\right) + \phi_{agg}\frac{\rho_{agg}}{\rho_f}} \right)$$

$$\frac{\sigma_{nf}}{\sigma_f} MH - HH' = 0, \tag{2.16}$$

$$\left(\frac{k_{nf}}{k_f} \right) \left(\frac{1}{\left(1 - \phi_{agg}\right) + \phi_{agg}\frac{(\rho C_p)_{agg}}{(\rho C_p)_f}} \right) \theta'' - \Pr \theta' H = 0, \tag{2.17}$$

The modified boundary conditions for the proposed work are as follows:

$$\left. \begin{array}{l} F(0) = -1 = \omega, \left(\dfrac{1}{\left(1 - \phi_{agg}\right) + \phi_{agg}\frac{\rho_{agg}}{\rho_f}} \right) \dfrac{1}{\Pr} M_e \dfrac{k_{nf}}{k_f}\theta'(0) + H(0) = 0, G(0) - 1 = 0 = \theta(0), \\[4mm] (F(\infty), H(\infty), G(\infty)) \to 0, \theta(\infty) \to 1. \end{array} \right\} \tag{2.18}$$

Dimensionless parameters for the proposed work are as follows:

$$\Pr = \frac{\mu_f C_p}{k_f}, \omega = \frac{C_0}{\Omega}, M = \frac{\sigma_f |B|^2}{\rho_f \Omega}, \mathrm{Re} = \frac{r^2 \Omega}{\nu_f}, M_e = \frac{C_f(T_\infty - T_m)}{\lambda + C_s(T_m - T_0)}.$$

3. Results and discussions

This sector explores the effect of a variety of substantial parameters on individual fields. The reduced equations are solved using the RKF-45 and the shooting process to better comprehend the model's behavior. To completely validate the suggested model's insight, the method is we repeated with different parameter values. When it comes to modeling and making predictions about complicated systems, artificial neural networks (ANNs) are an extremely useful tool. ANNs are especially helpful in this scenario because of their ability to understand the underlying patterns and correlations in the data, which enables more accurate forecasts of the fluid flow and the thermal behavior. The network can learn the complicated correlations among the input parameters and the output variables by being trained on a collection of data. ANNs are able to handle enormous volumes of data and are able to find complicated patterns that may not be obvious using standard mathematical modeling approaches. In addition, ANNs are capable of managing vast amounts of data. Fig. 2.1 shows the structure of ANN. The comparison analysis helps in ensuring that the data offered are accurate. By comparing to previously published research, the present numerical technique has been shown to be correct for particular reduced scenarios (see Table 2.2).

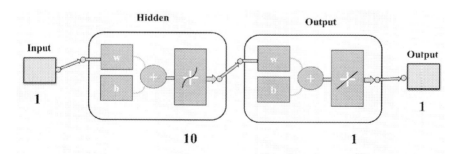

Figure 2.1 Structure of ANN.

Table 2.2 Comparison of the $F'(0)$ and $-\theta'(0)$ values for some reduced cases.

	$F'(0)$	$-\theta'(0)$
Ref. [30]	0.510233	
Ref. [31]	0.5102	0.9337
Ref. [32]	0.51023262	0.93387794
Present results	0.51023263	0.93387795

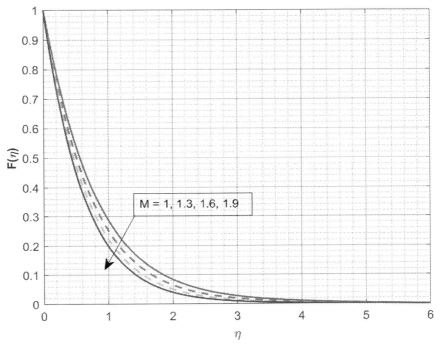

Figure 2.2 Impact of M on the $F(\eta)$.

Figs. 2.2 and 2.3 explore the impact of M on the $F(\eta)$ and $G(\eta)$, respectively. Here, rise in M values decline both the velocity profiles. The presence of a UMF in NF stream across a stretched disk with NP aggregation may modify the behavior of the NPs and change the flow properties of the liquid. Here, the magnetic forces acting on the NPs may lead them to align or agglomerate in certain directions. This may result in an increase in the liquid's effective viscosity. Consequently, the liquid's velocity field is influenced in both the axial and radial directions. Increased viscosity may lower flow velocity, especially in locations near the disk where particles are prone to concentrate. Furthermore, particle alignment may reduce the effective permeability of the liquid, resulting in a drop-in flow rate. Furthermore, the magnetic field may interact with the disk gyration, resulting in the formation of an azimuthal Lorentz force that opposes the liquid motion's direction. This force may lower flow velocity even more and change the stream pattern in both directions.

The consequence of ω on the $F(\eta)$ and $G(\eta)$ is portrayed in Figs. 2.4 and 2.5. Here, the increment in ω increases $F(\eta)$ but decays the $G(\eta)$.

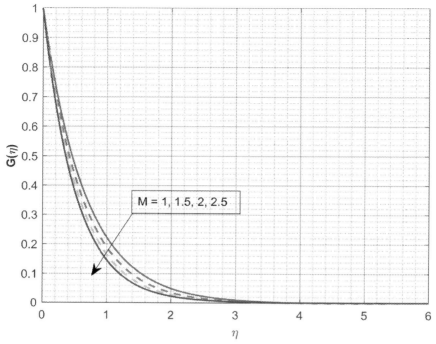

Figure 2.3 Impact of M on the $G(\eta)$.

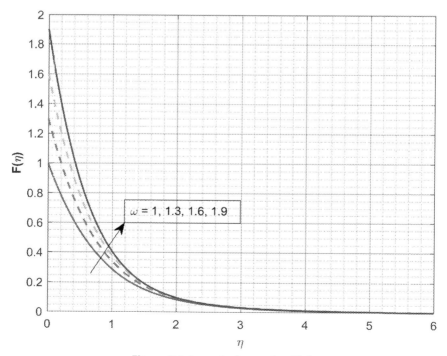

Figure 2.4 Impact of ω on the $F(\eta)$.

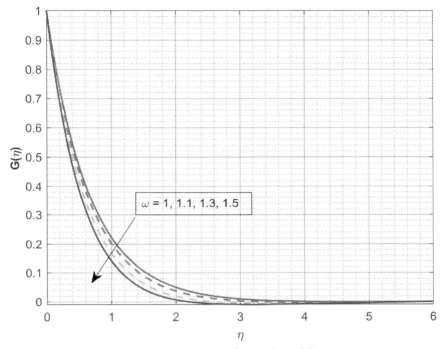

Figure 2.5 Impact of ω on the $G(\eta)$.

The rate at which the disk is stretched as well as the influence that this has on the flow of liquid is referred to as the stretching strength parameter. The stretching effect causes the flow velocity of the NF to rise in the radial direction when the ω is raised. The liquid particles travel outward as a result of the disk being stretched, which might raise the $F(\eta)$. However, in the tangential direction, a drop-in flow velocity may result from an upsurge in the ω. The stretching effect works perpendicular to the tangential direction; hence, it does not affect tangential flow velocity. Additionally, a rise in the liquid's effective viscosity owing to ω may result in a reduction in the $G(\eta)$ as a result of viscous forces. The interaction among the NPs and the liquid flow might further affect the flow behavior when NP aggregation is present. Changes in the liquid's effective viscosity brought on by the presence of NP aggregation may further alter the flow behavior.

Fig. 2.6 displays the impact of M_e over a $F(\eta)$. Here, $F(\eta)$ decreases for increased M_e. It is possible that increasing the values of M will result in a reduction in the $F(\eta)$. This is because melting may lead to a drop in the

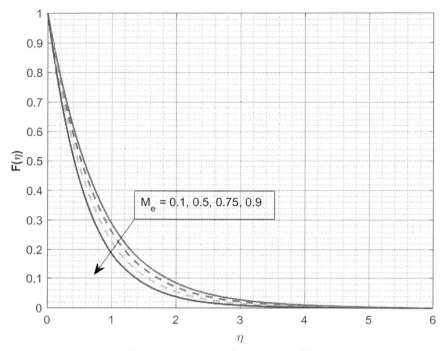

Figure 2.6 Impact of M_e over a $F(\eta)$.

effective viscosity of the liquid, which can then induce a reduction in the $F(\eta)$ owing to a decline in the viscous forces acting on the liquid.

Fig. 2.7 displays the impact of M_e on $\theta(\eta)$. Here, $\theta(\eta)$ increases for increased M_e. A rise in the M_e might lead to an upsurge in the $\theta(\eta)$. This is due to the fact that melting may lead to an upsurge in the heat transport rate, which in turn might affect the temperature throughout the liquid and change the thermal pattern. Moreover, the latent heat that is produced during the melting process is responsible for the escalation in the rate at which heat is transported. The quantity of latent heat that is released also rises as the M_e is raised, which eventually results in a superior heat transmission rate within the liquid. This may cause the $\theta(\eta)$ to rise.

Figs. 2.8 and 2.9 illustrate the designed model of convergence of validation, testing, and training progressions versus epochs indexes on $F(\eta)$ and $\theta(\eta)$. The best performance can be seen in epochs 317 and 343, with MSE values of $6.6009e - 10$ and $3.8806e - 09$, respectively.

Figs. 2.10 and 2.11 depict the convergence efficiency, correctness, and accuracy in solving the specified flow model. The step size and Mu is the

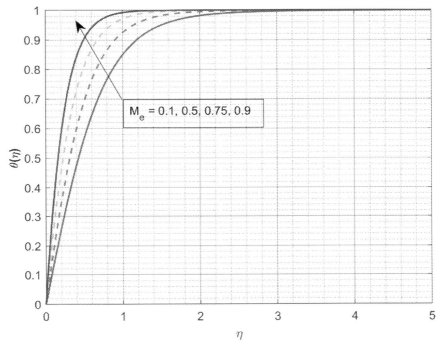

Figure 2.7 Impact of M_e over a $\theta(\eta)$.

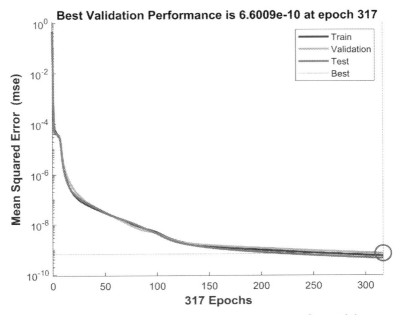

Figure 2.8 An examination of the MSE's outcomes about $F(\eta)$.

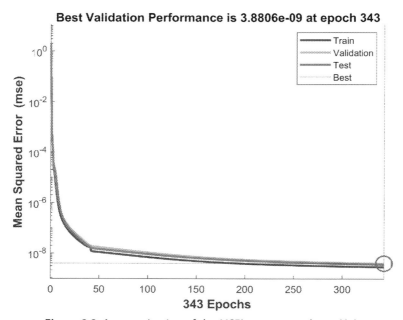

Figure 2.9 An examination of the MSE's outcomes about $\theta(\eta)$.

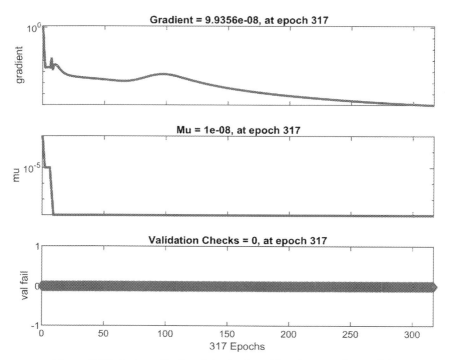

Figure 2.10 An investigation into the transitional states about $F(\eta)$.

Artificial intelligence—based Levenberg—Marquardt approach

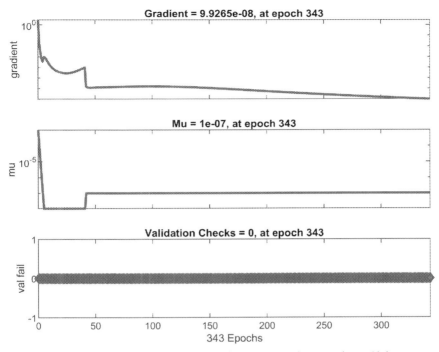

Figure 2.11 An investigation into the transitional states about $\theta(\eta)$.

gradient in these instances, as well as during training gradient in verdict one more vector. The matching gradient values for the anticipated flow are shown on the graph. The graph also shows that gradient and Mu values drop with increasing epoch size. Greater network testing and training results in outcomes that are more convergent for the lowest Mu and gradient values.

The error dynamics examination yields error histograms (Figs. 2.12 and 2.13) and fitness curves (Figs. 2.14 and 2.15). In addition to illuminating the error box of reference for the planned flow, a thorough examination of the error histogram will make it abundantly clear how many error values are far higher than the zero axis. Figs. 2.16 and 2.17 show the regression scrutiny for validation, training, testing, and total data. All over the operation, the output and target values are considered to be connected with a regression value of $R = 1$.

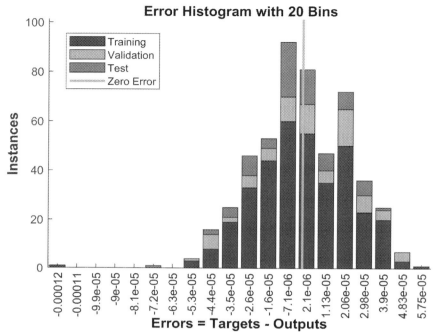

Figure 2.12 Analysis of error histograms on $F(\eta)$.

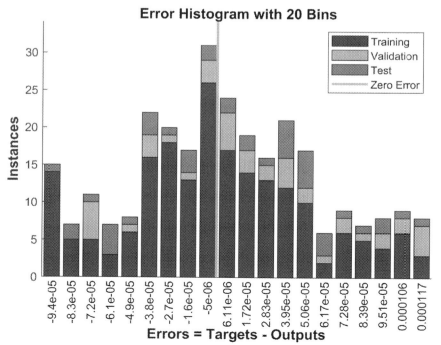

Figure 2.13 Analysis of error histograms on $\theta(\eta)$.

Artificial intelligence—based Levenberg—Marquardt approach

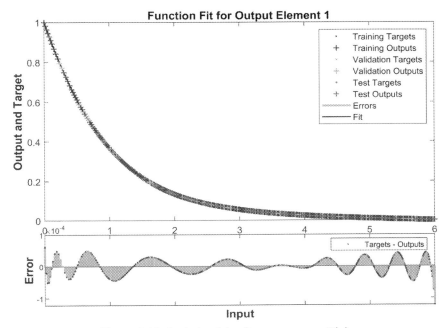

Figure 2.14 Analysis of the fitness curve on $F(\eta)$.

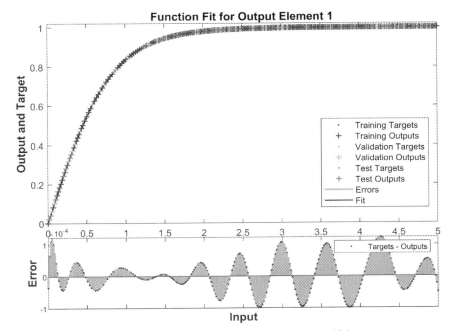

Figure 2.15 Analysis of the fitness curve on $\theta(\eta)$.

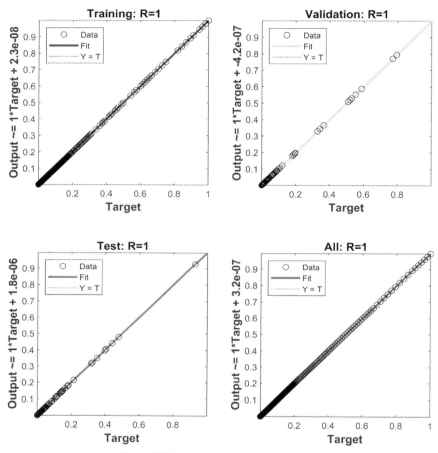

Figure 2.16 Regression analysis on $F(\eta)$.

4. Conclusions

This study examines the NP aggregation impact on the stream of NF through a disk. The current research employs the Krieger–Dougherty and Maxwell–Bruggeman models to investigate NP aggregation. These models accurately depict thermal conductivity and viscosity. By applying suitable transformations, the governing PDEs are transformed into ODEs. Subsequently, the system that was obtained is solved numerically through the implementation of the shooting technique and the utilization of the RKF-45 method. The major outcomes are that the rise in magnetic

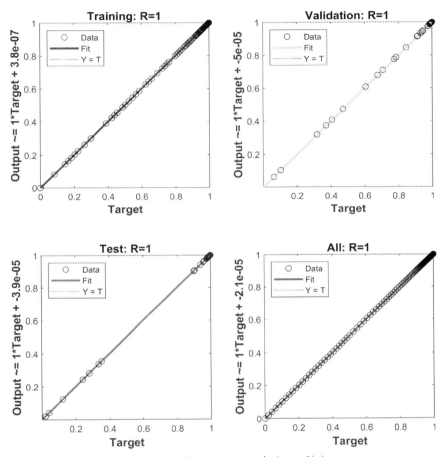

Figure 2.17 Regression analysis on $\theta(\eta)$.

parameter values declines the velocity field in both directions. The rise in stretching strength parameter values increases the velocity field in the radial direction but decreases in the tangential direction. The upsurge in melting parameter decreases the velocity field but improves the thermal field. The graph also demonstrates that the values of gradient and Mu drop as the epoch advances from one point to the next. More extensive testing and training of the network may lead to more convergence of the results for the lowest possible Mu and gradient value.

Nomenclature

σ Electrical conductivity
M_e Melting parameter
T_∞ Ambient temperature
k Thermal conductivity
M Magnetic parameter
T_0 Solid temperature
ρC_p Heat capacitance
B Magnetic field
(u, v, w) Velocity components
Ω angular velocity
Pr Prandtl number
ω Stretching strength parameter
T Temperature
λ Latent heat of the fluid
(r, ϕ, z) directions
α Thermal diffusivity
$F(\eta), G(\eta), H(\eta)$ Dimensionless velocity profiles
ρ Density
C_0 Stretching constant
ν Dynamic viscosity
T_m Temperature of the melting surface
C_s Heat capacity of the solid surface
η Dimensionless variable
μ Dynamic viscosity
$\theta(\eta)$ Dimensionless thermal profile
Re Local Reynolds number.
ϕ Volume fraction
Nu Nusselt number

Subscripts
f fluid
agg aggregate
nf nanofluid
s solid nanoparticle

References

[1] K. Sharma, N. Vijay, F. Mabood, I.A. Badruddin, Numerical simulation of heat and mass transfer in magnetic nanofluid flow by a rotating disk with variable fluid properties, Int. Commun. Heat Mass Tran. 133 (2022) 105977, https://doi.org/10.1016/j.icheatmasstransfer.2022.105977.

[2] R.J. Punith Gowda, R. Naveen Kumar, A.M. Jyothi, B.C. Prasannakumara, K.S. Nisar, KKL correlation for simulation of nanofluid flow over a stretching sheet considering magnetic dipole and chemical reaction, Z. Angew. Math. Mech. 101 (11) (2021) e202000372, https://doi.org/10.1002/zamm.202000372.

[3] T. Gul, Kashifullah, M. Bilal, W. Alghamdi, M.I. Asjad, T. Abdeljawad, Hybrid nano-fluid flow within the conical gap between the cone and the surface of a rotating disk, Sci. Rep. 11 (1) (2021) 1, https://doi.org/10.1038/s41598-020-80750-y.

[4] A. Tassaddiq, et al., Heat and mass transfer together with hybrid nanofluid flow over a rotating disk, AIP Adv. 10 (5) (2020) 055317, https://doi.org/10.1063/5.0010181.

[5] A. Hajizadeh, N.A. Shah, S.I.A. Shah, I.L. Animasaun, M. Rahimi-Gorji, I.M. Alarifi, Free convection flow of nanofluids between two vertical plates with damped thermal flux, J. Mol. Liq. 289 (2019) 110964, https://doi.org/10.1016/j.molliq.2019.110964.

[6] F. Wang, et al., The effects of nanoparticle aggregation and radiation on the flow of nanofluid between the gap of a disk and cone, Case Stud. Therm. Eng. 33 (2022) 101930, https://doi.org/10.1016/j.csite.2022.101930.

[7] P. Rana, B. Mahanthesh, K. Thriveni, T. Muhammad, Significance of aggregation of nanoparticles, activation energy, and Hall current to enhance the heat transfer phenomena in a nanofluid: a sensitivity analysis, Waves Random Complex Media (2022) 1−23, https://doi.org/10.1080/17455030.2022.2065043.

[8] Z. Mahmood, U. Khan, Nanoparticles aggregation effects on unsteady stagnation point flow of hydrogen oxide-based nanofluids, Eur. Phys. J. Plus 137 (6) (2022) 750, https://doi.org/10.1140/epjp/s13360-022-02917-y.

[9] Y. Yu, et al., Nanoparticle aggregation and thermophoretic particle deposition process in the flow of micropolar nanofluid over a stretching sheet, Nanomaterials 12 (6) (2022) 977, https://doi.org/10.3390/nano12060977.

[10] J. Chen, C.Y. Zhao, B.X. Wang, Effect of nanoparticle aggregation on the thermal radiation properties of nanofluids: an experimental and theoretical study, Int. J. Heat Mass Transf. 154 (Jun. 2020) 119690, https://doi.org/10.1016/j.ijheatmasstransfer.2020.119690.

[11] K. Singh, A.K. Pandey, M. Kumar, Numerical solution of micropolar fluid flow via stretchable surface with chemical reaction and melting heat transfer using Keller-Box method, Propuls. Power Res. 10 (2) (2021) 194−207, https://doi.org/10.1016/j.jppr.2020.11.006.

[12] T. Muhammad, H. Waqas, U. Farooq, M.S. Alqarni, Numerical simulation for melting heat transport in nanofluids due to quadratic stretching plate with nonlinear thermal radiation, Case Stud. Therm. Eng. 27 (2021) 101300, https://doi.org/10.1016/j.csite.2021.101300.

[13] K. Singh, A.K. Pandey, M. Kumar, Melting heat transfer assessment on magnetic nanofluid flow past a porous stretching cylinder, J. Egypt Math. Soc. 29 (1) (2021) 1, https://doi.org/10.1186/s42787-020-00109-0.

[14] T. Hayat, F. Shah, A. Alsaedi, B. Ahmad, Entropy optimized dissipative flow of effective Prandtl number with melting heat transport and Joule heating, Int. Commun. Heat Mass Transf. 111 (Feb. 2020) 104454, https://doi.org/10.1016/j.icheatmasstransfer.2019.104454.

[15] M. Jawad, Z. Khan, E. Bonyah, R. Jan, Analysis of hybrid nanofluid stagnation point flow over a stretching surface with melting heat transfer, Math. Probl. Eng. 2022 (2022) e9469164, https://doi.org/10.1155/2022/9469164.

[16] M. Turkyilmazoglu, Flow and heat over a rotating disk subject to a uniform horizontal magnetic field, Z. Naturforsch. 77 (4) (2022) 329−337, https://doi.org/10.1515/zna-2021-0350.

[17] P. Pal, K. Kumar, Role of uniform horizontal magnetic field on convective flow, Eur. Phys. J. B 85 (6) (2012) 201, https://doi.org/10.1140/epjb/e2012-30048-8.

[18] B.C. Prasannakumara, R.J.P. Gowda, Heat and mass transfer analysis of radiative fluid flow under the influence of uniform horizontal magnetic field and thermophoretic particle deposition, Waves Random Complex Media (2022) 1−12, https://doi.org/10.1080/17455030.2022.2096943.

[19] T. Tagawa, Effect of the direction of uniform horizontal magnetic field on the linear stability of natural convection in a long vertical rectangular enclosure, Symmetry 12 (10) (2020) 1689, https://doi.org/10.3390/sym12101689.

[20] F. Wang, et al., Aspects of uniform horizontal magnetic field and nanoparticle aggregation in the flow of nanofluid with melting heat transfer, Nanomaterials 12 (6) (2022) 1000, https://doi.org/10.3390/nano12061000.

[21] K. Naganthran, M. Mustafa, A. Mushtaq, R. Nazar, Dual solutions for fluid flow over a stretching/shrinking rotating disk subject to variable fluid properties, Phys. Stat. Mech. Appl. 556 (2020) 124773, https://doi.org/10.1016/j.physa.2020.124773.

[22] R. Naveen Kumar, H.B. Mallikarjuna, N. Tigalappa, R.J. Punith Gowda, D. Umrao Sarwe, Carbon nanotubes suspended dusty nanofluid flow over stretching porous rotating disk with non-uniform heat source/sink, Int. J. Comput. Methods Eng. Sci. Mech. 23 (2) (2022) 119–128, https://doi.org/10.1080/15502287.2021.1920645.

[23] S.-S. Zhou, M. Bilal, M.A. Khan, T. Muhammad, Numerical analysis of thermal radiative Maxwell nanofluid flow over-stretching porous rotating disk, Micromachines 12 (5) (2021) 540, https://doi.org/10.3390/mi12050540.

[24] A. Bhandari, Unsteady flow and heat transfer of the ferrofluid between two shrinking disks under the influence of magnetic field, Pramana 95 (2) (2021) 89, https://doi.org/10.1007/s12043-021-02107-y.

[25] A. Rauf, A. Mushtaq, N.A. Shah, T. Botmart, Heat transfer and hybrid ferrofluid flow over a nonlinearly stretchable rotating disk under the influence of an alternating magnetic field, Sci. Rep. 12 (1) (2022) 17548, https://doi.org/10.1038/s41598-022-21784-2.

[26] J. Mackolil, B. Mahanthesh, Sensitivity analysis of Marangoni convection in TiO2–EG nanoliquid with nanoparticle aggregation and temperature-dependent surface tension, J. Therm. Anal. Calorim. 143 (2021) 2085–2098, https://doi.org/10.1007/s10973-020-09642-7.

[27] R. Ellahi, M. Hassan, A. Zeeshan, "Aggregation effects on water base Al2O3–nanofluid over permeable wedge in mixed convection, Asia-Pac. J. Chem. Eng. 11 (2) (2016) 179–186, https://doi.org/10.1002/apj.1954.

[28] N. Acharya, K. Das, P.K. Kundu, Effects of aggregation kinetics on nanoscale colloidal solution inside a rotating channel, J. Therm. Anal. Calorim. 138 (1) (2019) 461–477, https://doi.org/10.1007/s10973-019-08126-7.

[29] L.T. Benos, E.G. Karvelas, I.E. Sarris, Crucial effect of aggregations in CNT-water nanofluid magnetohydrodynamic natural convection, Therm. Sci. Eng. Prog. 11 (2019) 263–271, https://doi.org/10.1016/j.tsep.2019.04.007.

[30] N. Kelson, A. Desseaux, Note on porous rotating disk flow, ANZIAM J. 42 (2000) C837–C855, https://doi.org/10.21914/anziamj.v42i0.624.

[31] N. Bachok, A. Ishak, I. Pop, Flow and heat transfer over a rotating porous disk in a nanofluid, Phys. B Condens. Matter 406 (9) (2011) 1767–1772, https://doi.org/10.1016/j.physb.2011.02.024.

[32] M. Turkyilmazoglu, Nanofluid flow and heat transfer due to a rotating disk, Comput. Fluids 94 (2014) 139–146, https://doi.org/10.1016/j.compfluid.2014.02.009.

CHAPTER THREE

Applications of nanofluids in refrigeration and air-conditioning

Jahar Sarkar
Department of Mechanical Engineering, Indian Institute of Technology (BHU) Varanasi, UP, India

Highlights

- Different roles of nanofluids in refrigeration systems are discussed with proper diagrams.
- Preparation, characterization, properties, heat transfer, and pressure drop are discussed.
- Performances of various systems using nanofluids in different roles are discussed in detail.
- Some application challenges and scopes in refrigeration and air-conditioning are discussed.

1. Introduction

Considering the energy security and environmental issues, there is a need for quality enhancement of the air-conditioning, heat pump, and refrigeration devices. This may be achieved by modifying either the system and system components or the properties of different fluids associated with the system. In recent years, many engineering applications have been attracted by the mono and hybrid nanofluids due to their superior thermo-physical, transport, mechanical, and other properties. Hence, the nanofluids (suspension of nano-sized particles in base fluids) may be utilized in refrigeration and heat pump applications to enrich the performance. For refrigeration and heat pump systems, the base fluid can be refrigerant, secondary fluid, lubricant, absorbent, etc. (organic or inorganic fluids), whereas the nanoparticles may be different metal, ceramic, carbon azeotrope, nitride, and carbide nanomaterials. The specific heat capacity, density, and thermal conductivity of various nanoparticles are listed in Table 3.1. Based on the

Table 3.1 Thermophysical characteristics of some used nanoparticles.

Materials	Density (kg/m^3)	Specific heat (J/kgK)	Thermal conductivity (W/mK)
Silver (Ag)	10,500	235	429
Copper (Cu)	8927	385	401
Aluminum (Al)	2700	900	237
Gold (Au)	19,300	129	318
Silicon (Si)	2330	710	149
Sodium (Na)	970	1230	142
Iron (Fe)	7874	451	80
Alumina (Al_2O_3)	3880	773	40
Titanium dioxide (TiO_2)	4156	686	8.95
Silica (SiO_2)	3970	740	1.4
Copper oxide (CuO)	6500	535	20
Zirconia (ZrO_2)	5670	550	3.7
Cerium oxide (CeO_2)	7215	670	6.9
Magnesia (MgO)	3580	910	50
Zink oxide (ZnO)	5675	494	23.4
Ferric oxide (Fe_2O_3)	5250	650	6
Iron oxide (Fe_3O_4)	5810	670	80
Graphene oxide (GO)	1910	710	2000
Silicon carbide (SiC)	3160	1340	350
Titanium carbide (TiC)	4930	700	25
Aluminum nitride (AlN)	3260	770	180
Diamond	3500	520	2200
Graphite	2270	710	119
Carbon nanotube	2600	410	3007
Graphene	2200	790	5000
Carbon quantum dot	1003	–	–
MXene	1500	–	–

types of nanoparticle suspension, the nanofluids are broadly classified as (i) mono nanofluids dispersing the similar nanoparticles, (ii) hybrid nanofluids dispersing the dissimilar nanoparticles mixture, and (iii) hybrid nanofluids dispersing the composite nanoparticles. This hybridization may be binary, ternary, or tetra, depending on the types of nanoparticles dispersing in nanofluids. Apart from the thermal conductivity enrichment, the following slip phenomena are also very important for the nanofluid heat transfer enrichment: Brownian phenomena, sedimentation, gravitational movement, clustering of nanoparticles, nanoparticle-surrounded layering, and thermophoresis [1]. These phenomena give relative velocity, resulting in this enhancement. Hence the nanofluid has emerged as a suitable candidate in different roles for the said applications due to their enhanced thermophysical, mechanical, and electrical properties [1].

This chapter focuses on the utilization of hybrid and mono nanofluids in different roles in air-conditioning, heat pump, and refrigeration devices to enhance the performance [2], such as (i) nanoparticle-dispersed secondary fluids (as liquids in the secondary circuit of evaporator, as a coolant in the condenser, and as a heating fluid in the generator of heat driven system), (ii) nanoparticle-dispersed refrigerants (called as nanorefrigerants), (iii) nanoparticle-dispersed lubricants (called as nanolubricants), (iv) nanoparticle-dispersed absorbents (called as nanoabsorbents), (v) nanoparticle-dispersed phase-change materials (called as nano-PCM) in the cold storage system, and (vi) nanoparticle-dispersed fluids in nonconventional systems. Finally, various challenges and future research and development scopes are discussed as well. As per the research paper publication trend, shown in Fig. 3.1, the application of nanofluids in different roles has emerged day-by-day in refrigeration, heat pump, and air-conditioning fields.

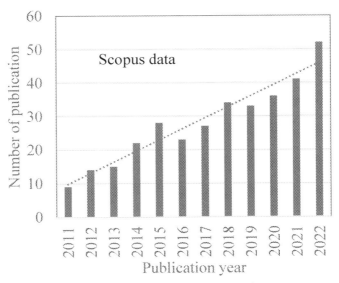

Figure 3.1 Publication trend of air-conditioning and refrigeration research using nanofluids.

2. Nanofluids as secondary fluid

The secondary fluid is used in the refrigeration and heat pump systems to effectively transfer heat between the system component and the heat source or sink. Water is widely utilized as a coolant (secondary liquid) to reject heat from various refrigeration system components (condenser, absorber, hot end of the thermoelectric cooler, etc.) to ambient air/cooling tower, river, ocean, etc. Similarly, water and various brines are used to transfer the cooling effect from the evaporator to application protocols (display cabins in the supermarket, food industries or air coolers for air-conditioning systems, etc.). The third option is the transfer of heat energy from any energy source, such as a solar collector, for heat-driven refrigeration systems (vapor absorption system, vapor adsorption system, ejector refrigeration system, etc.). Examples of above three applications of secondary fluid for refrigeration and heat pump systems are illustrated in Fig. 3.2.

The thermophysical and transport properties (particularly thermal conductivity) strongly influence the thermal characteristics of the heat exchange devices (contender, evaporator, generator, absorber, etc.) used in the refrigerating device and hence the refrigerant operating temperature for a given source and sink temperatures, which will ultimately affect the system performance (e.g., COP). As the heat conductivity of secondary fluids is significantly low with respect to the nanoparticle, the use of secondary fluid-based nanoparticle-dispersed nanofluids (having improved thermophysical and transport properties as compared to base fluid) can improve the refrigeration system performance. This section elaborates on the effect of using mono and hybrid nanoliquids as secondary medium in the said systems.

The basic and most important issue of both mono and hybrid nanofluids is their synthesis process for use in a refrigeration system, as the stability of

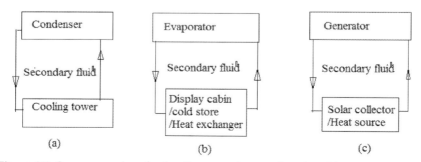

Figure 3.2 Some examples of using the secondary medium in refrigeration systems.

Applications of nanofluids in refrigeration and air-conditioning

prepared nanofluids is affected by it. Synthesis of nanofluids needs some special techniques for a stable and homogeneous solution having the least sedimentation and agglomeration problems. Nanofluids are prepared by dissolving suitable nanoparticles with the base mediums such as brine and water. There are usually two methods for nanofluid preparation: the one-step method and the two-step method [3]. Both preparation methods may involve chemical technique or mechanical technique or both. In the one-step process, nanoparticles are made and dispersed simultaneously in a base fluid. In the two-step process, we have to make the desired nanoparticles first and then disperse them in the base medium to prepare nanofluids. Fig. 3.3 provides the flowchart of both single-step and two-step techniques for the nanofluid preparation. The two-step process is in-general preferred over the single-step process because the simultaneous preparation and mixing of different types of nanoparticles using the one-step method are much more challenging. In both methods, the mixture is ultra-sonicated to get stable nanofluids. The sonication time and speed are very crucial as both will affect not only the stability and homogeneity but also the thermodynamic and transport characteristics of prepared nanofluids. For the preparation of hybrid (bi-hybrid or

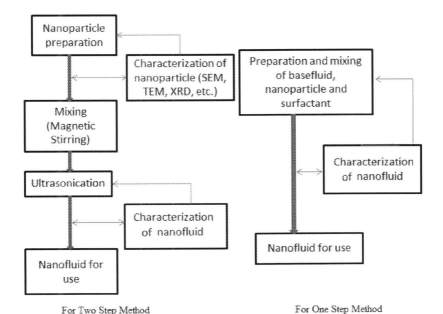

Figure 3.3 Flowchart for producing nanofluids (both one-step and two-step techniques).

tri-hybrid or tetra-hybrid) nanofluids, both single-step and two-step processes can be adopted [4]. However, researchers have mostly used the two-step process to prepare hybrid nanofluids. Various surfactants are used to synthesize stable homogeneous nanofluids, although stability duration depends on surfactant and nanofluid combination and synthesis method, sonication time, temperature, etc.

Stability analysis of same-type and hybrid nanofluids is very important as the enhanced stability leads to desired property improvement as well as an increase in operational life. Stable and homogeneous nanofluids indicate that the nanoparticles are properly and uniformly spaced in the base medium. Various stability analysis techniques can be used for synthesized nanofluids, such as the visual sedimentation method, zeta potential method, spectral characterization, and matching of pH values with the isoelectric point. Magnetic stirring, ultrasonication, surfactant addition, and surface modification are commonly utilized to enrich the stability of prepared mono or hybrid nanofluid. Choosing suitable sonication time, sonication frequency, and sonication temperature is very important for nanofluid preparation. Choosing a suitable amount of surfactant is also very important as the addition of a lower surfactant amount yields unstable nanofluids and the addition of a more surfactant amount leads to property deterioration. If we are able to select a suitable surfactant along with volume fraction, we may achieve long-term stability for the prepared nanofluid. The homogeneity of prepared mono and hybrid nanofluids can be checked by measuring the property at different positions of total nanofluid volume.

Four thermophysical and transport parameters (viscosity, density, thermal conductivity, and specific heat capacity) strongly influence the heat transfer behavior of nanofluids, used as the secondary fluid. The density enriches and specific heat capacity demolishes in general by using nanofluids as the nanoparticle density is higher and the nanoparticle specific heat capacity is lower than the base medium. For both mono and hybrid nanofluids, the density (ρ_{nf}) and specific heat capacity $(c_{p,nf})$ are theoretically calculated based on the mass balance and energy balance, respectively.

$$\rho_{nf} = \sum \phi_{np}\rho_{np} + \left(1 - \sum \phi_{np}\right)\rho_{bf} \tag{3.1}$$

$$\rho_{nf}c_{p,nf} = \sum \phi_{np}\rho_{np}c_{p,np} + \left(1 - \sum \phi_{np}\right)\rho_{bf}c_{p,bf} \tag{3.2}$$

where ϕ_{np}, ρ_{np}, ρ_{bf}, $c_{p,np}$, and $c_{p,bf}$ are nanoparticle volume fraction, nanoparticle density, base medium density, nanoparticle specific heat, and base fluid specific heat, respectively.

Both heat transport and flow friction of mono and hybrid nanofluids are influenced by the dynamic viscosity. The dynamic viscosity of the secondary fluid increases if we mix nanoparticles with them and also if we increase the percentage of nanoparticles in that medium. The following issues also influence the nanofluid viscosity: nanoparticle size, nanoparticle shape, medium temperature, pH value, the surfactant used, and synthesis method [5]. The coefficient of heat transport reduces and the flow friction enriches as the viscosity increases. Hence, the use of nanofluids is not favorable in terms of dynamic viscosity. Many theoretical models or correlations are available to predict the dynamic viscosity of mono nanofluids. For hybrid nanofluids also, many correlations are available but nanofluid specific. We have to give greater importance to the thermal conductivity of secondary fluid-based mono and hybrid nanofluids, which influence the heat transport performance. The heat transport coefficient and hence heat transport rate enriches by using nanofluids because the thermal conductivity enhances. Thermal conductivity depends on several factors, such as nanoparticle size, nanoparticle shape, nanoparticle volume fraction, medium temperature, pH value, surfactant, and synthesis method [6]. Many theoretical models or correlations are available to predict the thermal conductivity of mono-type nanofluids. For hybrid nanofluids also, many correlations are available but nanofluid-specific. The generalized correlations to predict viscosity (μ_{nf}) and thermal conductivity (k_{nf}) of mono/hybrid nanofluids having different nanoparticle morphologies are given by, respectively [7],

$$\mu_{nf} = \frac{1}{\phi} \sum_{i=1}^{n} \left(\phi_i \mu_{nf,i} \right) \text{ where } \phi = \sum_{i=1}^{n} \phi_i \tag{3.3}$$

$$k_{nf} = \frac{1}{\phi} \sum_{i=1}^{n} \left(\phi_i k_{nf,i} \right) \tag{3.4}$$

where, ϕ_i, $k_{nf,i}$, and $\mu_{nf,i}$ are nanoparticle volume concentration, thermal conductivity, and dynamic viscosity for ith nanoparticle type.

Due to enriched thermal conductivity as well as different slip mechanisms, the coefficient of heat transport as well as the rate of heat transport of mono/hybrid nanofluid used as secondary fluid increases. Both the pressure drop and heat transport enrich for mono/hybrid nanofluids; however, the pressure drop rise is desired to be negligible as compared to the heat transfer rise. Mono and hybrid nanofluids may be utilized as secondary mediums in various components, such as the evaporator, condenser, absorber, and generator [8]. Various types of heat-exchanging devices are used in refrigeration and heat pump systems, such as double-tube type, shell-tube type, and plate type. In the refrigeration and heat pump systems, the thermal conductance of the secondary medium (single-phase fluid) is significantly lower as compared to the primary refrigerant in the condenser, evaporator, absorber, and generator. As a result, the heat transport behavior of the secondary medium is much more sensitive to the performance of these components. This fact provides us with an emerging possibility to use mono/hybrid nanofluid as a secondary fluid in the heat pump and refrigeration systems. Many nanofluid-specific correlations have been developed to predict the Nusselt number or heat transport coefficient of mono nanofluid; however, a generalized correlation has been developed recently based on separation approach [9], which can be applicable for any base fluid and nanoparticle (with any shape) combinations. For hybrid nanofluids as secondary fluid, the mixing ratio of different nanoparticles may play an important role in heat transport and fluid friction behaviors [10].

The mono or hybrid nanofluid has been utilized as a cooling media in the hot side heat exchanger of the vapor compression refrigerating device to heat transfer to the cooling tower. By using mono or hybrid nanofluids as the coolant, the condenser size and pumping power reduce with the decrease in nanoparticle size and increment in nanoparticle volume concentration for a given cooling capacity. Sarkar [11] did the performance analyses of the transcritical CO_2 refrigerating cycle using nanofluid as a coolant in the shell-tube type gas cooler. They have found the cooling rate, heat exchanging effectiveness, and COP (coefficient of performance) improvements with negligible pumping energy increment. They reported the cooling COP improvement of up to 26% by using nanofluids as secondary fluid. Fig. 3.4 shows the COP improvement for different nanofluids, operating and design parameters of the gas cooler. As shown, the performance trends are similar, corresponding to the pressure in the gas cooler, mass flow rate of the nanofluid, and gas cooler heat transfer area; however, their deviation enriches with a rise in nanoparticle volume concentration. It has also been observed that the rise in nanoparticle volume concentration enhances the cooling COP; however, the increment of cooling COP is not significant for higher nanoparticle concentration due

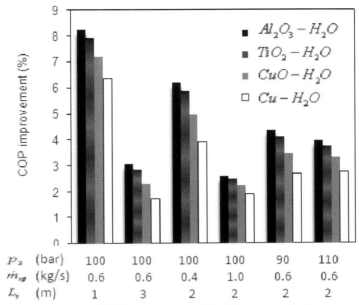

Figure 3.4 Comparison of COP improvement with different nanofluids (p_2 is gas cooler pressure, $\dot{m}_{nf}$ is nanofluid flow rate, and L_t is total gas cooler length) [11].

to the domination of the viscosity effect. The larger domination of negative effects due to the increase in viscosity may even degrade the said system capability by using said cooling medium [12]. Askari et al. [13] conducted an experiment on the counterflow cooling tower by utilizing two types of nanofluids. They found that both the efficacy and cooling range of the cooling tower increased by utilizing said fluids. The use of nanofluid also reduces the water loss in cooling towers, which leads to the reduction of water consumption. By utilizing nanofluids, both the condenser and cooling tower sizes reduce. The nanofluids can be utilized in the external cooling jacket of the condenser of the air-conditioning device. The experimental study reported the significant enhancement of COP and increased in performance with the enrichment in nanoparticle volume fraction. A maximum performance improvement of 29.6% has been reported [14]. In summary, the mono or hybrid nanofluid as a cooling media increases the capability of the discussed cooling devices, but the energy required for pumping may increase slightly.

The mono or hybrid nanofluid has also been used as the secondary refrigerant in the evaporator (in a secondary loop) to improve the performance.

Although, the overall improvement is influenced by the relative increase in heat transport rate and pumping energy rate. Whether the ratio of heat transport rate to pumping energy rate increases or decreases is dependent on the type of nanofluid used. Vasconcelos et al. [15] mentioned the superior cooling capability and COP of the system by using carbon nanotube (CNT)/water nanofluids as a secondary refrigerant. The use of nanofluid reduces the compressor power but increases the pump power and hence the total energy consumption is not affected substantially by using nanofluid. It may be noted that the optimum nanoparticle volume concentration has to be used in secondary fluid-based mono or hybrid nanofluids to enrich the capability of refrigeration or heat pump devices. The use of hybrid nanofluid affects not only the energy performance but also the exergy performance. The irreversibility of the plate evaporator can be reduced if we use nanofluidic secondary refrigerant. The nondimensional exergy destruction significantly reduces by utilizing brine-based hybrid nanofluids as a secondary refrigerant for different low-temperature applications. The use of brine-based hybrid nanofluids in the evaporator enhances the exergetic efficiency for the above-mentioned applications [16]. Due to improved heat transfer characteristics, the use of hybrid nanofluid as secondary fluid reduces the size of the evaporator and space requirement. The use of brine-based hybrid nanofluid as a secondary refrigerant also reduces primary refrigerant consumption. With the utilization of hybrid nanofluid, the running cost reduces, but the annual investment cost increases and the payback period depends on the nanoparticle used. Most of the hybrid nanofluids showed a payback period more than the evaporator life and therefore are not beneficial in the present condition [17]. However, the decreasing nanoparticle cost and nanofluid stability may decrease the payback period. By using nanofluid as a secondary medium, the evaporator heat transfer area and pressure drop decrease with a rise in nanoparticle concentration and a fall in nanoparticle size for a given cooling capacity. The refrigerant charge reduces due to the decrease in the area and hence the emission of CO_2 reduces. The cooling capacity, heat transfer performance, and COP increase by using mono or hybrid nanofluid, whereas the pumping power negligibly increases. The primary refrigerant consumption and hence its cost also reduce. The use of nanofluid in the secondary loop is also beneficial for the application heat exchanger (the cold energy taken from the evaporator is utilized through this heat exchanger). The TiO_2/water nanofluid was utilized in plate-type heat exchanger, and the experiment was performed by Tabari et al. [18]

for milk pasteurization. They have found that both heat transport rate (merit) and pressure drop (demerit) are higher than those of base fluid for all nanoparticle concentrations. It has been found that the higher flow rate of nanofluid enriches the heat transport coefficient and exergetic efficiency, which are merit; whereas, it enriches the pressure drop, pumping energy, the generation rate of entropy, and irreversibility, which are demerits [19]. The vapor compression system COP increases by utilizing hybrid nanofluids as a secondary working fluid and the COP increases with the enrichment in the nanofluid volumetric flow rate. The use of mono or hybrid nanofluids as the secondary working medium in chilled water-based air-conditioning systems reduces the compression ratio and hence compressor work [20].

The mono and hybrid nanofluids may be effectively used as the heat transfer fluid between the heat source (e.g., solar thermal collector) and vapor generator of the heat-driven refrigeration system. By using nanofluid in solar-driven refrigeration systems, both the COP and the exergy efficiency increase [21]. On the other hand, the solar collector area required per unit cooling capacity decreases, which leads to a decrease in collector cost and hence the total cost rate of the system [21]. It has been observed that thermodynamic performance rises with an increase in nanoparticle concertation [22]. The heat removal factor of solar collectors also improves by using nanofluid. However, no such difference in optimum operating parameters has been found by using nanofluid. Due to the lower value of optimal evaporator temperature by using nanofluid, the cooling capacity of the system also increases. However, the researchers have observed some drawbacks for utilizing mono or hybrid nanofluids due to their higher pumping power and cost in the real design.

Another important application of nono/hybrid nanofluids as secondary working fluid is with a thermoelectric cooler for heat transfer with the cold side or hot side. Some studies on this application have been carried out recently [23]. It has been observed that utilizing nanofluid in the thermoelectric refrigerator yields superior capability. The terminal temperature difference was found to be enriched with nanoparticle fraction for all studied nanofluids. It has been found from the conducted experiments that the temperature difference of the cooling cabin considerably improves by using nanofluids, even with small mass fractions of nanoparticles as compared to water. The mono and hybrid nanofluids can be employed as heat transfer fluid in many other refrigeration systems to improve performance.

3. Nanorefrigerants

The dispersion of nanoparticles in the refrigerants (leads to nanorefrigerants) may improve the thermophysical and tribological properties and hence heat transfer characteristics (single-phase heating or cooling, boiling, and condensation), which can ultimately enrich the exergetic and energetic capabilities of the refrigeration and heat pump systems. However, the preparation of nanorefrigerants is a key step as the system working fluid is not in the liquid state at ambient pressure. Similar to conventional nanofluids, a one-step technique or two-step technique can be utilized to prepare nanorefrigerants. Although widely used refrigerants are mostly in the vapor state under normal environmental conditions, as their normal boiling temperature is less than the temperature of ambient, and therefore the conventional nanofluid preparation method is unsuitable. For the refrigerant, which has normal boiling points higher than ambient temperature and hence is in a liquid state at the normal temperature of ambient, the commonly used nanofluid synthesis method has been utilized in the literature. Diao et al. [24] synthesized nanorefrigerant by dispersing Cu nanoparticles in the R141b refrigerant. They used the following steps: measurement, mixing, ultrasonication, and addition of suitable surfactant to stabilize the suspension. Similarly, Yang et al. [25] prepared R141b-based MWCNT nanorefrigerant using a two-step process for nanoparticle mass concentration of 0.1%, 0.2%, and 0.3%. They used Span-80 as a surfactant for the solution. Tazarv et al. [26] prepared the R141b-based nanorefrigerant by dispersing the required amount of TiO_2 nanoparticles with nanoparticle volume percentages ranging from 0.01% to 0.03% by a similar method. For the refrigerants having a lower normal boiling point than ambient (a gaseous state at the normal ambient condition), both the following methods have been adopted: (i) a cool environmental chamber has been used, where the temperature has been maintained lower the normal boiling temperature of the refrigerant. Hence, the refrigerant was in a liquid state in that chamber. Then the nanorefrigerant is synthesized by utilizing a single-step or two-step process as discussed within this chamber and (ii) nanolubricant is prepared first and then mixed with refrigerant. For example, Mahbubul et al. [27] utilized the first discussed technique. They used an orbital incubator shaker as a cold chamber, where a constant temperature of 15 C was maintained. Then they synthesized Al_2O_3/R134a nanorefrigerants by dispersing 30 nm-sized alumina nanoparticles in refrigerant with 5% nanoparticle volume percentage and sonicated in that chamber continuously for 24h duration at 240 rpm speed. The second method was also adopted by many investigators [2].

The characterization of nanorefrigerants includes (i) characterization of nanoparticles, which includes the estimations of their chemical behavior, size and shape, and also agglomeration size and (ii) stability analysis of dispersion. Various techniques for the characterization of nanoparticle have already been discussed in the previous section. Stability analysis is also very crucial for nanorefrigerants. Peng et al. [28] evaluated the experimental behavior of R141-TiO$_2$ nanorefrigerant for different particle concentrations, particle sizes, and surfactants. They observed that the larger nanoparticle size and concentration enrich the hydrodynamic diameter. Lin et al. [29] studied the stability of MWCNT dispersion in R141b refrigerant by adding the surfactant to the solution. They marked that good stability can be achieved by optimal surfactant concentration. Various stability analysis techniques can be used for nanorefrigerants, such as the visual sedimentation method, zeta potential analysis, spectral analysis, and matching of pH values with the isoelectric point. Magnetic stirring, ultrasonication, utilizing surfactant, and modification of surface are commonly used to improve the stability of prepared nanorefrigerants.

In general, the density increases, specific heat capacity decreases, dynamic viscosity increases, and thermal conductivity enriches by the addition of nanoparticles with refrigerants. By measuring the thermophysical behaviors of Al$_2$O$_3$/R-134a nanorefrigerant, Mahbubul et al. [27] observed that the density of nanorefrigerant increases by about 11%, whereas the specific heat capacity of nanorefrigerant decreases slightly. The volume concentration of nanoparticle and temperature has a strong influence on the specific heat capacity and density of nanorefrigerants. For nanorefrigerant, as the nanoparticle volume concentration rises and the operating temperature falls, the density enriches [2]. As evidence, Alawi et al. [30] estimated the density of CuO-R134a nanorefrigerant and concluded that as the nanoparticle volume concentration rises and the operating temperature falls, the density of prepared nanorefrigerant enriches. They also noticed that as the nanoparticle volume concentration or/and operating temperature rise, the specific heat of nanorefrigerants enriches [2]. It has also been found that the specific heat of nanorefrigerants is higher for nanoparticles with a larger diameter [2]. Several experiments (e.g., those by Mahbubul et al. [27] and Alawi et al. [30]) confirmed that as the nanoparticle volume percentage rises or/and the temperature falls, the dynamic viscosity of nanorefrigerant enriches. The viscosity is also strongly influenced by the nanoparticle shape. The nanoparticle having a larger ratio of its surface area to volume yields higher viscosity due to higher friction at the solid—fluid interface. The influences of various issues on the thermal conductivity of nanorefrigerants have been extensively examined [2], and the outcomes can be stated as follows: (i) the thermal conductivity of nanorefrigerant enriches with a rise in volume fraction of nanoparticle, (ii) the thermal conductivity of nanorefrigerant enriches with fall in the size of nanoparticle and enrich in aspect ratio for cylindrical shaped, and (iii) as the operating temperature rises, the thermal conductivity of nanorefrigerant enriches. The thermal conductivity enrichment is also influenced by nanoparticle clustering and aggregation [2]. As the thermal conductivity of nanorefrigerants is dependent on many parameters, the data-driven intelligent models (machine learning) can be very effective for the prediction [31].

The surface tension is also an essential property of nanorefrigerant as it influences the different phenomena of flow and pool boiling heat transfer behavior. In general, the surface tension of nanorefrigerants increases at higher concentrations of nanoparticles. It can also be noted that the surface tension falls with the addition of surfactant [2]. The surface tension of the nanorefrigerant enriches with an increment in nanoparticle volume percentage and size, whereas it falls with the increase in temperature as the nanoparticles are adsorbed at the liquid—vapor interface. Behabadi et al. [32] noticed experimentally that the surface tension of nanorefrigerant enriches with the addition of nanoparticles which enriches the heat transport.

The use of nanoparticles in refrigerants (nanorefrigerants) tailors the thermophysical and transport characteristics, leading to the improvement of both boiling and condensation heat transfer coefficients, as discussed. The following are some important facts that have been observed during the boiling and condensation of nanorefrigerants:

- Nanoparticles reduce the thickness of the boundary layer and therefore enhance the coefficient of heat transport during boiling and condensation.
- Nanoparticles are settled on the cooling surface in time of condensation leading to the enrichment of the coefficient of heat transport.
- The nanoparticle dispersion enriches the surface tension of refrigerant, which enriches the wettability and enriches the heat transport process.
- The maximum enrichment of the coefficient of boiling heat transport of nanorefrigerant can be observed for the optimum nanoparticle volume concentration.
- The boiling point temperature of refrigerant increases by dispersing nanoparticles.

- The evaporation rate of the nanorefrigerant may be enriched or reduced, which depends on the nanoparticle type and its volume concentration.

Overall, the use of nanoparticles with the refrigerant improves the coefficient of heat transport, which is influenced by the volume percentage nanoparticle. It has been found that the coefficient of heat transport of nanorefrigerants flowing inside the threaded tube is better as compared to that in a smoothed tube [33]. However, the enriched heat transport effect of nanorefrigerant inside the threaded tube is lower with respect to the smoothed tube for the large mass flow rate. It has also been observed that metallic nanoparticle-dispersed refrigerants show better heat transfer enhancement as compared to ceramic nanoparticle-dispersed refrigerants with lower vapor quality, whereas the effect of ceramic nanoparticles dispersed in nanorefrigerants is better with higher dryness fraction. In fact, the effect of dispersing nanoparticles in the refrigerants changes with vapor quality. In general, the coefficient of boiling heat transport enriches with a rise in nanoparticle volume concentration and vapor quality, whereas the coefficient of heat transport enhancement decreases with the rise in the flow rate of mass of nanorefrigerant [34]. It has also been found that spherical nanoparticle-dispersed refrigerant yields superior boiling heat transfer characteristics than the nonspherical-shaped nanoparticle-dispersed refrigerant may be because of aggregation, instabilities, and partial sedimentation of the nonspherical nanoparticles. On the other hand, the pressure drop of nanorefrigerant is higher with respect to base refrigerant for a specified flow rate of mass and the range of increment of the drop of pressure enriches with the increase in nanoparticle volume concentration. For the long-duration utilization of nanorefrigerant for flow boiling in the smooth tube, the nanoparticles deposit on the wall and hence the nanoparticle interaction near the wall promotes extra pool boiling, which leads to an increase in the coefficient of boiling heat transport; however, the pressure drop due to the fluid friction also increases [35].

The use of nanorefrigerants enhances the coefficients of both the boiling and condensation heat transports, which gives the advantages of compactness and light-weight refrigeration and heat pump devices and also compressor energy reduction, leading to the superior energy efficiency of the system. However, the nanolubricant–refrigerant mixture may degrade due to the long-period dispersion of nanoparticles. This effect is less at the lower nanoparticle concentration, greater mass fraction of lubricant, and reduced heating or/and cooling temperature [2]. Fig. 3.5 shows the layout of a vapor compression refrigerating device where nanorefrigerant is used to

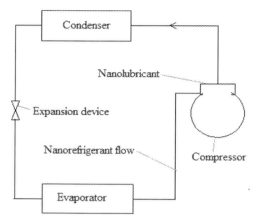

Figure 3.5 Use of nanorefrigerant and nanolubricant in vapor compression refrigeration cycle.

improve performance. The nanorefrigerant has been used in the refrigerator. The experiment illustrated that the performance of the refrigerator improved by 26.1% with TiO_2 nanoparticles of 0.1% mass percentage dispersed in R134a. The use of R410a-based nanorefrigerant illustrated that the ratio of energy efficiency of the refrigeration system can be enriched by 6%. The use of nanorefrigerant can also reduce energy consumption. Due to the improvement of heat transport characteristics in the evaporating equipment, the cooling capacity improves by using nanorefrigerant, which leads to the enrichment in the COP of the refrigeration system. It has been observed that the suction pressure, suction temperature, and discharge pressure decrease with the increase in nanoparticle volume concentration. Several experimental studies on refrigeration systems using various nanorefrigerants using different combinations of refrigerants and nanoparticles have been conducted and reported significant performance improvement [2]. By using the nanorefrigerant in the heat pump system also, the COP improves for both the cooling mode and heating mode [36]. It has been observed recently that the cryogenically treated nanopowder-dispersed refrigerant yields better performance improvement of refrigeration systems [37]. The COP of the refrigeration and heat pump systems using nanorefrigerant has been found to be improving with the rise in volumetric flow rate. A substantial temperature increase in the evaporator (4.69%—39.30%) and the temperature decrease in the condenser (3.0%—23.77%) have also been noticed for the refrigerating device in time investigation [38]. Recently, it has been observed that the nanoparticle dimension has no significant effect on the capability of the refrigerating device at the lower nanoparticle volume percentage and the overall refrigeration system performance deteriorates beyond a certain volume concentration of nanoparticle [39]. The discharge temperature of the compressor reduces with the utilization of nanorefrigerant in the refrigerating device, which will enrich the compressor life also [40]. The recent cost analysis of a heat pumping device with nanorefrigerant showed that the running cost decreases due to an increase in COP, but the initial system cost increases due to the cost of nanorefrigerant, resulting in a payback period that can be shortened under favorable conditions [36].

4. Nanolubricants

The lubricant is necessary for the compressor for the refrigeration system to decrease the losses because of friction and hence the power consumption, which will enrich the system's capability. Apart from the lubrication property, the interaction with the electrical circuit (related to the electrical conductivity) is also an important issue, particularly for the refrigeration system with a hermetic compressor. For the miscible lubricating oil, which flows with the refrigerant, the heat transfer characteristic is another important issue. The nanoparticle-dispersed lubricant is called nanolubricant, which can be prepared by dispersing the suitable nanoparticle in the lubricant or lubricating oil (as base fluid). One can see in Fig. 3.5, the miscible nanolubricant flows along with the refrigerant; otherwise, a separate flow circuit is used. The use of nanolubricant in the refrigeration device showed improvement in terms of lubrication, electrical and heat transport properties. As the lubricants are in liquid form at the ambient condition, the conventional nanofluids preparation processes (either single-step or two-step process), as discussed in Section 2, are used to prepare nanolubricants and also the similar characterization techniques are applicable to check the stability. The thermophysical properties of nanolubricants also have a similar trend as for nanofluids. The specific heat capacity of the nanolubricant increases linearly with the rise in temperature and decreases linearly with the rise in the mass fraction of the nanoparticle. Similar to the conventional refrigerant, the nanoparticle size does not alter the specific heat capacity of the nanolubricant. However, the mass fraction weighted model is better as compared to that model based on volume fraction to estimate the specific heat capacity of the nanolubricants [41]. Both thermal conductivity and dynamic viscosity increase by using nanolubricants as we have seen before. However, both the dynamic viscosity and thermal conductivity of nanolubricants increase with a rise in particle concentration and a decrease in temperature [42]. For the miscible nanolubricant, the thermophysical properties may influence the heat transport characteristics of the lubricant-refrigerant mixture in the refrigerating device.

The dispersion of nanoparticles in lubricant enriches the solubility of the lubricant in the refrigerant and also enhances the capability of heat transport. The solubility enhancement with refrigerant reduces the lubricant deposition on the condenser or other component walls, which can reduce the performance deterioration of the component and enhance its life [2]. Hence, the compressor life will increase. The improvement of solubility with refrigerant by the mixing of nanoparticles with the lubricant can also increase the oil return ratio. The friction coefficient also reduces by dispersing nanoparticles leading to a decrease in wear rate [2]. By dispersing the nanoparticle in the lubricant, the tribological performance enhances and hence, the capability of the compressor in the vapor compression refrigerating and heat pump systems can be improved with the use of nanolubricants. A disadvantage of using nanolubricant has been found in the case of rotatory compressors in refrigerating and heat pump systems, as the reduction of the system COP has been found as compared to base lubricant [2].

Various electrical properties, such as electrical conductivity, dielectric parameters, and breakdown behaviors, are also important issues of nanolubricant used in refrigeration and heat pump systems. There is a possibility of leakage of nanolubricant in the lubrication-related components during circulation through the compressor, which can come in contact with the electrical circuit. Furthermore, the mixture of refrigerant and lubricant has direct contact with the electrical motor in the case of a hermetic compressor in the refrigerating and heat pump devices. Hence the electrical properties are also very important for the nanolubricants. Researchers have seen that the utilization of nanoparticles in lubricating oil improves the electrical conductivity and dielectric properties, which enriches with the rise in nanoparticle fraction and temperature [2]. Hence, the breakdown strength improves, which increases the compressor's safety and durability by using nanolubricant [43].

Many studies showed that the fall of discharge temperature in compressor and compressor energy consumption would enrich the COP of refrigerating and heat pump devices by using nanolubricant [2]. For example, Sharif et al. [44] experimented with the capability enhancement of an air-conditioning device by utilizing SiO_2/PAG nanolubricant in the compressor. They reported that the average COP enriches by 10.5%. They also noticed that the COP is maximum at 0.05% nanoparticle volume percentage. Yang et al. [45] also utilized the graphene-dispersed lubricant in the household refrigerator. They noticed a fall in the temperature of compressor discharge, an improvement of cooling capacity by up to 4.7%, and a fall in energy consumed in the compressor by up to 20.3%. The utilization of a nanolubricant-refrigerant mixture in a vapor compression refrigerating device also increases the subcooling degree at the entry of the expansion device [46], which is the reason behind the improvement of COP using nanolubricant.

5. Nanoabsorbents

There are two important applications of liquid absorbent in the field of refrigerating and heat pumping and air-conditioning: (i) vapor absorption refrigeration system, where the absorbent is used to absorb the refrigerant vapor in the absorber and desorb in the generator and (ii) liquid desiccant-based air-conditioning device, where the desiccant (absorber) is used to absorb and desorb of water vapor. Fig. 3.6 shows the utilization of nanoabsorbent in the vapor absorption refrigerating device. The dispersion of nanoparticles in the liquid absorber (called a nanoabsorbent) may improve the absorption and desorption properties and hence enhance the capability of the system. The use of nanoabsorbent

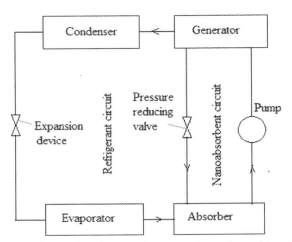

Figure 3.6 Use of nanoabsorbent in the vapor absorption refrigeration cycle.

enriches the absorption mass and heat transport capability, which leads to improvement of the capability of the absorption refrigerating device [2]. As the absorbents are in liquid form at the ambient condition, the conventional nanofluids preparation processes (either single-step or two-step process), as discussed in Section 2, are used to prepare the nanoabsorbents and also the similar characterization techniques are applicable to check the stability. For the nanoabsorbent also, both thermal conductivity and dynamic viscosity enrich; however, the viscosity can be reduced by adding the surfactant [47].

Zhang et al. [48] conducted an experiment on lithium bromide–water binary fluid by dispersing various nanoparticles and concluded that: with a rise in the rate of solution flow, a rise in the mass fraction of nanoparticle, and a decrease in nanoparticle size, the rate of water vapor absorption enriches. They also achieved a very good increment by adding metal nanoparticles. Wang et al. [49] also conducted a similar experiment with a lithium bromide–based nanoabsorbent and reported the enrichment of the coefficient of mass transport up to 41%. The cause behind the enrichment of the coefficient of mass transport is as follows: the nanoparticles randomly move and collide with each other in the absorbent (base fluid) due to the Brownian effect and hence the field disturbance caused by the nanoparticle movements causes the microconvection which enhances the mass transfer. With the use of nanoabsorbent, COP of the vapor-absorption device improves and it has been reported the COP improvement is up to 27% [50]. The vapor absorption rate of nanoabsorbents can be further improved by applying a magnetic field [51]. Modi et al. [52] compared two nanoabsorbents for a solar-powered absorption device and found better capability with lithium bromide absorbent. However, the exergy efficiency of vapor absorption decreases by using nanoabsorbent, and the largest part of the exergy destruction is in the rectifier and solution-side internal heat exchanger [53].

6. Miscellaneous applications

The nanoparticle dispersion in various base fluids has great application potential in the food industries, particularly for food processing, food packaging, and milk pasteurization [2]. The nanofluids can be used in the ice-making process, and it has been observed that the process of nucleation on the surface of metal can be enhanced by using nanofluids. When nanoparticles are dispersed in the process of ice formation, the degree of supercooling increases, which protects the ice adhesion on the cooling surfaces and also enhances the rate of nucleation. Hence the capability of ice-producing units will be increased by using nanofluid. Longo et al. [54] developed a modified milk dispenser utilizing nanofluidic technology. Their innovative milk dispenser is made by a tank covered by a serpentine tube jacket with the circulation of Al_2O_3—ethylene glycol nanofluid through it. Their experiment showed that the innovative milk disposer yields 63% —70% energy savings as compared to the conventional one. Their modified tank design using nanofluid saves energy consumption and enhances food safety.

Another important application of nanofluids is nanoparticle-enhanced PCM (phase-change material)-based cold storage in refrigerating, heat pumping, and air-conditioning devices. The cold storage capability can be improved by using nanofluid in two ways: (i) the heat transfer enhancement due to improved thermophysical and transport properties and (ii) the fast crystal formation during the solidification process of PCM [55]. The mixing of any additives (here, nanoparticles) in the base liquid reduces the freezing point. Both the nucleation time and supercooling degree decrease, while the nanoparticle concentration rises in the liquid and hence both absorption and desorption rates of latent heat are enhanced which reduces the power consumption. However, the nanoparticle volume concentration needs to be optimized by considering the trade-off between the performance and cost parameters. Solidification and melting times can be reduced by utilizing nanoparticle-enhanced PCM because of the enriched heat capacity rate, which offers smart cooling/heating in refrigerating, heat pumping, and air-conditioning devices [56]. Furthermore, the use of nanoparticle-enhanced PCM increases the chiller cooling capability and reduces the compressor energy requirement and hence the COP of the refrigerating device can be increased. This leads to the reduction of system irreversibility also. To increase the nucleation rate and dispersion stability, the gum acacia (GA) can be used as a nucleating agent along with nanoparticles [57], which will significantly reduce the degree of supercooling. Hence the nanoparticle-enhanced PCM has a strong energy-saving potential for cold thermal energy storage.

7. Challenges and future scope

The nanofluids can be utilized in refrigerating, heat pumping, and air-conditioning devices in different roles (secondary fluid, refrigerant, lubricant, absorbent, etc.) to improve the performance, which leads to significant energy-saving potential. However, it suffers from various challenges, which have to be overcome before commercial use. Various challenges, along with future research and development scopes, are listed here as follows:

- The main challenge with nanofluids is their stability for long-term use. Frequent changes in nanofluid will increase the running cost of the system, which is not compensated by energy-saving costs in the present scenario. Stability is essential to avoid fouling and sedimentation, which can impair the nanoparticle transport in the device. Hence major research and development effort is needed to enrich the stability of nanofluids.
- With the present market cost of nanoparticles, the payback period is considerably higher for nanofluids. Hence, a research effort is needed to decrease nanoparticle costs, so that the advantage of using nanofluids can be realized.
- Preparation and characterization of nanorefrigerants based on low normal boiling point refrigerants are more challenging issues, and therefore, more research efforts are required.
- In view of environmental concerns, the leakage of nanofluids may pollute the environment. Hence this issue has to be considered in the future.
- Various brine solutions are used as the secondary refrigerant in low-temperature applications, and there is an excellent opportunity to use nanobrines. Hence, extensive research and development efforts are needed for these brine-based nanofluids.
- The nanoparticle can clog the particle filters and expansion devices and deposit them on other component surfaces of refrigeration and heat pump systems, which can be dangerous to the device operation. Furthermore, the agglomeration of the nanoparticles may aggravate the friction and wear on the moving surface, leading to damage.

Nomenclature

c_p Specific heat capacity
k Thermal conductivity
ϕ Nanoparticle volume concentration
μ Dynamic viscosity
ρ Fluid density

Subscripts
bf base fluid
nf nanofluid
np nanoparticle

References

[1] R.V. Pinto, F.A.S. Fiorelli, Review of the mechanisms responsible for heat transfer enhancement using nanofluids, Appl. Therm. Eng. 108 (2016) 720–739.

[2] A. Bhattad, J. Sarkar, P. Ghosh, Improving the performance of refrigeration systems by using nanofluids: a comprehensive review, Renew. Sustain. Energy Rev. 82 (2018) 3656–3669.

[3] D.K. Devendiran, V.A. Amirtham, A review on preparation, characterization, properties and applications of nanofluids, Renew. Sustain. Energy Rev. 60 (2016) 21–40.

[4] V. Kumar, J. Sarkar, Research and development on composite nanofluids as next-generation heat transfer medium, J. Therm. Anal. Calorim. 137 (2019) 1133–1154.

[5] A.K. Sharma, A.K. Tiwari, A.R. Dixit, Rheological behaviour of nanofluids: a review, Renew. Sustain. Energy Rev. 53 (2016) 779–791.

[6] W.H. Azmi, K.V. Sharma, R. Mamat, G. Najafi, M.S. Mohamad, The enhancement of effective thermal conductivity and effective dynamic viscosity of nanofluids — a review, Renew. Sustain. Energy Rev. 53 (2016) 1046–1058.

[7] M. Sahu, J. Sarkar, Steady state energetic and exergetic performances of single phase natural circulation loop with hybrid nanofluids, J. Heat Tran. 141 (2019) 082401.

[8] M.T. Nitsas, I.P. Koronaki, Investigating the potential impact of nanofluids on the performance of condensers and evaporators — a general approach, Appl. Therm. Eng. 100 (2016) 577–585.

[9] S. Upadhyay, L. Chandra, J. Sarkar, A generalized Nusselt number correlation for nanofluids and look-up diagrams to select a heat transfer fluid for medium temperature solar thermal applications, Appl. Therm. Eng. 190 (2021) 116469.

[10] A. Bhattad, J. Sarkar, P. Ghosh, Experimentation on effect of particle ratio on hydro-thermal performance of plate heat exchanger using hybrid nanofluid, Appl. Therm. Eng. 162 (2019) 114309.

[11] J. Sarkar, Performance of nanofluid -cooled shell and tube gas cooler in transcritical CO_2 refrigeration systems, Appl. Therm. Eng. 31 (2011) 2541–2548.

[12] R.P. Yee, C.J.L. Hermes, A thermodynamic study of water-based nanosuspensions as secondary heat transfer fluids in refrigeration systems, Int. J. Refrig. 89 (2018) 104–111.

[13] S. Askari, R. Lotfi, A. Seifkordi, A.M. Rashidi, H. Koolivand, A novel approach for energy and water conservation in wet cooling towers by using MWNTs and nanoporous graphene nanofluids, Energy Convers. Manag. 109 (2016) 10–18.

[14] F. Ahmed, W.A. Khan, Efficiency enhancement of an air-conditioner utilizing nanofluids: an experimental study, Energy Rep. 7 (2021) 575–583.

[15] A.A. Vasconcelos, A.O. Gómez, E.P.B. Filho, J.A.R. Parise, Experimental evaluation of SWCNT-water nanofluid as a secondary fluid in a refrigeration system, Appl. Therm. Eng. 111 (2017) 1487–1492.

[16] A. Bhattad, J. Sarkar, P. Ghosh, Exergetic analysis of plate evaporator using hybrid nanofluids as secondary refrigerant for low-temperature applications, Int. J. Exergy 24 (2017) 1–20.

[17] A. Bhattad, J. Sarkar, P. Ghosh, Energy-economic analysis of plate evaporator using brine based hybrid nanofluids as secondary refrigerant, Int. J. Air-Cond. Refrig. 26 (2018) 1850003.

[18] Z.T. Tabari, S.Z. Heris, M. Moradi, M. Kahani, The study on application of TiO_2/water nanofluid in plate heat exchanger of milk pasteurization industries, Renew. Sustain. Energy Rev. 58 (2016) 1318–1326.

[19] A. Bhattad, J. Sarkar, P. Ghosh, Energetic and exergetic performances of plate heat exchanger using brine based hybrid nanofluid for milk chilling application, Heat Tran. Eng. 41 (2020) 522–535.

[20] M.S. Ahmed, A.M. Elsaid, Effect of hybrid and single nanofluids on the performance characteristics of chilled water air conditioning system, Appl. Therm. Eng. 163 (2019) 114398.

[21] F.A. Boyaghchi, M. Mahmoodnezhad, V. Sabeti, Exergoeconomic analysis and optimization of a solar driven dual evaporator vapor compression-absorption cascade refrigeration system using water/CuO nanofluid, J. Clean. Prod. 139 (2016) 970–985.

[22] E. Bellos, C. Tzivanidis, Performance analysis and optimization of an absorption chiller driven by nanofluid based solar flat plate collector, J. Clean. Prod. 174 (2018) 256–272.

[23] E. Cuce, T. Guclu, P.M. Cuce, Improving thermal performance of thermoelectric coolers (TECs) through a nanofluid driven water to air heat exchanger design: an experimental research, Energy Convers. Manag. 214 (2020) 112893.

[24] Y.H. Diao, C.Z. Li, Y.H. Zhao, Y. Liu, S. Wang, Experimental investigation on the pool boiling characteristics and critical heat flux of Cu-R141b nanorefrigerant under atmospheric pressure, Int. J. Heat Mass Transf. 89 (2015) 110–115.

[25] D. Yang, B. Sun, H. Li, X. Fan, Experimental study on the heat transfer and flow characteristics of nanorefrigerants inside a corrugated tube, Int. J. Refrig. 56 (2015) 213–223.

[26] S. Tazarv, M.S. Avval, F. Khalvati, E. Mirzaee, Z. Mansoori, Experimental investigation of saturated flow boiling heat transfer to TiO_2/R141b nanorefrigerant, Exp. Heat Tran. 29 (2016) 188–204.

[27] I.M. Mahbubul, A. Saadah, R. Saidur, M.A. Khairul, A. Kamyar, Thermal performance analysis of Al_2O_3/R-134a nanorefrigerant, Int. J. Heat Mass Transf. 85 (2015) 1034–1040.

[28] H. Peng, L. Lin, G. Ding, Influences of primary particle parameters and surfactant on aggregation behavior of nanoparticles in nanorefrigerant, Energy 89 (2015) 410–420.

[29] L. Lin, H. Peng, G. Ding, Dispersion stability of multi- walled carbon nanotubes in refrigerant with addition of surfactant, Appl. Therm. Eng. 91 (2015) 163–171.

[30] O.A. Alawi, N.A.C. Sidik, Influence of particle concentration and temperature on the thermophysical properties of CuO/R134a nanorefrigerant, Int. Commun. Heat Mass Transf. 58 (2014) 79–84.

[31] S. Zhang, Y. Li, Z. Xu, C. Liu, Z. Liu, Z. Ge, L. Ma, Experimental investigation and intelligent modeling of thermal conductivity of R141b based nanorefrigerants containing metallic oxide nanoparticles, Powder Technol. 395 (2022) 850–871.

[32] M.A. Akhavan-Behabadi, M.K. Sadoughi, M. Darzi, M. Fakoor-Pakdaman, Experimental study on heat transfer characteristics of R600a/POE/CuO nano-refrigerant flow condensation, Exp. Therm. Fluid Sci. 66 (2015) 46–52.

[33] D. Yang, B. Sun, H. Li, C. Zhang, Y. Liu, Comparative study on the heat transfer characteristics of nano-refrigerants inside a smooth tube and internal thread tube, Int. Commun. Heat Mass Transf. 113 (2017) 538–543.

[34] B. Sun, H. Wang, D. Yang, Effects of surface functionalization on the flow boiling heat transfer characteristics of MWCNT/R141b nanorefrigerants in smooth tube, Exp. Therm. Fluid Sci. 92 (2018) 162–173.

[35] P.S. Deokar, L. Cremaschi, Effect of nanoparticle additives on the refrigerant and lubricant mixtures heat transfer coefficient during in-tube single-phase heating and two-phase flow boiling, Int. J. Refrig. 110 (2020) 142–152.

[36] G. Kosmadakis, P. Neofytou, Investigating the effect of nanorefrigerants on a heat pump performance and cost-effectiveness, Therm. Sci. Eng. Prog. 13 (2019) 100371.

[37] D. Senthilkumar, Influence of cryogenic treatment on TiC nanopowder in R600a and R290 refrigerant used in vapor compression refrigeration system, Int. J. Air-Cond. Refrig. 27 (2019) 1950040.

[38] L. Kundan, K. Singh, Improved performance of a nanorefrigerant-based vapor compression refrigeration system: a new alternative, Proc. Inst. Mech. Eng. A: J. Power Energy 235 (2021) 106−123.

[39] P. Dey, B.K. Mandal, Performance enhancement of a shell-and-tube evaporator using Al_2O_3/R600a nanorefrigerant, Int. J. Heat Mass Tran. 170 (2021) 121015.

[40] D.K. Singh, S. Kumar, S. Kumar, R. Kumar, Potential of MWCNT/R134a nanorefrigerant on performance and energy consumption of vapor compression cycle: a domestic application, J. Braz. Soc. Mech. Sci. Eng. 43 (2022) 540.

[41] L. Lin, M.A. Kedzierski, Specific heat of aluminum-oxide nanolubricants, Int. J. Heat Mass Transf. 126 (2018) 1168−1176.

[42] M.Z. Sharif, W.H. Azmi, A.A.M. Redhwan, R. Mamat, Investigation of thermal conductivity and viscosity of Al_2O_3/PAG nanolubricant for application in automotive air conditioning system, Int. J. Refrig. 70 (2016) 93−102.

[43] D.F.M. Pico, J.A.R. Parise, E.P.B. Filho, Nanolubricants in refrigeration systems: a state-of-the-art review and latest developments, J. Braz. Soc. Mech. Sci. Eng. 45 (2023) 88.

[44] M.Z. Sharif, W.H. Azmi, A.A.M. Redhwan, R. Mamat, T.M. Yusof, Performance analysis of SiO_2/PAG nanolubricant in automotive air conditioning system, Int. J. Refrig. 75 (2017) 204−216.

[45] S. Yang, X. Cui, Y. Zhou, C. Chen, Study on the effect of graphene nanosheets refrigerant oil on domestic refrigerator performance, Int. J. Refrig. 110 (2020) 187−195.

[46] V. Nair, A.D. Parekh, P.R. Tailor, Experimental investigation of a vapour compression refrigeration system using R134/Nano-oil mixture, Int. J. Refrig. 112 (2020) 21−36.

[47] W. Jiang, K. Du, Y. Li, L. Yang, Experimental investigation on the influence of high temperature on viscosity, thermal conductivity and absorbance of ammonia−water nanofluids, Int. J. Refrig. 82 (2017) 189−198.

[48] L. Zhang, Z. Fua, Y. Liu, L. Jina, Q. Zhang, W. Hu, Experimental study on enhancement of falling film absorption process by adding various nanoparticles, Int. Commun. Heat Mass Transf. 92 (2018) 100−106.

[49] G. Wang, Q. Zhang, Z. Zeng, R. Xu, G. Xie, W. Chu, Investigation on mass transfer characteristics of the falling film absorption of LiBr aqueous solution added with nanoparticles, Int. J. Refrig. 89 (2018) 149−158.

[50] W. Jiang, S. Li, L. Yang, K. Du, Experimental investigation on performance of ammonia absorption refrigeration system with TiO_2 nanofluid, Int. J. Refrig. 98 (2019) 80−88.

[51] S. Wu, C.R. Ortiz, Experimental investigation of the effect of magnetic field on vapour absorption with LiBr−H_2O nanofluid, Appl. Therm. Eng. 193 (2020) 116640.

[52] N. Modi, B. Pandya, J. Patel, Comparative analysis of a solar-driven novel salt-based absorption chiller with the implementation of nanoparticles, Int. J. Energy Res. 43 (2019) 1563−1577.

[53] E.Y. Gurbuz, A. Kecebas, A. Sozen, Exergy and thermoeconomic analyses of the diffusion absorption refrigeration system with various nanoparticles and their different ratios as work fluid, Energy 248 (2022) 123573.

[54] G.A. Longo, G. Righetti, C. Zilio, Development of an innovative raw milk dispenser based on nanofluid technology, Int. J. Food Eng. 12 (2) (2016) 165−172.

[55] A.A.A. Attia, A.A. Altohamy, M.F.A. Rabbo, R.Y. Sakr, Comparative study on Al_2O_3 nanoparticle addition on cool storage system performance, Appl. Therm. Eng. 94 (2016) 449–457.

[56] A.S. Kumar, V. Kumaresan, R. Velraj, Solidification characteristics of water based graphene nanofluid PCM in a spherical capsule for cool thermal energy storage applications, Int. J. Refrig. 66 (2016) 73–83.

[57] P. Sundaram, A. Kalaisselvane, Cold thermal energy storage performance of graphene nanoplatelets–DI water nanofluid PCM using gum acacia in a spherical encapsulation, J. Therm. Anal. Calorim. 147 (2022) 14973–14985.

CHAPTER FOUR

Heat transfer enhancement with ferrofluids

Zouhaier Mehrez[1,2]

[1]Faculty of Sciences of Tunis, Laboratory of Energy, Heat and Mass Transfer (LETTM), Department of Physics, El Manar University, El Manar, Tunisia
[2]Gabes Preparatory Engineering Institute, Gabes, Tunisia

Highlights

- The stability criteria for ferrofluids were examined.
- The different preparation methods for magnetic nanoparticles were presented.
- The thermophysical properties of ferrofluids under the action of an external magnetic field were portrayed.
- The governing equations of FHD were provided.
- The recent advances on heat transfer enhancement in thermal systems using ferrofluids were reported in a state-of-the-art analysis.

1. Introduction

Heat transfer constitutes the predominant mode of energy exchange in numerous engineering applications, including microelectronics, electronic equipment, solar collectors, and heat exchangers. With the ongoing advancements in science and technology, these applications have evolved into smaller, more compact forms, thereby necessitating dependable thermal systems capable of effectively dissipating significant amounts of heat. To enhance heat transfer within these systems, various techniques are employed, with flow control being particularly noteworthy. Flow control hinges on two primary strategies: passive control (without energy addition): this approach involves making subtle modifications to the flow domain's geometry. Such adjustments might encompass the incorporation of fins and turbulators, the smoothing of sharp edges, and alterations to the nature of the exchange surface. Active control (with energy addition): in this strategy, external perturbations are applied to the flow, accompanied by an input of energy. These perturbations encompass actions such as suction, blowing,

Advanced Materials-Based Fluids for Thermal Systems
ISBN: 978-0-443-21576-6
https://doi.org/10.1016/B978-0-443-21576-6.00003-0

Copyright © 2024 Elsevier Inc.
All rights are reserved, including those for text and data mining, AI training, and similar technologies.

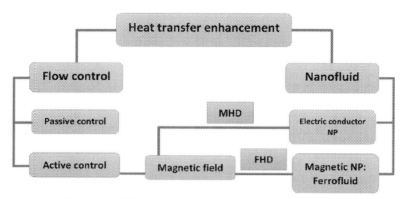

Figure 4.1 Different methods to enhance heat transfer.

the introduction of mechanical and acoustic vibrations, the application of electrical fields, and the utilization of magnetic fields (as illustrated in Fig. 4.1).

Despite the reliability exhibited by various heat transfer enhancement control strategies, their effectiveness remains constrained by system geometry, flow dynamics, and the type of coolant fluid used. To overcome these limitations, researchers have directed their focus over the past two decades toward augmenting the thermophysical properties of coolant fluids. This is achieved by incorporating nanoparticles (NPs) with high thermal conductivity into the base fluid, resulting in a stable colloidal suspension known as a nanofluid. This nanofluid showcases enhanced thermal characteristics in comparison to conventional fluids such as water, oil, and ethylene glycol. Interestingly, a synergistic approach can be employed by harnessing both flow control through magnetic fields and utilizing nanofluids. Particularly, when the nanofluid possesses high electrical conductivity due to metallic nanoparticles, the fluid behavior is altered in the presence of a magnetic field due to the emergence of the magnetohydrodynamic (MHD) phenomenon. Additionally, if the nanoparticles are magnetic (MNPs), the resulting composite is referred to as a ferrofluid. The enhanced magnetic properties of ferrofluids enable interaction with magnetic fields, subsequently inducing alterations in the flow structure through the phenomenon known as ferrohydrodynamics (FHD). This manipulation of flow structure can be finely controlled by adjusting the magnetic source's position and intensity. Consequently, by manipulating these magnetic field parameters, significant enhancements in heat transfer within a ferrofluid flow can be achieved.

The main objective of this chapter is to demonstrate the effects of using ferrofluid in the presence of a magnetic field on heat transfer enhancement. Firstly, the chapter presents the methods used for preparing ferrofluids and the different formulations used to model their thermophysical characteristics. Next, it discusses various mathematical models that deal with FHD in ferrofluid flows. The final section of the chapter provides clarification on heat transfer enhancement in FHD flows through numerical simulations and experimental studies.

2. Preparation of ferrofluid
2.1 Stability of ferrofluids

Ferrofluids can be defined as colloidal suspensions containing ferromagnetic nanoparticles, typically around 10 nm in size, dispersed within a carrier fluid. These suspensions acquire magnetic properties when subjected to an external magnetic field, all the while maintaining their colloidal stability. Ensuring the stability of a ferrofluid is imperative, whether a magnetic field is present or not. The root cause of colloidal instability within a ferrofluid is the tendency of nanoparticles to aggregate, resulting in sedimentation. Consequently, achieving a uniform colloidal suspension necessitates precise control over nanoparticle interactions with the base fluid during ferrofluid synthesis. This stability accounts for the monophasic nature of ferrofluids, which, in turn, leads to uniform flow and deformation patterns under the influence of a magnetic field. On another note, suspended nanoparticles are subject to diverse internal and external forces including gravity, magnetic fields, and Van der Waals forces, among others. These forces have a tendency to induce nanoparticle aggregation, ultimately destabilizing the ferrofluid. However, the presence of thermal agitation acts as a stabilizing factor for the colloidal stability of ferrofluids.

In the process of creating a stable ferrofluid, it becomes imperative to achieve a balance between various energies: the interparticle magnetic interaction energy (E_{PP}), the magnetic interaction energy between particles and an external field (E_{PF}), and the potential energy stemming from gravity (E_{Pot}). This equilibrium is achieved by countering these energies with the energy resulting from thermal agitation (E_{TA}). Utilizing these criteria, it becomes possible to theoretically determine the nanoparticle diameter (d) values that confer stability upon the ferrofluid. Table 4.1, as presented by Petit [1], provides a visual representation of nanoparticle diameters (d) corresponding to different stability conditions. An illustrative case within

Table 4 1 Stability condition of ferrofluids [1].

Stability condition	Nanoparticle diameter formula	Nanoparticle diameter for the soft ferrite
Interparticle magnetic interaction	$d \leq \sqrt[3]{\frac{144 k_B T}{\pi \mu_0 M^2}}$	11.5 nm
Interaction with extern magnetic field	$d \leq \sqrt[3]{\frac{6 k_B T}{\pi \mu_0 H M_S}}$	6.3 nm
Gravity interaction	$d \leq \sqrt[3]{\frac{6 k_B T}{\pi (\rho_P - \rho_f) g h}}$	12.7 nm

this table pertains to soft ferrite $(Mn,Zn)Fe_2O_4$, wherein the nanoparticle diameter (d) is calculated based on material density (ρ_P) of 4.9, saturation magnetization (M_S) of 0.4 T, temperature (T) of 273 K, magnetic field value (H) of 100 mT, and the use of pure water as the carrier fluid, with a maximum altitude (h) of 10 cm representing the height of the container holding the ferrofluid. In the expressions for "d," "k_B" denotes the Boltzmann constant, "μ_0" stands for the vacuum permeability, and "ρ_f" represents water density. Theoretical insights suggest that nanoparticles with diameters smaller than 10 nm are necessary to achieve stable ferrofluids, a goal that proves challenging to accomplish experimentally. To address this predicament, researchers have introduced surfactant coatings on nanoparticles to mitigate the effects of the aforementioned interactions, ultimately leading to the creation of stable colloidal suspensions.

The subsequent section outlines the process of nanoparticle preparation and their suspension within a carrier fluid.

2.2 Synthesis of ferrofluids

There exist two approaches for preparing ferrofluid:

- One-step method involves the vaporization of a solid material under a vacuum, followed by the direct condensation of its vapor into the liquid phase.
- Two-step method entails an initial phase of preparing a dry nanopowder, followed by its subsequent mixing and dispersion within the carrier liquid.

Magnetic nanoparticle (MNPs) synthesis can be accomplished through a variety of techniques, with three primary methods being distinguished: physical, chemical, and biological approaches. Each of these methods can be further subdivided into various strategies for preparation. Fig. 4.2 offers a comparative overview of the diverse approaches used to prepare MNPs [2].

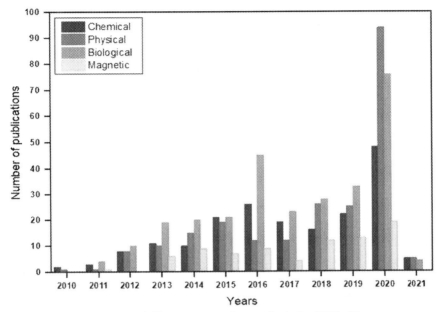

Figure 4.2 Different preparation methods for MNPs [2].

2.2.1 Physical preparation
2.2.1.1 Gas-phase deposition
This approach involves a collection of vacuum deposition techniques used for producing thin films:
- ***Vacuum evaporation:*** The material intended for deposition is vaporized within a sealed vacuum chamber. The vacuum environment enables the particles to directly contact the substrate, where they subsequently recondense into a solid state. This technique facilitates a streamlined one-step preparation of the ferrofluid, as it enables the direct condensation of MNPs into the carrier fluid (Fig. 4.3).
- ***Electron beam evaporation*** represents a type of physical vapor deposition. In this process, a high vacuum target anode is subjected to an electron beam emitted from a charged tungsten filament. The impact of the electron beam causes the molecules within the target material to transition into a gaseous state. These gaseous molecules subsequently condense into a solid form, coating the vacuum chamber with a thin layer of the anode material (Fig. 4.4).
- ***Sputtering*** is a phenomenon characterized by the detachment of particles from a cathode within a low-pressure environment. This technique

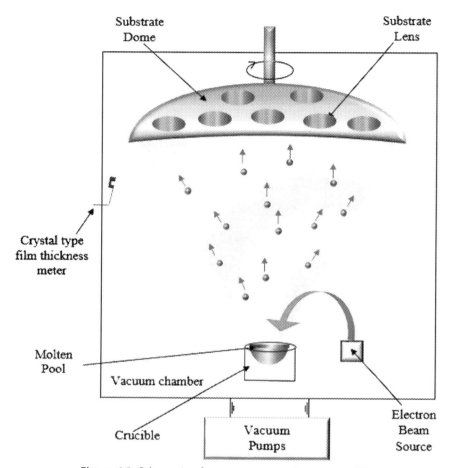

Figure 4.3 Schematic of vacuum evaporation process [3].

enables the creation of various materials through the condensation of metallic vapor originating from a solid target source onto a substrate (Fig. 4.5).

- **Pulsed laser deposition (PLD)**: Atoms and ions are vaporized through the impact of intense laser radiation within a vacuum chamber, leading to their transformation into a plasma plume. This plume is subsequently deposited onto a substrate (Fig. 4.6). Recently, Torres et al. [4] utilized this method to generate Fe_3O_4 nanoparticles. Employing a top-down approach, they initially employed the pulsed laser deposition (PLD) technique to synthesize nanoparticles from an α-Fe target. Subsequently,

Heat transfer enhancement with ferrofluids 67

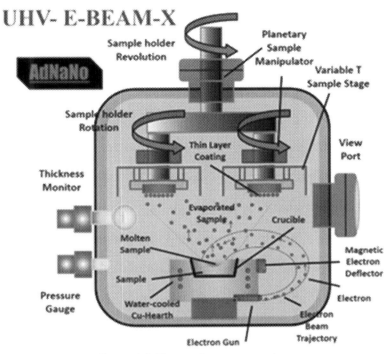

Figure 4.4 Electron beam evaporator.

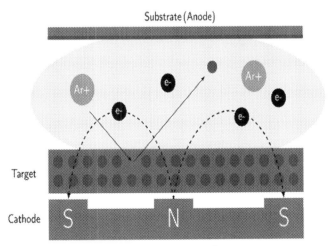

Figure 4.5 Sputtering technique.

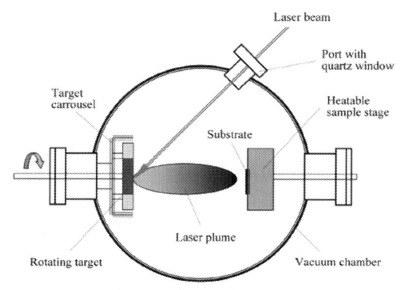

Figure 4.6 Schematic of PLD device.

adopting a bottom–up approach, they utilized a chemical method to produce both coated and pure SiO_2 nanoparticles in the second step.
- *Molecular beam epitaxy*: This technique involves directing one or more molecular beams toward a preselected substrate in order to achieve epitaxial growth (Fig. 4.7)
- *Deposition by electric arc*: Atoms and ions are vaporized by the influence of a robust electric current, generated through an electric discharge between two electrodes with a significant potential difference. This process dislodges metal particles, causing them to transition into a gaseous phase (Fig. 4.8).

2.2.1.2 Laser-induced pyrolysis

This method relies on the interaction between a CO_2 infrared laser beam and a stream of reagents within a controlled atmosphere reactor, where the two flows intersect. The energy transfer results in an increase in temperature within the reaction zone, leading to the dissociation of the precursors and the emergence of a flame. Within this flame, nanoparticles are generated without coming into contact with the reactor walls. The precursors can either be in gaseous or liquid form. In situations involving liquid precursors, an aerosol form of the precursor is introduced into the reactor (Fig. 4.9).

Heat transfer enhancement with ferrofluids

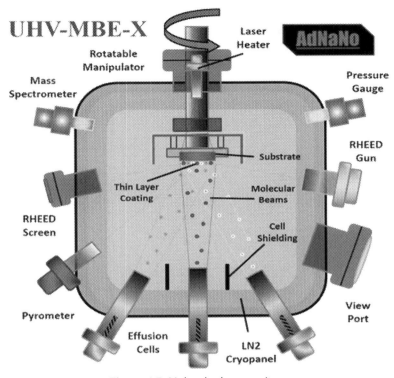

Figure 4.7 Molecular beam epitaxy.

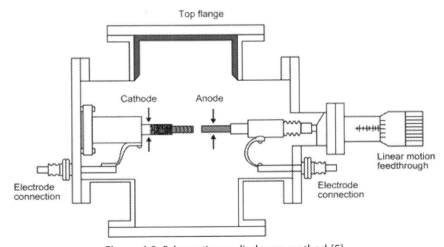

Figure 4.8 Schematic arc discharge method [5].

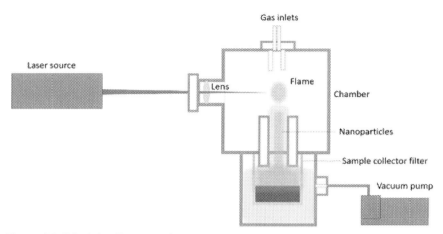

Figure 4.9 Principle of laser pyrolysis: Interaction between an infrared laser beam and a gaseous or liquid precursor [6].

Among the various techniques available for the synthesis of nanomaterials, laser pyrolysis stands out due to its remarkable flexibility and the wide array of compounds it can generate, encompassing diverse chemical compositions, morphologies, and degrees of crystallinity.

The stages of the procedure involve:
- inducing vibrational states within molecules through the absorption of infrared radiation;
- propagation of excitation through collisions, imparting the excited state to all molecules within the medium;
- dissociation of molecules resulting in the creation of a saturated vapor;
- homogeneous nucleation;
- growth of nanoparticles.

2.2.1.3 Powder balls milling

This is a mechanical method where the material is placed with the balls in a rotating drum. The grinding results from the friction and the shock created by the fall of the balls against the product and by the collision of the particles between them. The intensity of the grinding depends mainly on the speed of the drum, the size, and the material of the balls as well as the duration of the stay of the product in the chamber (Fig. 4.10).

2.2.1.4 Flame spray pyrolysis (aerosol)

Flame spray pyrolysis is a method that generates nanoparticles within a gaseous phase using a flame at elevated temperatures. The precursors are

Heat transfer enhancement with ferrofluids 71

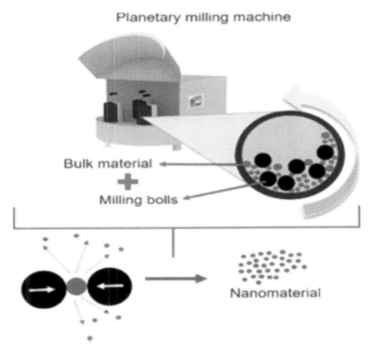

Figure 4.10 Synthesis process of nanomaterials by a planetary milling machine [2].

vaporized and subsequently subjected to combustion. The particle formation process can be divided into four stages: nucleation, condensation, coalescence, and coagulation. Initially, support particles are formed due to their relatively low vapor pressure. Subsequently, as the distance from the flame increases, nanoparticles are generated and swiftly deposit onto the surfaces of the support particles [7] (Fig. 4.11).

2.2.2 Chemical preparation
Various methodologies stem from this approach.

2.2.2.1 Precipitation
This approach is characterized as rapid reactive crystallization, functioning under notably high supersaturation conditions of chemical reagents where the product concentration greatly surpasses the equilibrium constant. This heightened supersaturation serves as the driving force behind precipitation, a fundamental concept that also dictates precipitation kinetics processes (including nucleation, growth, aggregation, and agglomeration). Various

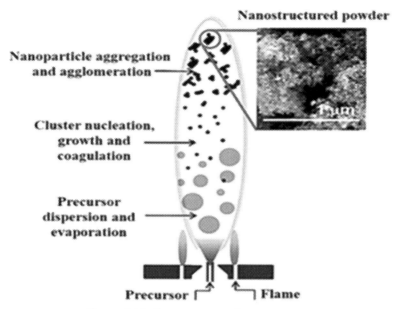

Figure 4.11 *Flame spray pyrolysis process* [7].

techniques employ this method for nanoparticle synthesis, including hydrothermal, coprecipitation, and sol–gel methods. Within the coprecipitation method, agglomeration and growth proceed simultaneously, with the nucleation process unfolding gradually, resulting in the formation of uniform nanoparticles. Venugopal et al. [8] harnessed this method to fabricate cadmium-doped Fe_3O_4 ferrofluids. Similarly, Victory et al. [9] employed coprecipitation to synthesize Fe–Mn ferrite coated with oleic acid. Babykutty et al. [10] employed this technique to create chromium-substituted magnetite ($Cr_xFe_{1-x}Fe_2O_4$) ferrofluids, notable for their optical and antibacterial properties. Fig. 4.12 provides an illustration of the ferrofluid fabrication process implemented by these researchers.

Hydrothermal synthesis encompasses diverse techniques for crystallizing substances within a sealed container, conducted in an aqueous solution under high vapor pressure. This method yields well-crystallized MNPs with substantial saturation magnetization. Furthermore, hydrothermal treatment enables the production of various iron oxide nanoparticles, including nanocubes and hollow spheres. Within this framework, Boštjancic et al. [11] successfully engineered alcohol-containing ferrimagnetic nanoplatelets (NPLs)

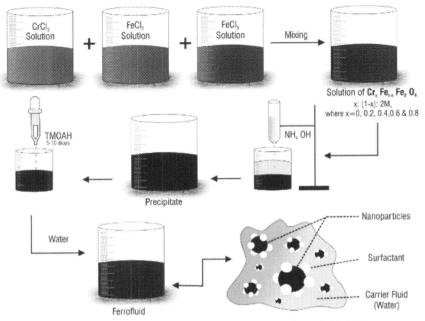

Figure 4.12 Synthesis process of $Cr_xFe_{1-x}Fe_2O_4$ ferrofluids [10].

of barium hexaferrite to create ferrofluids. This process is depicted in Fig. 4.13.

The sol—gel process involves generating a stable suspension (sol) from chemical precursors in a solution. Through interactions between the suspended species and the solvent, these "sols" undergo a transition into a three-dimensional solid network, spreading throughout the liquid medium. This state is referred to as the "freeze" phase. Subsequently, these gels are converted into amorphous dry materials either by removing the solvents within their gaseous or supercritical domains (resulting in aerogels) or by straightforward evaporation under atmospheric pressure (resulting in xerogels) (Fig. 4.14).

2.2.2.2 Thermal decomposition

This technique revolves around the thermal decomposition of a metal complex (precursor) at elevated temperatures, aided by the presence of a ligand (surfactant) in a solvent with a high boiling point. The resulting nanoparticles are coated with this ligand, ensuring their dispersion and stability within organic solvents. The ultimate characteristics of the

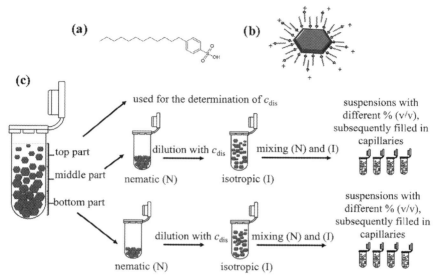

Figure 4.13 (A) DBSA molecular structure, (B) schematic diagram of DBSA double-layer around of NPL, and (C) schematic representation of the preparation sample [11].

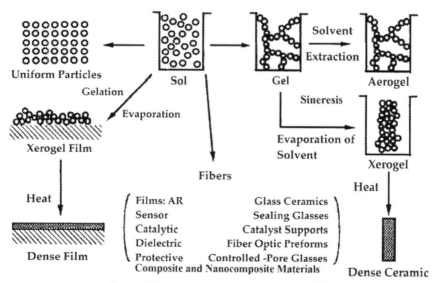

Figure 4.14 Schematic sol–gel process [12].

nanoparticles, such as size and shape, are intricately linked to the manipulation of germination and growth mechanisms through various experimental factors, including temperature, solvent type, ligand-to-precursor

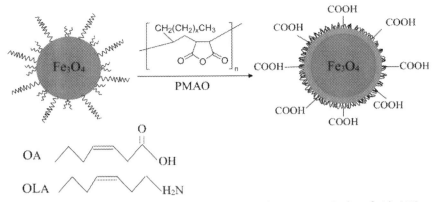

Figure 4.15 Schematic transferring mechanism of PMAO@Fe$_3$O$_4$ ferrofluids [13].

ratio, heating rate, and more. Vuong et al. [13] executed this method using PMAO (maleic anhydride-alt-1-octadecene) as a phase-transfer ligand to synthesize ferrofluids composed of highly stable magnetite nanoparticles. These nanoparticles are designed for applications in magnetic resonance imaging (MRI) and hyperthermia. Fig. 4.15 illustrates the schematic depiction of the phase-transfer mechanism for PMAO@Fe$_3$O$_4$ ferrofluids.

2.2.2.3 Microemulsion

The microemulsion technique employs water droplets as nanoreactors within an oil continuous phase, facilitated by surfactant molecules. Within this method, iron precursors can be precipitated into iron oxide specifically within the aqueous phase, confined within the micelles' central regions. The organic phase does not see iron oxide precipitation due to the inactivity of iron precursors in this context. The nanoparticle size can be tailored by regulating the dimensions of the water droplets. The surfactants responsible for micellization can also serve to disperse iron oxide nanoparticles. Fig. 4.16 presents an illustration of three microemulsion methods: reverse micelle, one microemulsion method, and two-microemulsion method [14]. Pemartin et al. [15] employed the oil-in-water reaction microemulsion technique to craft Mn−Zn ferrite MNPs. More recently, Huang et al. [16] employed the two microemulsion method to investigate the generation of a double-emulsion ferrofluid manipulable under a magnetic field.

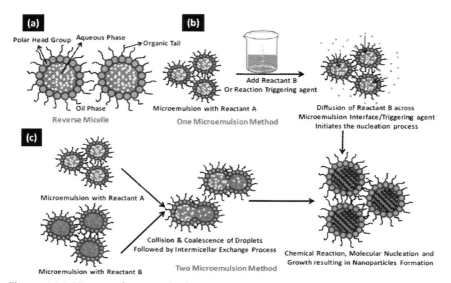

Figure 4.16 Microemulsion method. (A) System of reverse micellar, (B) method of one microemulsion, and (C) method of two microemulsion [14].

2.2.3 Biological preparation

Conventional techniques, encompassing both physical and chemical methods, exhibit diverse drawbacks such as high expenses and the production of hazardous and toxic substances. In pursuit of more economical and environmentally benign pathways for nanoparticle synthesis, researchers have turned toward microorganisms and plant extracts. This approach possesses a notable potential to yield nanomaterials with precise shapes and meticulously controlled structures.

The biological method for nanoparticle preparation involves three primary stages: selecting an appropriate solvent for synthesis, opting for a safe reducing agent for the experimenter, and choosing a nontoxic material to stabilize the nanoparticles.

Typically, bioreduction utilizing plant extracts involves combining an aqueous solution of the plant extract with an aqueous metal salt solution. This reaction, typically swift and usually completed at room temperature, leads to color changes during the reaction process, with the final precipitate color indicating the creation of the desired material. Fig. 4.17, as presented by Barizão et al. [2], offers an illustration of distinct stages in the synthesis of magnetite nanoparticles. The reducing agents utilized here are molecules derived from plant extracts.

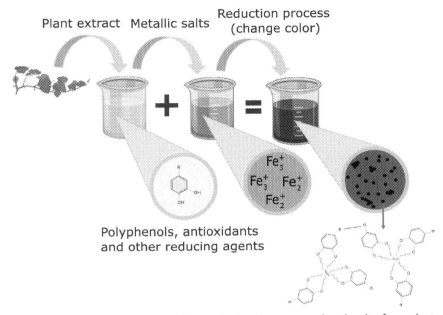

Figure 4.17 Magnetite nanoparticles synthesis using extracted molecules from plants as reducing agents [2].

3. Thermophysical properties of ferrofluids

Similar to a basic nanofluid, the act of incorporating MNPs into a carrier fluid results in heightened thermal conductivity and dynamic viscosity. This process also enhances various other thermal attributes of the fluid, including heat capacity, thermal diffusivity, and the coefficient of thermal expansion. The characteristics of these properties are contingent upon:
- volume fraction of nanoparticles;
- temperature
- carrier fluid nature;
- size of nanoparticles;
- shape of nanoparticles.

Within a ferrofluid devoid of an external magnetic field, the magnetic moments carried by the nanoparticles are distributed randomly, resulting in a total fluid magnetization of zero. When the ferrofluid is exposed to a magnetic field, it attains a nonzero magnetization. The magnetic moments of the particles align themselves with the applied field's direction. This magnetic response significantly influences its thermophysical characteristics.

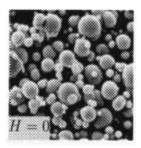 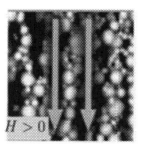

Figure 4.18 Structural behavior of ferrofluid particles under the action of a magnetic field. *Image obtained by electronic microscope [17].*

Consequently, alongside the factors already mentioned, these parameters are also contingent upon magnetic field attributes such as its nature (uniform, nonuniform, and variable), intensity, direction, and frequency (applicable in the case of alternating fields).

Fig. 4.18 depicts the response of magnetic particles within a ferrofluid when subjected to a magnetic field. As evident, in the presence of a magnetic field (H > 0) [17], the nanoparticles, acting as magnetic dipoles, align themselves in accordance with the field lines, forming chains of particles known as aggregates. Consequently, the entire fluid acquires a magnetic attribute and becomes polarized. This transformation leads to alterations in the fluid's homogeneity and isotropy. Consequently, the thermohydraulic attributes of the fluid can be fine-tuned based on the intensity and orientation of the applied magnetic field.

Mohammadfam et al. [18] conducted an experimental investigation into the rheological behaviors of Fe_3O_4/hydraulic oil magnetic nanofluids. Their findings indicated that elevating the volume fraction of nanoparticles leads to an increase in the kinematic viscosity of the oil. Additionally, the rheological characteristics of the ferrofluid shift from being Newtonian to non-Newtonian, exhibiting dilatant behavior. It was observed that the viscosity of the ferrofluid can increase for specific combinations of nanoparticle fraction and magnetic field intensity. Lei et al. [19] utilized a two-step method to prepare Fe_3O_4/water ferrofluids aiming for heat transfer enhancement. They explored various surfactants, nanoparticle volume fractions, and magnetic field orientations. The results revealed that the surfactant tetramethylammonium hydroxide (TMAH) provided the best ferrofluid stability. Across all cases, both viscosity and thermal conductivity rose with increasing nanoparticle volume fraction. Under parallel magnetic field direction, the heat transfer rate notably exceeded that under perpendicular magnetic field direction. However, the heat transfer rate in the absence of an applied magnetic field fell in between these values. Li et al. [20] delved into the rheological properties of a perfluoropolyether-based ferrofluid. They highlighted that these properties are contingent upon shear rate and magnetic field strength. The magnetoviscous effect substantially diminishes with higher shear rates. In a theoretical study, Bhandari et al. [21] revealed that the ferrofluid's viscosity decreases with increased magnetic core diameter and rises with surfactant thickness. This viscosity also relies on magnetic field characteristics like strength, steadiness, and frequency. Siebert et al. [22] explored the impact of introducing small cobalt ferrite nanoparticles into a water-based ferrofluid. They elucidated that the addition of these small nanoparticles to the base fluid augments the magnetoviscous effect. This effect arises from the heightened interaction of large particles within the ferrofluid, coupled with the stabilization of chain-like agglomerates facilitated by the presence of small nanoparticles.

This leads to a reduction in shear thinning when subjected to an external magnetic field.

In a ferrofluid, the aggregation phenomenon of nanoparticles triggered by an external magnetic field leads to an augmentation in heat diffusion and consequently thermal conductivity. This outcome hinges upon the nanoparticles' size and proportion as well as the intensity of the magnetic field. Researchers have demonstrated that thermal conductivity increases up to a maximum point at a critical magnetic field strength, after which it diminishes. This phenomenon is attributed, following Shima et al. [23], to the escalating aspect ratio of magnetic field-induced aggregates with the increasing field intensity. Beyond the critical field strength, the decrease in thermal conductivity is primarily ascribed to zippering transitions, which reduce the aspect ratio of the aggregates. Through their experimental study of kerosene-based magnetite nanofluids, these researchers also observed that this trend becomes more pronounced with larger particle sizes, ranging from 2.8 to 9.5 nm. This enlargement in particle size elevates dipolar moments and dipole—dipole interactions between nanoparticles, which induce aggregate formation under the influence of an applied magnetic field. In a separate study involving a ferrofluid based on Fe_3O_4 nanoparticles (6.7 nm) dispersed in kerosene oil, Philip et al. [24] confirmed the thermal conductivity behavior in the presence of a magnetic field. They expounded that the maximum enhancement occurs when the magnetic field strength aligns with the saturation magnetization of the ferrofluid. These same authors,

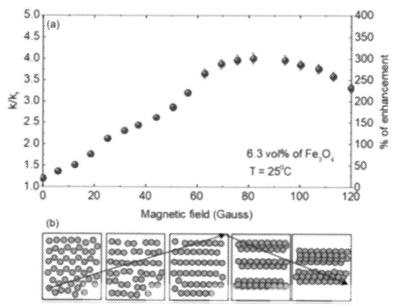

Figure 4.19 (A) The ratio of effective thermal conductivity (k/k$_f$) and its corresponding enhancement rate with respect to magnetic field intensity. (B) Schematic depiction illustrating the varying structures of chain-like formations within the ferrofluid for diverse magnetic field intensities. The arrow denotes the shift in thermal conductivity due to the strengthening of the magnetic field [25].

in Ref. [25], established that the enhancement in thermal conductivity due to a magnetic field is attributed to the formation of chainlike structures. These structures act as high-conductivity heat pathways when dipolar interaction energy surpasses thermal energy (refer to Fig. 4.19). Their findings showcased that adding 7.8 vol% of Fe$_3$O$_4$ enhances thermal conductivity by 23%, and this percentage can escalate to 300% in the presence of a magnetic field with 6.3 vol% of particles. In a research paper analyzing the thermal conductivity behavior of water-based hematite (Fe$_2$O$_3$) and magnetite (Fe$_3$O$_4$) ferrofluids, Karimi et al. [26] revealed the time-dependent characteristic of thermal conductivity under a uniform magnetic field. This trend results from the growth of chain-like cluster size over time, leading to a sharp increase in thermal conductivity until reaching a peak value at magnetization saturation (refer to Fig. 4.20). As indicated by the review papers of Kole et al. [27] and Alsaady et al. [28], a variety of researchers, through experimental studies, have validated that the enhancement in thermal conductivity is

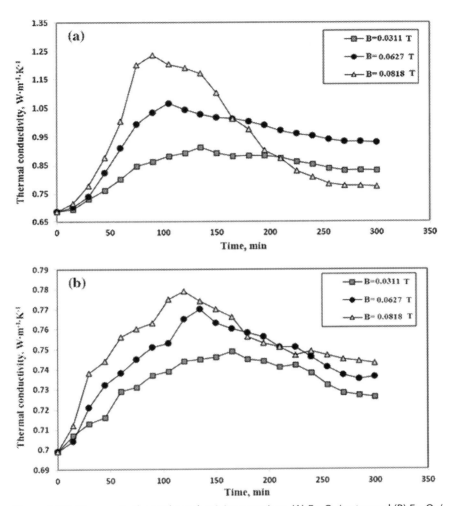

Figure 4.20 Changes in thermal conductivity over time. (A) Fe$_2$O$_3$/water and (B) Fe$_3$O$_4$/water [26].

achieved when heat flux and the direction of an applied magnetic field are parallel.

Despite the considerable volume of published research providing models for the thermophysical properties of nonmagnetic nanofluids, there is a limited number of articles addressing the same topic for ferrofluids. Within this context, Ajith et al. [29] embarked on preparing a stable disk-shaped MgFe$_2$O$_4$/water ferrofluid through a hydrothermal procedure. Their aim was to investigate and formulate models for its thermophysical properties in the presence of a magnetic field. Their findings unveiled notable enhancements in viscosity, density, and thermal conductivity, reaching up to 28.31%, 5.33%, and 13.92%, respectively. Drawing from their experimental data, the authors proposed correlations for the thermal conductivity, density, and viscosity of the magnesium ferrite ferrofluid both in the absence and presence of a magnetic field. These correlations are structured to account for the magnetic flux (M), which varies from 50 to 350 Gauss when the magnetic field is applied. These relationships are expressed as follows:

Effective thermal conductivity, measured at temperature $T = 298$K.

✔ Without magnetic field:

$$\frac{k_f}{k_b} = 1.0522\left[(1 + \varphi)^{0.2241}\right] \tag{4.1}$$

✔ With magnetic field:

$$\frac{k_f}{k_b} = 1.04522\left[(1 + \varphi)^{0.2365}\left(1 + \frac{M}{M_{max}}\right)^{0.053218}\right] \tag{4.2}$$

- Effective density, measured for a temperature ranging from 298 to 343 K.

  ✔ Without magnetic field:

$$\frac{\rho_f}{\rho_b} = 1.002\left[(1 + \varphi)^{0.265}\left(\frac{T}{T_{max}}\right)^{-0.0036}\right] \tag{4.3}$$

  ✔ With magnetic field:

$$\frac{\rho_f}{\rho_b} = 0.99246\left[(1 + \varphi)^{0.262}\left(\frac{T}{T_{max}}\right)^{-0.00615}\left(\frac{M}{M_{max}}\right)^{-0.0042}\right] \tag{4.4}$$

- Effective density, measured for a temperature ranging from 298 to 343 K.

  ✔ Without magnetic field:

$$\frac{\mu_f}{\mu_b} = 1.024\left[(1 + \varphi)^{1.115236}\left(1 + \frac{T}{T_{max}}\right)^{0.057065}\right] \tag{4.5}$$

Heat transfer enhancement with ferrofluids

With magnetic field:

$$\frac{\mu_f}{\mu_b} = 1.0314\left[(1+\varphi)^{1.0314}\left(1+\frac{T}{T_{max}}\right)^{0.04754}\left(1+\frac{M}{M_{max}}\right)^{0.0418}\right] \quad (4.6)$$

Within the equations mentioned above, the symbol φ denotes the volume fraction of nanoparticles, while the suffix "max" signifies the maximum value.

4. Mathematical formulation of FHD

FHD is a branch of physics that investigates the behavior of ferromagnetic fluids within the influence of an external magnetic field. In this study, the ferrofluid is treated as a homogeneous and isotropic continuum, analyzed at the mesoscopic scale. It is integrated with volumetric forces that influence fluid movement, including the forces of viscosity and gravity, in addition to the magnetic force arising from the interaction between the ferrofluid and the external magnetic field. The Navier–Stokes equation, which characterizes the motion of a ferrofluid, is expressed as follows:

$$\rho\left(\frac{\partial \vec{v}}{\partial t} + \left(\vec{v}.\overrightarrow{grad}\right)\vec{v}\right) = -\overrightarrow{grad}P + \eta\overrightarrow{\Delta}\vec{v} + \rho\vec{g} + \vec{f}_{mv} \quad (4.7)$$

where ρ is the density, η the dynamic viscosity, P pressure field in the fluid, and $\vec{v}$ the velocity field. The volumetric magnetic force $\vec{f}_{mv}$ is a conservative force, also called Kelvin force. It derivates from volumetric magnetic energy ε_m and it can be written as follows:

$$\vec{f}_{mv} = -\overrightarrow{grad}\varepsilon_m \quad (4.8)$$

where ε_m is equal to the difference between the volumetric magnetic energy of the particle and that of the vacuum, and it is expressed as:

$$\varepsilon_m = -\frac{1}{2}\vec{H}.\vec{B} - \left(-\frac{\mu_0}{2}H^2\right) \quad (4.9)$$

In Eq. (4.9), $\vec{H}$ denotes the magnetic excitation, $\vec{B}$ the magnetic field, and μ_0 the vacuum magnetic permeability. $\vec{H}$ is linked with the magnetization vector $\vec{M}$ by the following relation:

$$\vec{M} = \chi\vec{H} \quad (4.10)$$

With χ denotes the magnetic susceptibility of the ferrofluid.

In addition, $\overrightarrow{H}$, $\overrightarrow{B}$, and $\overrightarrow{M}$ are related by the following expression:

$$\overrightarrow{H} = \frac{\overrightarrow{B}}{\mu_0} - \overrightarrow{M} \tag{4.11}$$

By using Eqs. (4.10) and (4.11), $\overrightarrow{B}$ can be expressed as:

$$\overrightarrow{B} = \mu_0 \left(\overrightarrow{H} + \overrightarrow{M} \right) \tag{4.12}$$

$$\overrightarrow{B} = \mu_0 (1 + \chi) \overrightarrow{H} \tag{4.13}$$

Replacing the expression of $\overrightarrow{H}$ of Eq. (4.9) by those from Eq. (4.13), the volumetric magnetic energy is expressed as:

$$\varepsilon_m = - \frac{\mu_0 \, \chi}{2} H^2 \tag{4.14}$$

$$\varepsilon_m = - \frac{\mu_0 \, \chi}{2\mu_0^2 (1 + \chi)^2} B^2 \tag{4.15}$$

The characteristic paramagnetic of ferrofluid allows to use the fact that $\chi \ll 1$, and this can write ε_m from Eq. (4.15) as follows:

$$\varepsilon_m = - \frac{\chi}{2\mu_0} B^2 \tag{4.16}$$

Eqs. (4.9) and (4.16) give the following expression of volumetric magnetic force:

$$\overrightarrow{f}_{mv} = \frac{\chi}{2\mu_0} \overrightarrow{grad} B^2 \tag{4.17}$$

By using a gradient property, $\overrightarrow{f}_{mv}$ can be expressed as:

$$\overrightarrow{f}_{mv} = \frac{\chi}{\mu_0} B \overrightarrow{grad} B \tag{4.18}$$

Most researchers have expressed $\overrightarrow{f}_{mv}$ as function as magnetic magnetization module M by replacing in Eq. (4.18) B by $\frac{\mu_0}{\chi} M$ (by using Eqs. 4.10 and 4.13) giving so:

$$\overrightarrow{f}_{mv} = M \overrightarrow{grad} B = \mu_0 M \overrightarrow{grad} H \tag{4.19}$$

So, the expression of Navier–Stokes equation, taking into the FHD effect (Eq. 4.7), is as follows:

$$\rho \left(\frac{\partial \overrightarrow{u}}{\partial t} + \left(\overrightarrow{v} . \overrightarrow{grad} \right) \overrightarrow{v} \right) = - \overrightarrow{grad} P + \eta \overrightarrow{\Delta} \overrightarrow{v} + \rho \overrightarrow{g} + \mu_0 M \overrightarrow{grad} H$$

$$\tag{4.20}$$

In the case where the ferrofluid is electrically nonconducting and by using the Maxwell's equations, the conditions on the magnetic excitation and magnetic field are written as:

$$\overrightarrow{rot}\,\overrightarrow{H} = 0 \tag{4.21}$$

$$div\,\overrightarrow{B} = 0 \tag{4.22}$$

The following equation (mass conservation) indicates the noncompressibility of ferrofluid:

$$div\,\overrightarrow{v} = 0 \tag{4.23}$$

To determine the magnetization M in the equation for a numerical studies, researchers used the Langevin theory to express $M(H, T)$ as follows:

$$M = nmL(\xi) \text{ where } \xi = \frac{mH}{k_B T} \text{ and } L(\xi) = coth\,\xi - \xi^{-1} \tag{4.24}$$

With $m = M_d V$ denotes the magnetic moment of a single subdomain magnetic particle, n its number density, $V = \frac{\pi d^3}{6}$ is the particle volume, and M_d represents the domain magnetization of dispersed ferromagnetic material and k_B the Boltzmann constant [30].

Other researchers used the relation of Matsuki et al. [31] to calculate $M(H, T)$. It is expressed as:

$$M = KH(T_C - T) \tag{4.25}$$

Here K is a constant and T_C is the curie temperature [32].

In addition to the previously outlined magnetic volumetric force, a noteworthy phenomenon that arises within a ferrofluid under the influence of a magnetic field is termed the magnetocaloric effect. This effect pertains to a physical property of magnetic materials, wherein the material experiences heating or cooling around its Curie temperature due to the impact of an external magnetic field. Within a ferrofluid, this effect plays a role in the overall energy balance. The thermal power per unit volume stemming from the magnetocaloric effect is mathematically defined as follows [33]:

$$-\mu_0 T \frac{\partial M}{\partial T} \overrightarrow{v}.\overrightarrow{grad}H \tag{4.26}$$

The energy equation for a ferrofluid by considering the magnetocaloric effect can be written as:

$$\rho c_P \left(\frac{\partial T}{\partial t} + \overrightarrow{v}.\overrightarrow{grad}\,T \right) = k\nabla^2 T - \mu_0 T \frac{\partial M}{\partial T} \overrightarrow{v}.\overrightarrow{grad}H \tag{4.27}$$

In Eq. (4.27), c_P stands the heat capacitance of ferrofluid and k its thermal conductivity.

In conclusion, the equations governing FHD with the underlying assumptions of a homogenous, isotropic, incompressible, and nonconducting ferrofluid can be summarized as follows:

$$\begin{cases} \overrightarrow{rot}\overrightarrow{H} = 0 \quad (4.21); \quad div\,\overrightarrow{B} = 0 \quad (4.22); \quad div\,\overrightarrow{v} = 0 \quad (4.23) \\[2ex] \rho\left(\dfrac{\partial \overrightarrow{u}}{\partial t} + \left(\overrightarrow{v}.\overrightarrow{grad}\right)\overrightarrow{v}\right) = -\overrightarrow{grad}P + \eta\overrightarrow{\Delta}\overrightarrow{v} + \rho\overrightarrow{g} + \mu_0 M\overrightarrow{grad}H \quad (4.20) \\[2ex] \rho c_P\left(\dfrac{\partial T}{\partial t} + \overrightarrow{v}.\overrightarrow{grad}\,T\right) = k\nabla^2 T - \mu_0 T\dfrac{\partial M}{\partial T}\overrightarrow{v}.\overrightarrow{grad}H \quad (4.27) \end{cases}$$

5. Heat transfer enhancement using ferrofluids

Due to its enhanced magnetic and thermal properties, ferrofluid holds potential for applications involving thermal systems. When devoid of a magnetic field, ferrofluid functions like a typical nanofluid, augmenting convective thermal exchange due to its elevated thermal conductivity. However, in the presence of a magnetic field, the thermal efficiency of ferrofluid is further amplified (as discussed in Section 2), and the fluid flow can be manipulated through the influence of magnetic forces that are modulated by variations in magnetic field attributes.

Numerous experimental and numerical studies have been conducted to investigate the flow of Fe_3O_4/water ferrofluid in a horizontal channel under the influence of a nonuniform magnetic field. Mehrez et al. [34] observed the formation of a recirculation zone near the magnetic source located on the heated wall, which negatively affected the thermal boundary layer. This recirculation zone had a significant impact on heat transfer, leading to an enhancement of up to 86%, depending on the Reynolds number. The authors attributed the presence of the recirculation zone to the swirling effect of the Kelvin force induced by the magnetic field. Nassab et al. [35], in their numerical study of the same channel configuration, explained the existence of the recirculation zone by the variation in magnetization of the nanoparticles, which increases as the temperature decreases. Mehrez et al. [33] further investigated the forced convection problem of Fe_3O_4/water ferrofluid flow in a horizontal channel with four discrete thermal sources. They introduced four magnetic sources near each heat source, resulting in the

Heat transfer enhancement with ferrofluids 87

formation of recirculation zones at the thermal sources. This modification of the flow structure significantly influenced the temperature field, with the position of the magnetic sources playing a crucial role (see Fig. 4.21). By

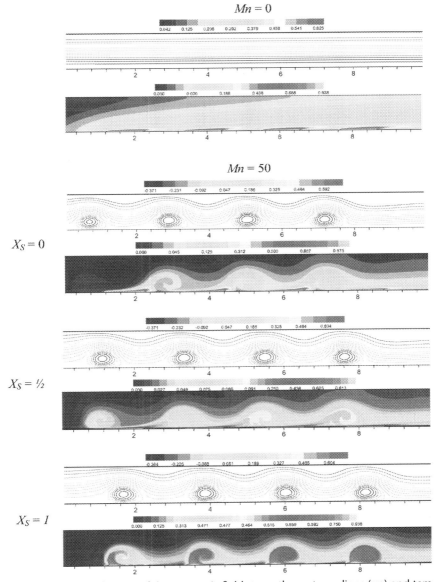

Figure 4.21 The influence of the magnetic field strength on streamlines (up) and temperature field (down) for different magnetic source positions (Mn = 0, i.e., without magnetic field). [33].

carefully selecting the positions of the magnetic sources, the volume fraction of nanoparticles, and the Reynolds number, heat transfer enhancement of up to 300% was achieved under the combined effect of the magnetic field and nanoparticles. Mousavi et al. [36] conducted a numerical study on the 3D flow of Fe_3O_4/water ferrofluid in a horizontal channel using a neodymium block magnet to create the magnetic field. They observed the formation of a swirling structure at the heated portion of the upper wall when the magnetic field was applied. The authors also highlighted the dominance of the FHD effect on heat transfer enhancement compared to the MHD effect, with a potential heat transfer enhancement of up to 100%.

Shyam et al. [37] conducted both numerical and experimental investigations on the flow and heat transfer characteristics of iron/water ferrofluid in a heated stainless steel tube under the influence of a constant and time-dependent magnetic field (MF). Using infrared thermogram visualization, they were able to determine that the reverse secondary flow (recirculation) was caused by the deposition of chain-like clusters of MNPs near the wall adjacent to the magnetic source. This deposition resulted in the formation of irregular spiked humps that disrupted the flow field and induced secondary flow, thereby enhancing the transport of energy and momentum (refer to Fig. 4.22). The shape and volume of these depositions were found to depend on the nature and frequency of the magnetic field. The maximum heat transfer enhancement achieved was 39%, which was obtained with an alternative magnetic field at a frequency of 0.1 Hz.

Mokhtari et al. [38] conducted a numerical study on the forced convection heat transfer of ferrofluid flow inside a tube with twisted tapes, focusing on the FHD effect. They confirmed that the swirling effect of the magnetic field had a significant impact on the temperature distribution and flow field, leading to a heat transfer enhancement of 30%. Abadeh et al. [39] conducted experimental investigations on the FHD problem of Fe_3O_4/water in a circular tube (see Fig. 4.23). They prepared a stable ferrofluid using a two-step method and ensured its stability by adding a surfactant and using a wave mixer. The study reported a maximum heat transfer enhancement of 11.96% and 14.8% with a constant and alternative magnetic field, respectively. Shafii et al. [40] performed an experimental study on the forced convection of ferrofluid flow in nonmagnetizable and magnetizable porous media within a copper circular tube. They prepared the ferrofluid through a chemical reaction between iron II and iron III ions in an aqueous ammonia solution to produce magnetite. The study found heat transfer enhancements

Heat transfer enhancement with ferrofluids

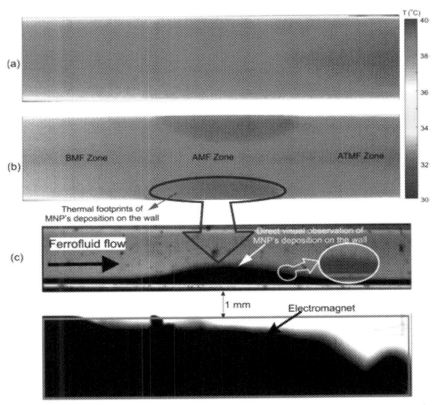

Figure 4.22 IR thermogram taken with close up IR lens for (A) B = 0, (B) B = 1080 G for Re = 66, (C) qualitative bright field visualization of ferrofluid flow in glass capillary. BMF: before MF, AMF: at MF, and ATMF: after MF [37].

of 6.39% and 9% in the cases of magnetizable and nonmagnetizable porous media, respectively.

The efficiency of the FHD effect in enhancing heat transfer in a 3D porous fin heat sink has been demonstrated in the numerical study conducted by Bezaatpour et al. [41] (see Fig. 4.24). The study revealed a remarkable 35% enhancement in heat transfer, while the presence of a magnetic field had no significant impact on the pressure drop. These findings hold significant importance for a wide range of applications, particularly in the field of electronics cooling.

Numerous studies have underscored the significance of ferrofluid flow subjected to magnetic fields in enhancing heat transfer across various industrial and engineering applications. To illustrate, in the context of heat

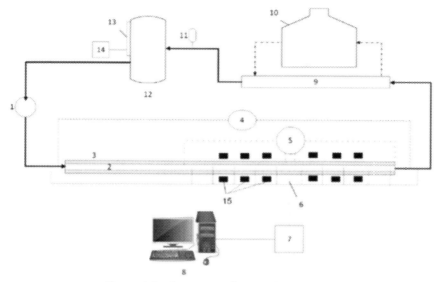

Figure 4.23 Experimental setup diagram [39].

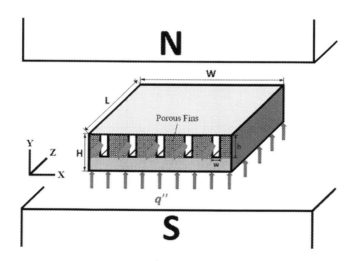

Figure 4.24 3D porous fin heat sink [41].

exchangers, Bezaatpour et al. [42] harnessed magnetite ferrofluid flow and an external magnetic field to amplify heat transfer within a compact fin-and-tube heat exchanger (as depicted in Fig. 4.25). The researchers noted the emergence of vortices induced by magnetic forces behind each tube,

Heat transfer enhancement with ferrofluids

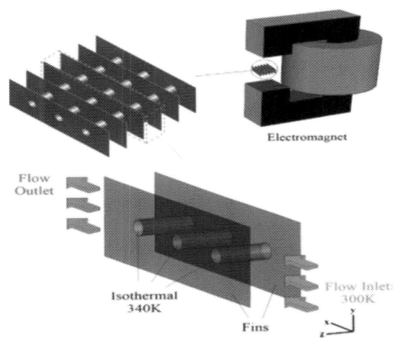

Figure 4.25 Schematic representation of the mini fin-and-tube compact heat exchanger placed in a C-shape electromagnet [42].

intensifying flow mixing and, consequently, enhancing heat transfer. Interestingly, they found that the effectiveness of the heat exchanger escalated at low Reynolds numbers upon the application of a magnetic field, leading to a maximum heat transfer enhancement of 52.4%. In a separate study, Hosseinizadeh et al. [43] conducted numerical investigations on a triple tube heat exchanger filled with Fe_3O_4/water and subjected to a uniform magnetic field (as shown in Fig. 4.26). The researchers identified the formation of a secondary flow within the inner tube in the presence of the magnetic field, contributing to increased flow mixing. The heat exchanger's performance index, defined as the normalized heat transfer rate per pumping power, was enhanced by up to 68% at a Reynolds number of 400 and a magnetic field intensity of 2400G. Zheng et al. [44] took an experimental approach to explore thermal performance and pressure drop in a plate heat exchanger. They prepared a ferrofluid consisting of 0.1 vol% Fe_3O_4 nanoparticles suspended in deionized water using a two-step method. Employing magnets arranged horizontally and vertically (as shown in Fig. 4.27) to create the

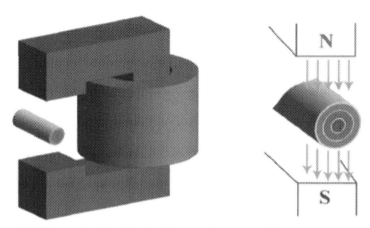

Figure 4.26 Schematic of triple tube heat exchanger and the magnet location [43].

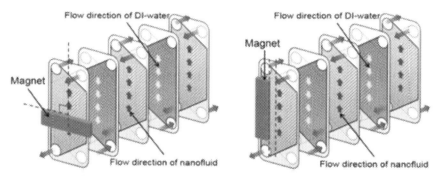

Figure 4.27 Schematic representation of vertical and horizontal arrangements of magnets in a plate heat exchanger [44].

magnetic field, the authors found that the optimal arrangement involved vertically placing magnets outside the sidewalls in a side-by-side configuration. This arrangement led to a remarkable 23.8% increase in the Nusselt number accompanied by a 10% reduction in pressure drop. These findings are particularly significant as they demonstrate the simultaneous enhancement of heat transfer and reduction in flow resistance.

Numerous researchers have recently investigated the improvement of thermal performance in renewable energy systems through the FHD effect. One such study conducted by Shojaeizadeh et al. [45] examined the flow of $Mn-Zn-Fe_2O_4$ nanoparticles in water within a flat-plate solar collector under a nonuniform magnetic field (see Fig. 4.28). Their findings

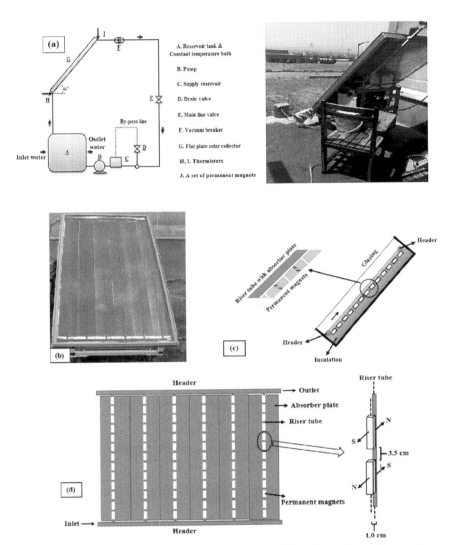

Figure 4.28 (A) Solar collector experimental setup; (B) flat-plate solar collector photograph; (C) permanent magnets locations; and (D) schematic of the collector backside without frame and insulation [45].

demonstrated that by using a volume fraction of nanoparticles of 0.8%, a fluid flow rate of 0.0083 kg/s, and a magnetic field intensity of 1T, the thermal efficiency can be enhanced by up to 52.15%. In another study, Razzaghi et al. [46] employed a numerical approach to investigate the performance of

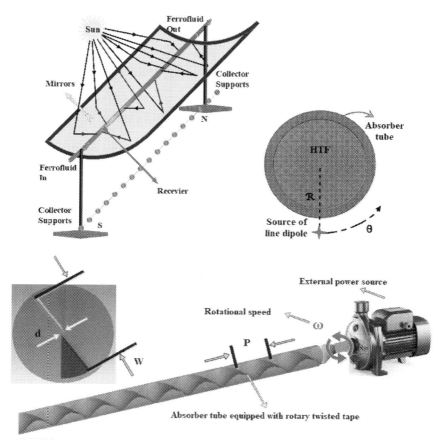

Figure 4.29 Schematic diagram of solar collector and twisted tape [46].

the LS-3 parabolic solar collector, which was filled with ferrofluid and subjected to a magnetic field and a rotary twisted tape (see Fig. 4.29). They discovered that the magnetic field increased heat transfer by up to 97%, while the twisted tape enhanced it by 305%. Furthermore, Xing et al. [47] conducted an experimental study on the enhancement of thermal performance in solar photothermal energy conversion. They utilized a hybrid Fe_3O_4-MWCNTs/water ferrofluid and an external magnetic field (see Fig. 4.30). The results showed that the presence of Fe_3O_4-MWCNTs in pure water at a concentration of 0.01 wt% increased solar photothermal conversion efficiency by up to 30%. Additionally, the efficiency further improved by 6% when subjected to a magnetic field.

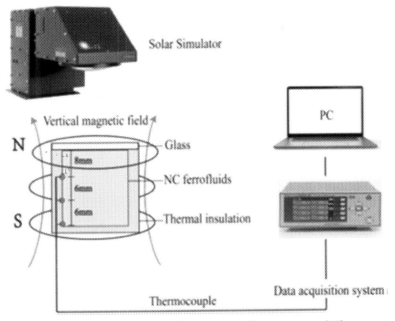

Figure 4.30 Photothermal conversion efficiency setup [47].

Across the various studies mentioned above, scholars have aimed to elucidate a shared outcome, highlighting the pronounced influence of the volumetric magnetic force, commonly known as the Kelvin force, and expressed in Eq. (4.19), on both the flow dynamics and thermal properties of ferrofluids. This force brings about alterations in momentum mechanisms and diffusion phenomena, thereby influencing the energy transfer process and subsequently impacting the enhancement of heat transfer. Notably, there were instances where the rate of enhancement exhibited remarkable strength under specific combinations of flow parameters, magnetic field characteristics, and geometrical configurations. Such outcomes bear substantial advantages for diverse engineering and industrial applications, as elaborated upon in this section.

6. Conclusion

Ferrofluids remain a subject of enduring fascination for researchers due to their inherent complexity, showcasing diverse physical phenomena that manifest across various applications, particularly when influenced by a magnetic field. These phenomena, including interactions between MNPs and base fluid molecules, MNPs and magnetic fields, Brownian motion, clustering, and agglomeration, constitute primary drivers behind the enhanced heat transfer observed across a broad spectrum of engineering and industrial contexts.

The first part of this book chapter delves into the stability of ferrofluids, beginning with an exploration of the physical phenomena that accompany the suspension of MNPs within a carrier fluid. Following this, various techniques for preparing ferrofluids are presented and elucidated using existing literature. Despite the extensive body of research in this area, the preparation and characterization of stable ferrofluids continue to be a pertinent topic. Achieving optimal quality, ensuring robust suspension properties suitable for industrial applications, remains a subject of ongoing importance.

Stable ferrofluids exhibit distinct thermophysical properties compared to their base fluids, and these properties undergo significant transformations when subjected to an external magnetic field. This aspect is the focal point of the second part, which extensively examines the thermal and rheological characteristics of ferrofluids, encompassing their alterations in response to magnetic field attributes like nature, direction, and intensity. Within this section, the interplay between viscosity and magnetic fields is elucidated, along with the significance of the chain-like structures that manifest within ferrofluids. These structures play a crucial role in enhancing thermal conductivity when influenced by a magnetic field.

The third part of this work provides an extensive exploration of the equations that govern the movement and energy transfers of a ferrofluid within a hydraulic circuit, offering detailed explanations and analysis. When a magnetic field is present, an additional term is introduced into the momentum equation, representing the volumetric magnetic force, known as the Kelvin force, which arises from the magnetic field's influence. This force, stemming from a potential referred to as magnetic energy, plays a pivotal role in inducing significant modifications in both the flow and temperature fields. Furthermore, an extra term is integrated into the energy equation to account for the magnetocaloric effect. Concluding this part, the ultimate equations for FHD and energy are presented, tailored specifically to the scenario of incompressible and nonconducting ferrofluids. The FHD formulation is a result of the intricate interconnection between Navier—Stokes and Maxwell's equations. The precise calculation details and the provided equations stand to offer valuable guidance for researchers aiming to conduct precise simulations of FHD, thereby enhancing the accuracy and precision of their modeling efforts.

The concluding section provides an overview of the current state of the field and highlights recent advancements in the realm of heat transfer enhancement within ferrofluid flow under the influence of an external magnetic field. The studies presented collectively underscore the profound alterations observed in flow structure, notably the emergence of vortices induced by the magnetic force. The authors of these studies have endeavored to elucidate the underlying physical mechanisms responsible for these observed modifications, which are pivotal in driving the notable enhancement of heat transfer. Furthermore, these investigations demonstrate how the application of ferrofluids in conjunction with a magnetic field can significantly enhance the thermal performance across a range of engineering applications, such as fin heat sinks, thermal exchangers, solar collectors, and solar photothermal conversion systems.

Given the wide array of issues covered across its various sections, this chapter holds the potential to captivate a diverse audience, including researchers in fields like engineering, physics, chemistry, applied mathematics, and numerical analysis.

Nomenclature

$\vec{B}$ Magnetic field
c_p Thermal capacity (J/Kg K)
d Particle diameter (m)
$\vec{f}_{mv}$ Volumetric magnetic force
g Gravitational acceleration (m/s^2)
h Heat exchange coefficient (W/m^{2}K)
$\vec{H}$ Magnetic excitation
k Thermal conductivity (W/mK)
k_B Boltzmann constant
$\vec{M}$ Magnetization
P Pressure (Pa)
t Time (s)
T Temperature (K)
$\vec{v}$ Velocity
ε_m Volumetric magnetic energy

Greek symbols

φ Nanoparticles volume fraction
η Dynamic viscosity (Kg/ms)
μ_0 Vacuum permeability ($4\pi \times 10^{-7}$ H/m)
ρ Density (Kg/m^3)
χ Magnetic susceptibility

Subscripts

b Base fluid
f Ferrofluid
p Particles
max Maximum

References

[1] M. Petit, Contribution à l'étude des systèmes de refroidissement basés sur le couplage magnétothermique dans les ferrofluides à faible température de Curie: mise en place d'outils de caractérisation et de modélisation, Doctorate Thesis, 2006.

[2] A.C. de Lima Barizão, J. Pinto de Oliveira, R.F. Gonçalves, S. Túlio Cassini, Nano-magnetic approach applied to microalgae biomass harvesting: advances, gaps, and perspectives, Environ. Sci. Pollut. Res. 28 (2021) 44795−44811, https://doi.org/10.1007/s11356-021-15260-z.

[3] M. Urbina, A. Rinaldi, S. Cuesta-Lopez, A. Sobetkii, A.E. Slobozeanu, P. Szakalos, Y. Qin, M. Prakasam, R.-R. Piticescu, C. Ducros, A. Largeteau, The methodologies and strategies for the development of novel material systems and coatings for applications in extreme environments - a critical review, Manuf. Rev. 5 (2018), https://doi.org/10.1051/mfreview/2018006.

[4] W.S. Torres, A.S. Alcantara, R.D. Bini, M.B. Alvim, M.C. Santos, L.F. Cotica, D.L. Rocco, Top-down and bottom-up approaches to obtain magnetic nanoparticle of Fe$_3$O$_4$ compound: pulsed laser deposition and chemical route, Mater. Chem. Phys. 290 (October 15, 2022) 126511.

[5] M. Javed Ansari, M.M. Kadhim, B.A. Hussein, H.A. Lafta, E. Kianfar, Synthesis and stability of magnetic nanoparticles, BioNanoScience 12 (2022) 627−638, https://doi.org/10.1007/s12668-022-00947-5.

[6] S. Wang, L. Gao, Chapter 7 - laser-driven nanomaterials and laser-enabled nanofabrication for industrial applications, in: Industrial Applications of Nanomaterials Micro and Nano Technologies, 2019, pp. 181−203, https://doi.org/10.1016/B978-0-12-815749-7.00007-4.

[7] M. Gautam, J.O. Kim, C.S. Yong, Fabrication of aerosol-based nanoparticles and their applications in biomedical fields, J. Pharm. Investig. 51 (2021) 361−375, https://doi.org/10.1007/s40005-021-00523-1.

[8] R. Venugopal, H. Thomas, R. Sini, Optical characterisation of cadmium doped Fe3O4 ferrofluids by co-precipitation method, Mater. Today: Proc. 25 (Part 2) (2020) A1−A5.

[9] M. Victory, R.P. Pant, S. Phanjoubam, Synthesis and characterization of oleic acid coated Fe−Mn ferrite based ferrofluid, Mater. Chem. Phys. 240 (2020) 122210.

[10] B. Babukutty, D. Ponnamma, S.S. Nair, J. Jose, S.G. Bhat, S. Thomas, Structural influence of chromium substituted magnetite ferrofluids on the optical and antibacterial characteristics, Mater. Today Commun. 34 (2023) 105439.

[11] P. Hribar Bostjancic, Z. Gregorin, N. Sebastián, N. Osterman, D. Lisjak, A. Mertelj, Isotropic to nematic transition in alcohol ferrofluids of barium hexaferrite nanoplatelets, J. Mol. Liq. 348 (2022) 118038.

[12] G. Valverde Aguilar, Sol-Gel Method, Book, 2018, https://doi.org/10.5772/intechopen.82487.

[13] T.K.O. Vuong, T.T. Le, H.D. Do, X.T. Nguyen, N.X. Ca, V.T. Thu, Le T. Lu, D.L. Tran, Materials Chemistry and Physics, vol. 245, 2020 122762.

[14] A. Das, A.K. Ganguli, Design of diverse nanostructures by hydrothermal and microemulsion routes for electrochemical water splitting, RSC Adv. 8 (2018) 25065−25078, https://doi.org/10.1039/C8RA04133D.

[15] P. Kelly, C. Solans, J. Alvarez-Quintana, Margarita Sanchez-Dominguez, Synthesis of Mn−Zn ferrite nanoparticles by the oil-in-water microemulsion reaction method, Colloids Surf. A Physicochem. Eng. Asp. 451 (June 1, 2014) 161−171.

[16] X. Huang, M. Saadat, M. Ali Bijarchi, M. Behshad Shafii, Ferrofluid double emulsion generation and manipulation under magnetic fields, Chem. Eng. Sci. 270 (2023) 118519, https://doi.org/10.1016/j.ces.2023.118519.

[17] H. Sleiman, Systèmes de suspension semi-active à base de fluide magnétorhéologique pour l'automobile, Doctorate Thesis, Arts et Métiers ParisTech, 2010.

[18] Y. Mohammadfam, S.Z. Heris, L. Khazini, Experimental Investigation of Fe_3O_4/hydraulic oil magnetic nanofluids rheological properties and performance in the presence of magnetic field, Tribol. Int. 142 (2020) 105995.

[19] J. Lei, Z. Luo, Q. Shan, X. Huang, F. Li, Effect of surfactants on the stability, rheological properties, and thermal conductivity of Fe_3O_4 nanofluids, Powder Technol. 399 (2022) 117197.

[20] Z. Li, J. Yao, D. Li, Research on the rheological properties of a perfluoropolyether based ferrofluid, J. Magn. Magn Mater. 424 (February 15, 2017) 33−38.

[21] A. Bhandari, Effect of the diameter of magnetic core and surfactant thickness on the viscosity of ferrofluid, J. Magn. Magn Mater. 548 (2022) 168975.

[22] E. Siebert, V. Dupuis, S. Neveu, S. Odenbach, Rheological investigations on the theoretical predicted "Poisoning" effect in bidisperse ferrofluids, J. Magn. Magn. Mater. 374 (January 15, 2015) 44−49.

[23] P.D. Shima, J. Philip, Tuning of thermal conductivity and rheology of nanofluids using an external stimulus, J. Phys. Chem. C 115 (2011) 20097−20104.

[24] J. Philip, P.D. Shima, B. Raj, Enhancement of thermal conductivity in magnetite based nanofluid due to chainlike structures, Appl. Phys. Lett. 91 (2007) 203108.

[25] J. Philip, P.D. Shima, B. Raj, Evidence for enhanced thermal conduction through percolating structures in nanofluids, Nanotechnology 19 (2008) 305706.

[26] A. Karimi, M. Goharkhah, M. Ashjaee, M. Behshad Shafii, Thermal conductivity of Fe_2O_3 and Fe_3O_4 magnetic nanofluids under the influence of magnetic field, Int. J. Thermophys. 36 (2015) 2720−2739, https://doi.org/10.1007/s10765-015-1977-1.

[27] M. Kole, S. Khandekar, Engineering applications of ferrofluids: a review, J. Magn. Magn. Mater. 537 (2021) 168222.

[28] M. Alsaady, R. Fu, B. Li, R. Boukhanouf, Y. Yan, Thermo-physical properties and thermo-magnetic convection of ferrofluid, Appl. Therm. Eng. 88 (2015) 14−21.

[29] K. Ajith, Archana Sumohan Pillai, I.V. Muthu Vijayan Enoch, A. Brusly Solomon, Effect of magnetic field on the thermophysical properties of low-density ferrofluid with disk-shaped $MgFe_2O_4$ nanoparticles, Colloids Surf. A Physicochem. Eng. Asp. 613 (2021) 126083.

[30] I. Mark, Shliomis, ferrohydrodynamics: retrospective and issues, in: Chapter in Lecture Notes in Physics, January 2008, pp. 85−111, https://doi.org/10.1007/3-540-45646-5_5.

[31] H. Matsuki, K. Yamasawa, K. Murakami, Experimental considerations on a new automatic cooling device using temperature sensitive magnetic fluid, IEEE Trans. Magn. 13 (5) (1977) 1143−1145.

[32] A. Jarray, Z. Mehrez, A. El Cafsi, Effect of magnetic field on the mixed convection Fe_3O_4/water ferrofluid flow in a horizontal porous channel, Pramana - J. Phys. 94 (2020) 156, https://doi.org/10.1007/s12043-020-02015-7.

[33] Z. Mehrez, A. El Cafsi, Forced convection Fe_3O_4/water nanofluid flow through a horizontal channel under the influence of a non-uniform magnetic field, Eur. Phys. J. Plus 136 (2021) 451, https://doi.org/10.1140/epjp/s13360-021-01410-2.

[34] Z. Mehrez, A. El Cafsi, Heat exchange enhancement of ferrofluid flow into rectangular channel in the presence of a magnetic field, Appl. Math. Comput. 391 (2021) 125634.

[35] W. Nessab, H. Kahalerras, B. Fersadou, D. Hammoudi, Numerical investigation of ferrofluid jet flow and convective heat transfer under the influence of magnetic sources, Appl. Therm. Eng. 150 (March 5, 2019) 271−284.

[36] S. Morteza Mousavi, A. Ali Rabienataj Darzi, M. Li, Modelling and simulation of flow and heat transfer of ferrofluid under magnetic field of neodymium block magnet, Appl. Math. Model. 103 (2022) 238−260.

[37] S. Shyam, B. Mehta, P.K. Mondal, S. Wongwises, Investigation into the thermohydrodynamics of ferrofluid flow under the influence of constant and alternating magnetic field by InfraRed Thermography, Int. J. Heat Mass Tran. 135 (2019) 1233−1247.

[38] M. Mokhtaria, S. Haririb, M. Barzegar Gerdroodbaryc, R. Yeganehd, Effect of non-uniform magnetic field on heat transfer of swirling ferrofluid flow inside tube with twisted tapes, Chem. Eng. Process: Process Intensif. 117 (2017) 70−79.

[39] A. Abadeh, M. Sardarabadi, M. Abedi, M. Pourramezan, M. Passandideh-Fard, M.J. Maghrebi, Experimental characterization of magnetic field effects on heat transfer coefficient and pressure drop for a ferrofluid flow in a circular tube, J. Mol. Liq. 299 (2020) 112206.

[40] M. Behshad Shafii, M. Keshavarz, Experimental study of internal forced convection of ferrofluid flow in non-magnetizable/magnetizable porous media, Exp. Therm. Fluid Sci. 96 (2018) 441−450.

[41] M. Bezaatpour, M. Goharkhah, Effect of magnetic field on the hydrodynamic and heat transfer of magnetite ferrofluid flow in a porous fin heat sink, J. Magn. Magn. Mater. 476 (2019) 506−515.

[42] M. Bezaatpoura, H. Rostamzadeh, Heat transfer enhancement of a fin-and-tube compact heat exchanger by employing magnetite ferrofluid flow and an external magnetic field, Appl. Therm. Eng. 164 (2020) 114462.

[43] S.E. Hosseinizadeh, S. Majidi, M. Goharkhah, J. Ali, Energy and exergy analysis of ferrofluid flow in a triple tube heat exchanger under the influence of an external magnetic field, Therm. Sci. Eng. Prog. 25 (2021) 101019.

[44] D. Zheng, J. Yang, J. Wang, S. Kabelac, B. Sundén, Analyses of thermal performance and pressure drop in a plate heat exchanger filled with ferrofluids under a magnetic field, Fuel 293 (2021) 120432.

[45] E. Shojaeizadeh, F. Veysi, K. Zareinia, A.M. Mansouri, Thermal efficiency of a ferrofluid-based flat-plate solar collector under the effect of non-uniform magnetic field, Appl. Therm. Eng. 201 (2022) 117726.

[46] M.J. Pour Razzaghi, M. Asadollahzadeh, M.R. Tajbakhsh, R. Mohammadzadeh, M.Z.M. Abad, E. Nadimi, Investigation of a temperature-sensitive ferrofluid to predict heat transfer and irreversibilities in LS-3 solar collector under line dipole magnetic field and a rotary twisted tape, Int. J. Therm. Sci. 185 (2023) 108104.

[47] M. Xing, C. Jia, H. Chen, R. Wang, L. Wang, Enhanced solar photo-thermal conversion performance by Fe3O4 decorated MWCNTs ferrofluid, Sol. Energy Mater. Sol. Cells 242 (2022) 111787.

CHAPTER FIVE

Nanofluids—Magnetic field interaction for heat transfer enhancement

Brahim Fersadou, Walid Nessab and Henda Kahalerras
Faculty of Mechanical and Process Engineering, Houari Boumediene University of Sciences and Technology (USTHB), Algiers, Algeria

Highlights

- Simulation methods of nanofluids and their interaction with magnetic field are given.
- MHD forced convection in a corrugated duct with elliptical porous blocks is studied.
- MHD effect on convective nanofluid flow in two interacting cavities is examined.
- FHD mixed convection of a ferrofluid in an open corrugated cavity is explored.
- Effects of Ha, blocks' size and orientation, Da, Mn, and R_{int} are analyzed.

1. Introduction

Any energy transformation from one form to another (mechanical, chemical, or electrical) is accompanied by more or less significant heat dissipation. This heat dissipation must be removed in most industrial systems to maintain efficient operation. Several passive or active cooling techniques, such as finned heat sinks equipped with fans [1–4], are employed for this purpose. As a result of the ongoing development of technology, new and improved cooling methods have been adopted to ensure the proper functioning of systems.

2. Nanofluids

Heat transfer fluids like water, frequently used for cooling or heating purposes, have low thermal conductivities that restrict their ability to

Advanced Materials-Based Fluids for Thermal Systems
ISBN: 978-0-443-21576-6
https://doi.org/10.1016/B978-0-443-21576-6.00007-8

remove heat from a system. Improving their thermophysical properties arose by suspending solid particles with outstanding thermal capabilities and a nanometric size. Choi [5] coined "nanofluids" to designate this new class of fluids. Nanoparticles can be made of any material with high thermal conductivity. Among the classic materials, we find pure metals (aluminum, copper, titanium, gold, silver, and iron), metal oxides (alumina, cupric oxide, and zinc oxide), various ceramics (carbides, nitrides, and sulfides), carbon nanotubes, fullerenes, diamond, and polymers. The geometry and average size of these small pieces of material also vary and can be adapted according to how they are manufactured (sphere, cylinder, disc, tube, and filament). Maxwell's work [6] has demonstrated the effectiveness of such a strategy for enhancing heat conduction. Experiments with particulate heat transfer fluids have shown that some issues, such as particle sedimentation, higher pressure drop, and increased thermal resistance at the walls, can occur when utilizing them. To obtain stable nanofluids with relatively little particle settling and minimal changes in chemical composition, several techniques must be used when preparing the nanofluids. These methods can focus on altering the suspension's pH, using surfactants that act as dispersants or ultrasonic vibrations [7]. The primary motivation behind implementing these strategies is to redesign the surface properties of nanoparticles to enhance their capacity to spread evenly in liquids and essentially prevent particle aggregation [8]. Utilizing nanofluids can enhance heat transfer and energy efficiency in numerous thermal systems, including:

- Cooling electronic systems like integrated circuits [9] or oscillatory heat pipes [10].
- Cooling of transmission motors [11] and lubricating moving mechanical parts [12].
- Cooling of high-power electronic devices and controlled-energy weapons in military systems.
- Enhancing heat transfer in flow boiling systems [13,14].
- In biomedicine, nanofluids could generate a higher temperature around tumors to destroy cancer cells without harming the surrounding healthy cells [15].
- Nanofluids can increase the transfer of heat from solar collectors to depository containers and energy density in the renewable energy industry.

2.1 Nanofluids simulation approaches

Most research works dealing with nanofluid convective heat transfer problems have used either homogeneous or nonhomogeneous approaches. In the nonhomogeneous model, several factors, including gravity, Brownian motion, and sedimentation, are responsible for the nonzero velocity between the carrier liquid and the nanoparticles. However, in the homogeneous model, the nanoparticles are highly diluted, and the slip rate between phases is negligible. This last strategy is more straightforward and more computationally effective.

2.1.1 Homogeneous approach

Even though nanofluids are combinations of solid and liquid, the approach typically adopted in most investigations treats the nanofluid as a homogeneous solution. Indeed, the nanoparticles in suspension are supposed to be transported at the same velocity as the base liquid due to their small size and volume fraction. Additionally, by taking into account the local thermal equilibrium, it is possible to assume that the liquid—solid particle combination operates like a traditional homogeneous fluid, with characteristics that can be estimated as functions of those of the components. It ought to be emphasized that this assumption isn't always accurate [16].

2.1.2 Nonhomogeneous approach

The nonhomogeneous technique may be accurate in the nanofluid study since it accounts for the motion between the solid and fluid. Various multiphase theories have been put forth and employed to adequately explain and forecast the flow and complex fluids' behavior. Mirmasoumi and Behzadmehr [17] conducted a study using a two-phase mixture model to scrutinize the laminar mixed convection of a nanofluid composed of water as base fluid and Al_2O_3 as nanoparticles. Their findings revealed that the concentration of nanoparticles is greater at the bottom of the tube and near the wall. Bianco et al. [18] utilized single and two-phase models to conduct a numerical simulation of turbulent forced convection in a circular conduit filled with Al_2O_3-water nanofluid. Akbarinia and Laur [19] employed a two-phase mixture model to investigate nanofluid flow in a curved circular duct. The outcomes showed that the spatial distribution of nanoparticles remains uniform. Buongiorno [20] utilized an approach in which both thermophoresis and Brownian motion are taken into account and treated the nanofluid as a mixture of two components: carrier liquid and nanoparticles.

Brownian motion is the name of the phenomenon that describes the random movement of particles submerged in a fluid. In a liquid, the molecules are close to each other but can still move. The moving molecules will collide with the suspended particles. The sum of the forces of these collisions on a particle allows it to acquire a certain amount of momentum and moves randomly; this movement is called Brownian motion. The larger the particles, the weaker the Brownian motion. According to several authors, Brownian motion significantly influences a nanofluid's transport properties. Koo and Kleinstreuer [21] constructed a thermal conductivity model for nanofluids. It considers nanoparticle volume fraction and dimensions, particles and base liquid properties, and temperature changes. The model shows that Brownian motion has a more significant impact on higher temperatures, confirmed through experiments. Another model for thermal conductivity has been developed by Prasher et al. [22]. It connects the convection caused by nanoparticles' Brownian motion with the Maxwell−Garnett conduction model to precisely catch the effects of different factors, such as particle size, the type of base fluid used, thermal resistance between particles and liquid, and temperature. To assess the impact of particle Brownian motion on thermal transport in nanofluids, Yang [23] established equations for thermal conductivity based on the kinetic theory with relaxation time approximations.

The Brownian motion is quantified as illustrated in Eq. (5.1):

$$D_B \overrightarrow{\nabla} \phi . \overrightarrow{\nabla} T$$
$$D_B = \frac{k_B \, T}{3\pi\mu d_p} \tag{5.1}$$

where D_B is the coefficient of Brownian diffusion.

The thermophoresis effect is the other phenomenon related to Brownian motion. The fluid absorbs heat energy when a temperature gradient is present in the flow domain close to a heated wall. The liquid molecules are then more thermally excited and, therefore, can have a higher velocity than those on the cold side. The nanoparticles are then pushed firmly by the liquid molecules on the hot side. The thermophoresis phenomenon was foremost examined by Tyndall [24] when he noticed that aerosol particles in a chamber moved away from a heated surface. The thermophoresis effect is modeled using Eq. (5.2):

$$\frac{D_T}{T_o} \overrightarrow{\nabla} T . \overrightarrow{\nabla} T$$
$$D_T = 0.26 \frac{k_{bf}}{2k_{bf} + k_p} \frac{\mu}{\rho_f} \phi \tag{5.2}$$

where D_T is the thermophoretic coefficient.

3. Magnetic nanofluids

Magnetic nanofluids do not exist naturally, so they have to be synthesized. The first attempt at magnetic nanofluids or ferrofluids was made by Gozin Knight in the 18th century, who prepared a fluid of iron filings. However, real ferrofluid synthesis was not achieved until 1963 by NASA chemist Steven Papell who was tasked with controlling the liquid fuel in a rocket in space. The lack of gravity allowed it to float in the tank, so it was difficult to pump it efficiently into the rocket engine. Papell planned to convert nonmagnetic rocket fuel into a fuel with magnetic properties, so he mixed magnetite Fe_3O_4 powder with paraffin (carrier liquid) in the presence of oleic acid (surfactant) so that he could control the flow of the fuel with a magnetic field. Steven Papell has paved the way for developing ferrofluid technology in various fields through his extensive scientific investigation, notably in enhancing heat exchange under a magnetic field. The most commonly used nanoparticles for the preparation of ferrofluids are magnetite, iron, or cobalt. The magnetic properties of these nanoparticles allow their motion to be controlled by an externally operated magnetic field. For ferrofluids to operate accurately in thermal systems, they must be stable, that is, the separation distance between the particles must be maintained regardless of the applied forces. Thus, some research studies focus on preventing these issues from arising during operation. Berger et al. [25] outlined the process of creating a reliable ferrofluid suitable for use in science or engineering laboratories at a beginner level. Pop et al. [26] proposed an optimal particle size distribution function that can help maintain the stability of magnetic fluids over a long period when placed in low-gradient magnetic fields. They employed a measurement apparatus established on Gouy's method to determine the magnetization curves and eliminate any clusters or aggregates that may affect the stability. A device including Hall-type microsensors and a differential instrument for measuring magnetic structure was employed by Bolshakova et al. [27] to track the motion of magnetic particles over time when subjected to a nonuniform magnetic field. Huang et al. [28] found that phenolic antioxidants significantly enhance the magnetic stability of the α-olefinic hydrocarbon synthetic oil-based iron-nitride magnetic fluid. Guo et al. [29] used a set of surfactants consisting of oleic acid, polyethylene glycol 4000 (PEG 4000), agar, and oleic acid to enhance the stability of a water-Fe_3O_4 magnetic fluid. The sequential addition of these surfactants produced magnetic nanoparticles with a strong saturation magnetization and a highly stable ferrofluid even under magnetic and gravitational fields. Using different exploration methods, a study by Kamali et al. [30] examined a ferrofluid based on $CoFe_2O_4$ for its structural and magnetization characteristics. They concluded that the ferrofluid remained stable under ambient conditions. Yu et al. [31] examined the progress made in enhancing the stability of thermal nanofluids. They focused on comprehending the mechanisms behind the dispersion behavior of these fluids, techniques for stabilizing them, and methods for characterizing their dispersion behavior. Lysenko et al. [32] produced magnetite colloids with a high concentration in kerosene–magnetic fluids. They tested various organic acids to see which best stabilized the magnetite nanoparticles and constructed precipitation curves for each stabilizer. They also introduced a new parameter to predict precipitants' coagulation effect, allowing them to approximate the precipitation curves with a single curve. Experiments have shown that the saturation polarization of materials falls with temperature. Its decay is increasingly rapid as one approaches the Curie point [33], noted T_c. There is, therefore, a direct link between magnetic properties and the thermal field. Each material has its own Curie temperature. Iron, for example, loses its spontaneous magnetization at $770°C$. Depolarization will not be prevalent at the usual temperatures for power electronics (from ambient to $200°C$). Conversely, there are materials whose Curie transition occurs at low temperatures; for example, Manganese–Zinc ferrites lose their polarization at $150°C$.

4. Ferrohydrodynamics

Ferrohydrodynamics, or FHD, is an emerging field of science that studies the flow of incompressible magnetized fluids. The most common fluids are colloidal ferrofluids composed of tiny magnetic particles dispersed in a base liquid. The forces in these types of fluids result from a magnetic field acting on the magnetizable material without any electric current.

In the theory of electromagnetism, the force per unit of volume acting on a magnetized material sample with magnetization M in a variable field H, called Kelvin force, is given by Eq. (5.3) according to Zebib [34] and Bashtovoy et al. [35]:

$$\mu_0(M \cdot \nabla)H \tag{5.3}$$

where μ_0 is the void's magnetic permeability.

If the magnetization of an element is always in the same direction as the magnetic field, Eq. (5.4) is utilized instead of Eq. (5.3):

$$\mu_0 \left(M \cdot \nabla \right) H = \left(\mu_0 M/H \right) \left(H \cdot \nabla \right) H \tag{5.4}$$

Magnetization is described as the volumetric density of the magnetic moment. Each of the nanoparticles in the ferrofluid has its magnetic moment. Without a magnetic field, the nanoparticles are randomly oriented. However, the dipole moments struggle to align with the applied field due to thermal agitation (thermophoresis effect) for low intensities. The magnetization characteristic generally relies on the magnetic field's strength and temperature. Matsuki et al. [36] performed an experimental investigation to remove waste heat from machines using an automatic cooling device containing a magnetic fluid. They found that the magnetization decreases linearly with increasing temperature following the relationship given by Eq. (5.5):

$$M = K' \left(T_c - T \right) H \tag{5.5}$$

T_c is the so-called Curie temperature, the value above which the material loses its magnetization.

The magnetocaloric effect consists of an alteration in the temperature of a magnetic material when exposed to a magnetic field. The magnetic moments of the atoms making up the material align themselves with the field lines, which causes a variation in the magnetic enthalpy of the material. The latter allows the magnetic material's temperature to rise. The German physicist Emil Warburg discovered the magnetocaloric effect in iron in 1881, which opened the door for further research in this context. A substantial magnetic entropy modification has been uncovered by Pecharsky and Geschneidner [37] in Gd_5 (Si_2Ge_2) when exposed to a magnetic field alteration. It surpasses the reversible magnetocaloric effect in any other magnetic material by at least twice the magnitude. In a study by Tzirtzilakis [38], a mathematical model was introduced to describe blood flow under a magnetic field. The model considers blood's magnetization and electrical conductivity, the magnetocaloric effect, and the Joule heating. An analogous model was utilized by Sheikholeslami and Ganji [39] when studying the convective heat exchange of a ferrofluid in a semiannular cavity. Hesham et al. [40] examined the magnetocaloric effect and magnetothermal features of $ErFe_2$. They employed the molecular field approach to computing the magnetization and the magnetic impact on the entropy and heat capacity.

The improvement of the magnetocaloric effect through adjusting phase transition temperatures on Ni_2FeGa Heusler glass-covered microwires was analyzed by Hennel et al. [41]. By strategically incorporating structural and magnetic transitions, they increased the magnetic entropy variation by over three times.

The magnetocaloric effect causes a thermal power per unit volume which is expressed by Eq. (5.6):

$$\mu_0 \, T \frac{\partial M}{\partial T} \left(\overrightarrow{V} \cdot \overrightarrow{\nabla} H \right) \tag{5.6}$$

5. Magnetohydrodynamics

Magnetohydrodynamics, or MHD, is a branch of hydrodynamics that describes the behavior of an electrical conductor fluid under the action of an electromagnetic field [42]. The Swedish physicist Hannes Alfvén [43] was the first to use the term magnetohydrodynamics in 1942. The connection of the magnetic field with these moving fluids generates an electric current. The latter's appearance under a magnetic field leads to the creation of a force called Lorentz force that affects fluid motion.

There are two types of MHD [44]:
- Ideal MHD when the fluid conducts the current without any electrical resistance. It is simulated to a perfect conductor.
- Resistive MHD corresponding to the case of a fluid with nonzero resistance.

Another way of distinguishing these two types of MHD is to consider the magnetic Reynolds number, which is calculated using Eq. (5.7):

$$R_m = \mu_0 \, \sigma \, v \, L \tag{5.7}$$

If R_m is more significant than 1, it is ideal MHD; otherwise, it is resistive MHD.

The set of equations describing the MHD combines Maxwell's equations of electromagnetism and Navier–Stokes' equations of fluid motion. At quiescent, Ohm's law states that the electric current through a conductor is directly proportional to the electric field $\overrightarrow{J} = \sigma \, \overrightarrow{E}$. When the fluid is flowing at a given velocity under magnetic and electric fields, the generalized form of Ohm's law is written as stated in Eq. (5.8):

$$\overrightarrow{J} = \sigma \left(\overrightarrow{E} + \overrightarrow{V} \times \overrightarrow{B} \right) \tag{5.8}$$

The electromagnetic force or Lorentz force is a force per unit of volume defined as the sum of two forces, electric and magnetic. The first is the Coulomb force depending on the electric field, while the second is directly related to the magnetic field and is called the Laplace force. Its expression is given by Eq. (5.9):

$$q\vec{E} + \vec{J} \times \vec{B} \tag{5.9}$$

Joule heat dissipation due to the electron movement (electric current) through a fluid is written as displayed in Eq. (5.10):

$$\frac{J^2}{\sigma} = \sigma\left(\vec{E} + \vec{V} \times \vec{B}\right)^2 \tag{5.10}$$

6. Applications
6.1 MHD forced convection in a corrugated channel with elliptical porous blocks

The physical domain under study and depicted in Fig. 5.1 is a two-dimensional corrugated channel with three porous elliptic blocks and a narrowing at the entrance. An MWCNT-water nanofluid enters with a constant velocity U_i and a cold temperature T_c. A thermal insulation is used on the side walls of the channel, while the lower and upper are kept at a warm temperature T_h. A uniform transverse magnetic field is applied.

The density, heat capacity, electrical conductivity, thermal conductivity, and viscosity of (MWCNT-water) nanofluid are given by Eqs. (5.11)–(5.15):

$$\rho_{nf} = (1-\phi)\rho_{bf} + \phi\rho_p \tag{5.11}$$

$$(\rho C_p)_{nf} = (1-\phi)(\rho C_p)_{bf} + \phi(\rho C_p)_p \tag{5.12}$$

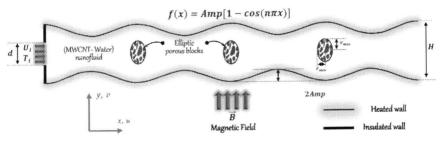

Figure 5.1 Physical domain.

$$\frac{\sigma_{nf}}{\sigma_{bf}} = 1 + \frac{3(\gamma - 1)\phi}{(\gamma + 2) - (\gamma - 1)\phi}; \quad \gamma = \frac{\sigma_p}{\sigma_{bf}} \tag{5.13}$$

$$\frac{k_{nf}}{k_{bf}} = \frac{(1 - \phi) + 2\phi \dfrac{k_p}{k_p - k_{bf}} \ln \dfrac{k_p + k_{bf}}{2k_{bf}}}{(1 - \phi) + 2\phi \dfrac{k_{bf}}{k_p - k_{bf}} \ln \dfrac{k_p + k_{bf}}{2k_{bf}}} \tag{5.14}$$

$$\mu_{nf} = \frac{\mu}{(1 - \phi)^{2.5}} \tag{5.15}$$

The fluid flow is steady-state, two-dimensional, and laminar. The porous medium is isotropic, homogeneous, and saturated by an incompressible and Newtonian nanofluid that is locally in thermal equilibrium with the solid matrix. The nanoparticles, smaller than the pores of the porous medium, are brought into suspension in the base fluid using surfactants, preventing their aggregation and deposition. The viscous dissipation is negligible.

The homogeneous approach is chosen to describe the MHD forced convection of nanofluid. Eqs. (5.16)–(5.18) of continuity, motion, and energy, in dimensionless form, are:

$$\frac{\partial U}{\partial X} + \frac{\partial V}{\partial Y} = 0 \tag{5.16}$$

$$\frac{1}{\varepsilon^2} \left[U \frac{\partial U}{\partial X} + V \frac{\partial U}{\partial Y} \right] = -\frac{\partial P}{\partial X} + \frac{\rho_{bf}}{\rho_{nf}} \frac{\mu_{nf}}{\mu} \frac{R_\mu}{Re} \left(\frac{\partial^2 U}{\partial X^2} + \frac{\partial^2 U}{\partial Y^2} \right)$$
$$- \frac{\rho_{bf}}{\rho_{nf}} \frac{\mu_{nf}}{\mu} \frac{1}{Re\,Da} U - \frac{C}{\sqrt{Da}} \left| \vec{V} \right| U - \frac{Ha^2}{Re} \frac{\sigma_{nf}}{\sigma_{bf}} \frac{\rho_{bf}}{\rho_{nf}} U \tag{5.17a}$$

$$\frac{1}{\varepsilon^2} \left[U \frac{\partial V}{\partial X} + V \frac{\partial V}{\partial Y} \right] = -\frac{\partial P}{\partial Y} + \frac{\rho_{bf}}{\rho_{nf}} \frac{\mu_{nf}}{\mu} \frac{R_\mu}{Re} \left(\frac{\partial^2 V}{\partial X^2} + \frac{\partial^2 V}{\partial Y^2} \right)$$
$$- \frac{\rho_{bf}}{\rho_{nf}} \frac{\mu_{nf}}{\mu} \frac{1}{Re\,Da} V - \frac{C}{\sqrt{Da}} \left| \vec{V} \right| V \tag{5.17b}$$

$$U \frac{\partial \theta}{\partial X} + V \frac{\partial \theta}{\partial Y} = \frac{(\rho C_p)_{bf}}{(\rho C_p)_{nf}} \frac{R_k}{RePr} \frac{k_{nf}}{k_{bf}} \left[\frac{\partial^2 \theta}{\partial X^2} + \frac{\partial^2 \theta}{\partial Y^2} \right] + Ec \frac{Ha^2}{Re} \frac{\sigma_{nf}}{\sigma_{bf}} \frac{(\rho C_p)_{bf}}{(\rho C_p)_{nf}} U^2 \tag{5.18}$$

The associated boundary conditions are stated in Eqs. (5.19)—(5.21):

Inlet

$$\begin{cases} U = V = 0, \dfrac{\partial\theta}{\partial X} = 0 \quad \text{Wall} \\ U = 1, V = 0, \theta = 0 \quad \text{Jet} \end{cases} \tag{5.19}$$

Corrugated walls

$$U = V = 0, \theta = 1 \tag{5.20}$$

Exit

$$\frac{\partial U}{\partial X} = V = \frac{\partial\theta}{\partial X} = 0 \tag{5.21}$$

The permeability configurations and sizes of the porous blocks were varied to conduct the parametric study. Details are given in Tables 5.1 and 5.2.

Fig. 5.2 illustrates the evolution of the streamlines and isotherms for different permeability configurations at a porous block orientation angle of $0°$. The resulting flow structures are consistent with the channel's overall shape and the block permeability degree. The confined nanofluid jet generates recirculation cells just at the channel's entrance, followed by an acceleration of the flow at the ripple necks, whose intensity varies according to the permeability of the blocks. A flow asymmetry is present at the first corrugation for configurations C_1, C_2, and C_4, where the first block is either at high

Table 5.1 Configurations of porous blocks' permeability.

Porous blocks arrangement	Porous blocks permeability		
	First block	Second block	Third block
C_1	$Da = 10^{-1}$	$Da = 10^{-1}$	$Da = 10^{-1}$
C_2	$Da = 10^{-3}$	$Da = 10^{-3}$	$Da = 10^{-3}$
C_3	$Da = 10^{-6}$	$Da = 10^{-6}$	$Da = 10^{-6}$
C_4	$Da = 10^{-1}$	$Da = 10^{-3}$	$Da = 10^{-6}$
C_5	$Da = 10^{-6}$	$Da = 10^{-3}$	$Da = 10^{-1}$

Table 5.2 Sizes of the elliptical porous blocks.

	r_{min}	r_{max}
Size 1 (S_1)	0.13	0.2
Size 2 (S_2)	0.2	0.3
Size 3 (S_3)	0.26	0.4

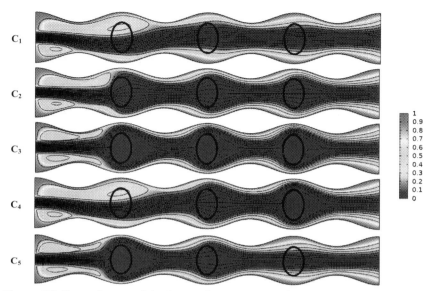

Figure 5.2 Streamlines and isotherms for various configurations of porous blocks' permeability: size S_2 and Ha = 5

or medium permeability, allowing the nanofluid to flow along a trajectory controlled by the confined jet. For configurations C_3 and C_5, the first block permeability being very low, it acts as a solid front that delays the flow at the central part of the channel and reduces the elongation of the nanofluid jet. From a thermal standpoint, when the permeability of a porous block is reduced, there is a decrease in the boundary layers' thickness and, therefore, an improvement in the heat exchange in the surrounding space because of the acceleration of the fluid flow at the solid walls of the channel.

In Fig. 5.3, the structures of the dynamic and thermal fields are plotted by intensifying the MHD effect controlled by the Hartmann number. It can be noticed that the magnetic field delays the fluid flow in the central part of the channel and accelerates it near the walls, eliminating any recirculation zones produced by the jet and decreasing the thermal boundary layers' thickness near the corrugated walls. Applying a magnetic field produces an added resistance on the nanofluid flow similar to the local inertial resistances produced by the moderate permeability blocks.

To emphasize the contribution of the MHD effect on wall cooling and entropy generation in the channel, the evolution of Nu_{mlow}, Nu_{mup}, and S with Ha is shown in Fig. 5.4. Overall, applying a uniform magnetic field and

Nanofluids—Magnetic field interaction for heat transfer enhancement 113

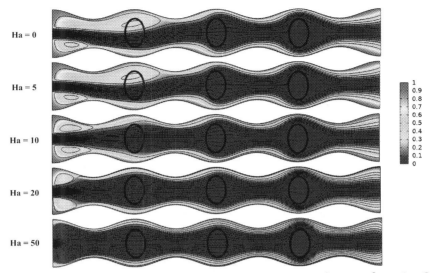

Figure 5.3 Isotherms and streamlines for various Hartmann numbers: configuration C_4 and size S_2.

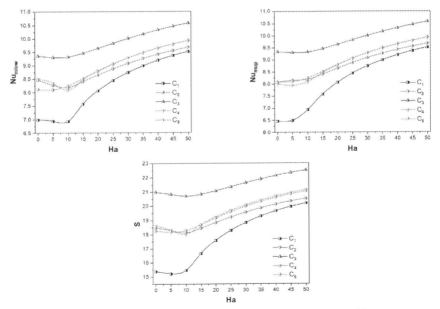

Figure 5.4 Mean Nusselt numbers and total entropy generation versus Hartmann number for various configurations of porous blocks' permeability at size S_2.

increasing its strength benefits the heat transfer rate but is unfavorable to entropy production. The best wall cooling is obtained with configuration C_3 corresponding to the situation where the permeability of the three elliptical blocks is similar and of low value. However, it led to the highest rate of disorder. The C_1 configuration generated the least thermodynamic disorder because, due to the porous blocks' high permeability, the dynamic and thermal field structures were not modified, resulting in a low heat exchange at the hot walls. As for C_4 and C_5, using porous blocks with different permeabilities is an acceptable solution since medium cooling rates and moderate thermodynamic disorder are obtained.

Figs. 5.5 and 5.6 depict the streamlines and isotherms at different block sizes and orientations for the C_4 configuration. The deviations of the streamlines and the reduction of the thermal boundary layer thicknesses become more pronounced with increasing the block size. Each orientation provides a particular deflection of the nanofluid leading to local diminution of the thermal boundary layers.

Fig. 5.7 displays the evolution of the mean Nusselt numbers and the total entropy production with the orientation angle for different sizes of elliptical blocks. The heat transfer rates increase significantly with increasing block size. Each block acts as a deflector that deviates the flow according to size, orientation, and permeability (see Figs. 5.5 and 5.6). The closer the block is to the channel's parietal boundaries, the better the heat exchange. From a thermodynamic point of view, increasing the blocks' size causes an increase in entropy generation. So moderate block size with a vertical orientation should be chosen to ensure suitable cooling with an acceptable irreversibility rate.

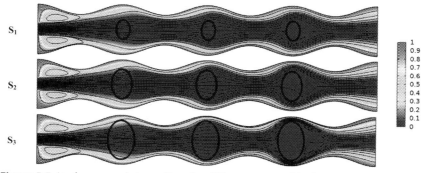

Figure 5.5 Isotherms and streamlines for different porous blocks size: configuration C_1 and Ha = 10.

Nanofluids—Magnetic field interaction for heat transfer enhancement 115

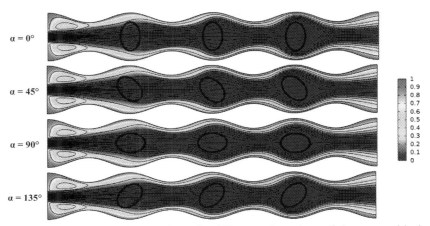

Figure 5.6 Isotherms and streamlines for different orientations of the porous blocks: configuration C_4, size S_2, and Ha = 10.

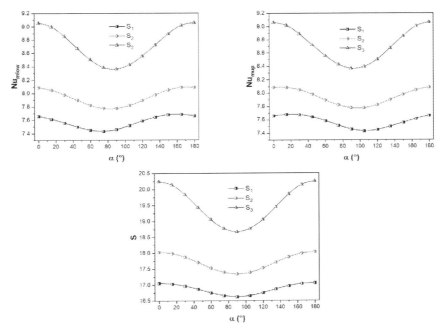

Figure 5.7 Mean Nusselt numbers and total entropy generation versus orientation angle for various porous blocks' size at configuration C_4 and Ha = 10.

6.2 MHD mixed convection in an open cavity

The physical domain under investigation is an enclosure of aspect ratio l/H = 5 divided into two open rectangular cavities, of opening width d, by a thin wall. Each cavity's right and left walls alternately contain cooled or heated sources of width s that are kept at heat flux densities q_h and q_c, respectively, and adiabatic regions of width w/2. The organization of these zones is depicted in Fig. 5.8, with the enclosure's lower and upper sides being adiabatic. A nanofluid, made of Cu nanoparticles and water, penetrates the system with constant velocity U_i, temperature T_i, and nanoparticles volume fraction ϕ_i. An external magnetic field is applied uniformly in the transverse direction. The separation between the two cavities holds a porous part of width d, which location changes. It is in contact with the right wall in case 1, whereas in cases 2 and 3, it is located in the center and on the left. This geometrical configuration is similar to the one considered in the study performed by Fersadou et al. [45], with the difference being that the partition wall was entirely solid.

Eqs. (5.22)–(5.25) provide the nanofluid density, heat capacity, thermal expansion coefficient, and electrical conductivity:

$$\rho_{nf} = (1-\phi)\rho_{bf} + \phi\rho_p \quad (5.22)$$

$$(\rho C_p)_{nf} = (1-\phi)(\rho C_p)_{bf} + \phi(\rho C_p)_p \quad (5.23)$$

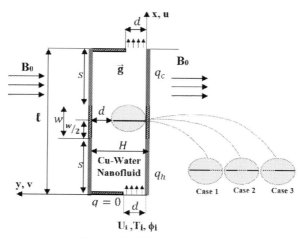

Figure 5.8 Physical domain.

$$(\rho\beta)_{nf} = (1 - \phi)(\rho\beta)_{bf} + \phi(\rho\beta)_p \tag{5.24}$$

$$\frac{\sigma_{nf}}{\sigma_{bf}} = 1 + \frac{3(\gamma - 1)\phi}{(\gamma + 2) - (\gamma - 1)\phi}; \quad \gamma = \frac{\sigma_p}{\sigma_{bf}} \tag{5.25}$$

The Cu-water mixture's viscosity and thermal conductivity are computed using the Corcione model [46], as shown in Eqs. (5.26)−(5.27):

$$\mu_{nf} = \frac{\mu}{1 - 34.87 \left(d_p / d_{bf}\right)^{-0.3} \phi^{1.03}} \tag{5.26}$$

$$\frac{k_{nf}}{k_{bf}} = 1 + 4.4 Re_B^{0.4}\, Pr^{0.6} \left[\frac{T}{T_{fr}}\right]^{10} \left[\frac{k_p}{k_{bf}}\right]^{0.03} \phi^{0.66} \quad Re_B = \frac{\rho_{bf}\, d_p}{\mu}\, \frac{2 k_B T}{\pi \mu d_p^2} \tag{5.27}$$

The fluid flow is steady-state, two-dimensional, and laminar. The nanofluid is electrically conductive, Newtonian, incompressible, and regarded as a nonhomogeneous mixture of two constituents (nanoparticles and base fluid) in which no chemical reaction occurs between these two wholly diluted components. The base fluid and nanoparticles are locally in thermal equilibrium, and viscous dissipation is neglected.

The nonhomogeneous approach is adopted to describe nanofluid MHD mixed convection. Eqs. (5.28)−(5.31) of continuity, motion, energy, and nanoparticles volume fraction in the dimensionless form are written as follows:

$$\frac{\partial U}{\partial X} + \frac{\partial V}{\partial Y} = 0 \tag{5.28}$$

$$U\frac{\partial U}{\partial X} + V\frac{\partial U}{\partial Y} = -\frac{\partial P}{\partial X} + \frac{\rho_{bf}}{\rho_{nf}}\frac{1}{Re}\left[\frac{\partial}{\partial X}\left(\frac{\mu_{nf}}{\mu}\frac{\partial U}{\partial X}\right) + \frac{\partial}{\partial Y}\left(\frac{\mu_{nf}}{\mu}\frac{\partial U}{\partial Y}\right)\right]$$
$$+ Ri\frac{(\rho\beta)_{nf}}{(\rho\beta)_{bf}}\frac{\rho_{bf}}{\rho_{nf}}\theta - \frac{Ha^2}{Re}\frac{\sigma_{nf}}{\sigma_{bf}}\frac{\rho_{bf}}{\rho_{nf}}U$$

$$\tag{5.29a}$$

$$U\frac{\partial V}{\partial X} + V\frac{\partial V}{\partial Y} = -\frac{\partial P}{\partial Y} + \frac{\rho_{bf}}{\rho_{nf}}\frac{1}{Re}\left[\frac{\partial}{\partial X}\left(\frac{\mu_{nf}}{\mu}\frac{\partial V}{\partial X}\right) + \frac{\partial}{\partial Y}\left(\frac{\mu_{nf}}{\mu}\frac{\partial V}{\partial Y}\right)\right]$$

$$\tag{5.29b}$$

$$U\frac{\partial\theta}{\partial X}+V\frac{\partial\theta}{\partial Y}=\frac{(\rho C_p)_{bf}}{(\rho C_p)_{nf}}\frac{1}{RePr}\left[\frac{\partial}{\partial X}\left(\frac{k_{nf}}{k_{bf}}\frac{\partial\theta}{\partial X}\right)+\frac{\partial}{\partial Y}\left(\frac{k_{nf}}{k_{bf}}\frac{\partial\theta}{\partial Y}\right)\right]$$

$$+\frac{(\rho C_p)_{bf}}{(\rho C_p)_{nf}}\frac{1}{RePr}\left\{N_B\left[\frac{\partial\phi}{\partial X}\frac{\partial\theta}{\partial X}+\frac{\partial\phi}{\partial Y}\frac{\partial\theta}{\partial Y}\right]\right.$$

$$\left.+N_T\left[\left(\frac{\partial\theta}{\partial X}\right)^2+\left(\frac{\partial\theta}{\partial Y}\right)^2\right]\right\}+Ec\frac{Ha^2}{Re}\frac{\sigma_{nf}}{\sigma_{bf}}\frac{(\rho C_p)_{bf}}{(\rho C_p)_{nf}}U^2$$

$$(5.30)$$

$$U\frac{\partial\phi}{\partial X}+V\frac{\partial\phi}{\partial Y}=\frac{1}{RePr}\left[\frac{\partial}{\partial X}\left(\frac{1}{Le}\frac{\partial\phi}{\partial X}\right)+\frac{\partial}{\partial Y}\left(\frac{1}{Le}\frac{\partial\phi}{\partial Y}\right)\right]$$

$$+\frac{1}{RePr}\left[\frac{\partial}{\partial X}\left(\frac{1}{Le}\frac{N_T}{N_B}\frac{\partial\theta}{\partial X}\right)+\frac{\partial}{\partial Y}\left(\frac{1}{Le}\frac{N_T}{N_B}\frac{\partial\theta}{\partial Y}\right)\right]$$

$$(5.31)$$

The related boundary conditions are specified in Eqs. (5.32)−(5.35):

Inlet

$$\begin{cases} U=V=0,\dfrac{\partial\theta}{\partial X}=\dfrac{\partial\phi}{\partial X}=0 & \text{Wall} \\[2mm] U=1,V=0,\theta=0,\phi=1 & \text{Jet} \end{cases}$$

$$(5.32)$$

Right wall

$$\begin{cases} U=V=0,\dfrac{\partial\theta}{\partial Y}=-\dfrac{k_{bf}}{k_{nf}}\dfrac{\partial\phi}{\partial Y},\dfrac{\partial\phi}{\partial Y}=-\dfrac{N_T}{N_B}\dfrac{\partial\theta}{\partial Y} & 0\leq X\leq S \\[3mm] U=V=0,\dfrac{\partial\theta}{\partial Y}=\dfrac{\partial\phi}{\partial Y}=0 & S<X\leq(S+W) \\[3mm] U=V=0,\dfrac{\partial\theta}{\partial Y}=\dfrac{q_c}{q_h}\dfrac{k_{bf}}{k_{nf}}\dfrac{\partial\phi}{\partial Y},\dfrac{\partial\phi}{\partial Y}=-\dfrac{N_T}{N_B}\dfrac{\partial\theta}{\partial Y} & (S+W)<X\leq \\ & \qquad(2S+W) \end{cases}$$

$$(5.33)$$

Left wall

$$\begin{cases} U=V=0,\dfrac{\partial\theta}{\partial Y}=-\dfrac{q_c}{q_h}\dfrac{k_{bf}}{k_{nf}}\dfrac{\partial\phi}{\partial Y},\dfrac{\partial\phi}{\partial Y}=-\dfrac{N_T}{N_B}\dfrac{\partial\theta}{\partial Y} & 0\leq X\leq S \\[3mm] U=V=0,\dfrac{\partial\theta}{\partial Y}=\dfrac{\partial\phi}{\partial Y}=0 & S<X\leq(S+W) \\[3mm] U=V=0,\dfrac{\partial\theta}{\partial Y}=\dfrac{k_{bf}}{k_{nf}}\dfrac{\partial\phi}{\partial Y},\dfrac{\partial\phi}{\partial Y}=-\dfrac{N_T}{N_B}\dfrac{\partial\theta}{\partial Y} & (S+W)<X\leq(2S+W) \end{cases}$$

$$(5.34)$$

Exit

$$\begin{cases} U = V = 0, \dfrac{\partial \theta}{\partial X} = \dfrac{\partial \phi}{\partial X} = 0 \quad \text{wall} \\[2ex] \dfrac{\partial U}{\partial X} = V = \dfrac{\partial \theta}{\partial X} = \dfrac{\partial \phi}{\partial X} = 0 \quad \text{jet} \end{cases} \tag{5.35}$$

To highlight the effects of MHD on the thermal and dynamic fields, the isotherms and streamlines are depicted in Fig. 5.9 for the three configurations of the porous partition at various Hartmann number values. Without a magnetic field and for all three cases, the nanofluid motion in the first cavity is characterized by an up-flow controlled by the jet originating from the entrance and a down-flow near the cold wall. Between these two flows, a recirculation zone is developed. For the second cavity, the flow is controlled by the arrangement of the separating wall's porous portion and its permeability (see Fig. 5.10). In case 3, the dynamic distribution is similar to the first cavity's. In contrast, in case 1 and case 2, the nanofluid penetrates through several outlets, and consequently, the magnitude of the reverse flow is attenuated. Applying the magnetic field causes substantial alterations in streamlines patterns, where there is competition between the ascending flow and the eddies whose size will lower as the Hartmann number rises until they vanish entirely at Ha = 50. This behavior is because a magnetic field produces a supplementary resistance in the nanofluid. The force of Lorentz provides this resistance and operates vertically and in the opposing direction to the buoyancy force and primary nanofluid movement. For more increased magnetic field strengths, the ascending flow caused by the force of Lorentz in the vicinity of the cooled sources will outweigh the descending flow provoked by the buoyancy force. This results in the complete removal of the reverse flow.

Regarding the evolution of the mean Nusselt numbers per cavity with the number of Hartmann and for various permeable wall arrangements, case 1 is thermally the most efficient in the first cavity due to the heat discharge from the heated source to the second cavity through the porous portion. As for the effect of the magnetic field intensity, its growth results in two tendencies; the first is marked by the decrease in heat exchange up to a critical Hartmann number between 20 and 25. This behavior occurs due to the slowing down of the nanofluid movement around the hot source and the contraction of the recirculation cell close to the cool one. Above the Ha critical value, a second trend arises where the mean Nusselt number increases due to the Lorentz force's benefit over the fluid. Looking at the

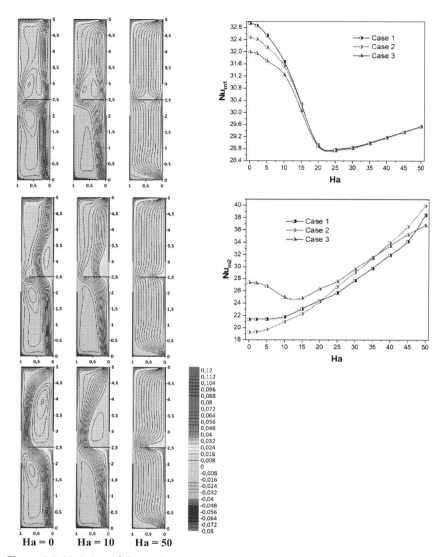

Figure 5.9 Variation of the isotherms, streamlines and Nu$_m$ with Ha for the three cases: Ri = 10 and Da = 10^{-3}.

Nu$_{m2}$ curves, the identical evolutionary tendency obtained formerly is encountered in case 3, the most thermally efficient for the second cavity, with a shift occurring at a critical Hartmann number around the value of 15. In contrast, for case 1 and case 2, the strengthening of the magnetic field is always advantageous for heat transfer.

Nanofluids—Magnetic field interaction for heat transfer enhancement

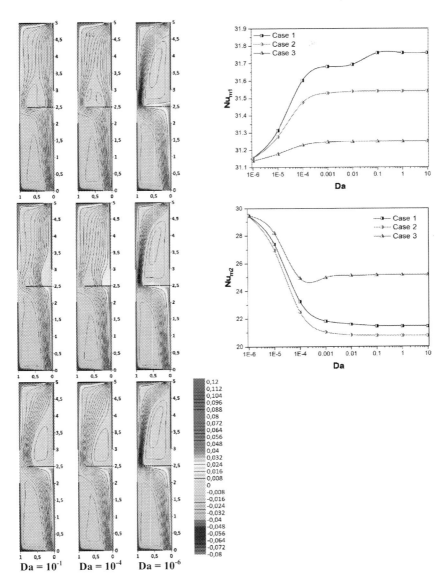

Figure 5.10 Variation of the isotherms, streamlines and Nu_m with Da for the three cases: $Ri = Ha = 10$.

From Fig. 5.10, for all three investigated cases, the growth in Darcy number of the porous portion improves the heat transfer at the first cavity. In contrast, degradation is noted in the second cavity. Reducing the Darcy number for case 3 leads to substantial flow deviations toward the second cavity's hot wall, translating into a better heat exchange at this level. This behavior affects the heat exchange of the first cavity, where the nanofluid flow is removed from the hot source. Increasing Da amplifies the difference in heat transfer rates between the different configurations. Case 1 is the most thermally favorable in the first cavity, and case 3 is in the second.

Conversely to the Hartmann number impact, Fig. 5.11 demonstrates that the augmentation in the Richardson number, reflecting the buoyancy strength, donates to the growth of the recirculation zone size, pushing the nanofluid flow upwards near the heated sources with increasing velocities to maintain the flow rate constant. This reorganization makes heat transfer more efficient for the various partition combinations in the two cavities. Additionally, it is apparent in the isotherms that the thermal boundary layers around the heated sources decrease in size as the Ri value grows.

6.3 FHD mixed convection of a ferrofluid flow

The chosen geometry is a vertical two-dimensional open cavity having an $L/H = 5$ aspect ratio. A (Fe_3O_4-water) ferrofluid enters the channel from a narrowing of width d with a constant velocity U_i and a uniform temperature T_i. At the outlet, the flow is separated into two jets by a deflector of dimension d. The two corrugated right and left walls are maintained at a warm temperature, whereas the rest are thermally isolated. Four sources create a nonuniform magnetic field, arranged as shown in Fig. 5.12, through which either identical or different electric currents pass. The axial and transverse components of the magnetic field are expressed as indicated in Eq. (5.36):

$$H_x(x, y) = \sum_{i=1}^{i=4} \frac{I_{r.l}}{2\pi} \frac{(y - b_i)}{(x - a_i)^2 + (y - b_i)^2} \tag{5.36a}$$

$$H_y(x, y) = \sum_{i=1}^{i=4} -\frac{I_{r.l}}{2\pi} \frac{(x - a_i)}{(x - a_i)^2 + (y - b_i)^2} \tag{5.36b}$$

where a_i and b_i are the coordinates in the axial and transverse directions giving the location of each magnetic source. I_r and I_l are the electric currents across the right and left ones.

The fluid flow is steady-state, laminar, and two-dimensional. The Newtonian and incompressible ferrofluid is treated as a homogeneous mixture with negligible slip velocity between the liquid and solid phases. The interactions between the Fe_3O_4 nanoparticles are suppressed by coating their

Nanofluids—Magnetic field interaction for heat transfer enhancement

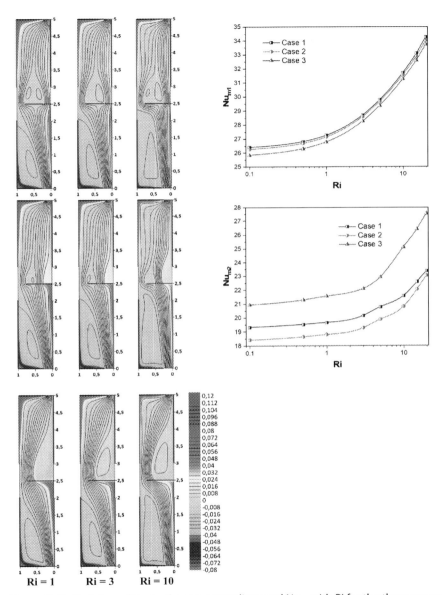

Figure 5.11 Variation of the isotherms, streamlines and Nu_m with Ri for the three cases: Ha = 10 and Da = 10^{-3}.

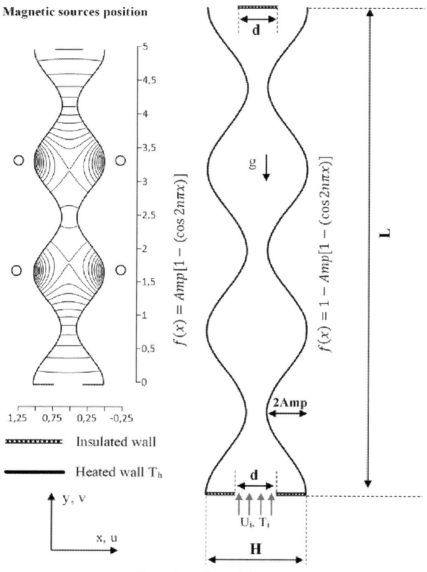

Figure 5.12 Physical domain.

surface with surfactants to prevent aggregation and sedimentation. All thermophysical properties are constant except the density in the expression of the floating force for which the approximation of Boussinesq is used. Viscous dissipation is negligible, and there are no internal heat sources.

The homogeneous methodology is employed to characterize the nanofluid FHD mixed convection. The dimensionless form of continuity, motion, and energy equations is given in Eqs. (5.37)–(5.39):

$$\frac{\partial U}{\partial X} + \frac{\partial V}{\partial Y} = 0 \tag{5.37}$$

$$U\frac{\partial U}{\partial X} + V\frac{\partial V}{\partial Y} = -\frac{\partial P}{\partial X} + \frac{1}{Re}\frac{\rho_{bf}}{\rho_{nf}}\frac{\mu_{nf}}{\mu}\left(\frac{\partial^2 U}{\partial X^2} + \frac{\partial^2 U}{\partial Y^2}\right) + Ri\frac{(\rho\beta)_{nf}}{(\rho\beta)_{bf}}\frac{\rho_{bf}}{\rho_{nf}}\theta$$
$$+ \frac{\rho_{bf}}{\rho_{nf}}M_n(Z_1 - \theta - Z_2)H\frac{\partial H}{\partial X} \tag{5.38a}$$

$$U\frac{\partial V}{\partial X} + V\frac{\partial V}{\partial Y} = -\frac{\partial P}{\partial Y} + \frac{1}{Re}\frac{\rho_{bf}}{\rho_{nf}}\frac{\mu_{nf}}{\mu}\left(\frac{\partial^2 V}{\partial X^2} + \frac{\partial^2 V}{\partial Y^2}\right)$$
$$+ \frac{\rho_{bf}}{\rho_{nf}}M_n(Z_1 - \theta - Z_2)H\frac{\partial H}{\partial Y} \tag{5.38b}$$

$$U\frac{\partial \theta}{\partial X} + V\frac{\partial \theta}{\partial Y} = \frac{1}{Re\,Pr}\frac{k_{nf}}{k_{bf}}\frac{(\rho C_p)_{bf}}{(\rho C_p)_{nf}}\left(\frac{\partial^2 \theta}{\partial X^2} + \frac{\partial^2 \theta}{\partial Y^2}\right)$$
$$+ M_n Ec\frac{(\rho C_p)_{bf}}{(\rho C_p)_{nf}}(\theta + Z_1)H\left(U\frac{\partial H}{\partial X} + V\frac{\partial H}{\partial Y}\right) \tag{5.39}$$

The corresponding boundary conditions are specified in Eqs. (5.40)–(5.42):

Inlet

$$\begin{cases} U = V = 0, \dfrac{\partial \theta}{\partial X} = 0 \quad \text{Wall} \\[2mm] U = 1, V = 0, \theta = 0 \quad \text{Jet} \end{cases} \tag{5.40}$$

Corrugated walls

$$U = V = 0, \theta = 1 \tag{5.41}$$

Exit

$$\begin{cases} U = V = 0, \dfrac{\partial \theta}{\partial X} = 0 \quad \text{Wall} \\[2mm] \dfrac{\partial U}{\partial X} = V = \dfrac{\partial \theta}{\partial X} = 0 \quad \text{Jet} \end{cases} \tag{5.42}$$

To emphasize the influence of the ferrohydrodynamic phenomenon and the impact of the current intensity in the sources on the thermal and dynamic fields, this section is devoted to the effect of the magnetic number Mn and the intensity ratio R_{int}. For this purpose, the number of wall corrugations has been fixed at n = 3 and the Richardson number at Ri = 1, considering two amplitudes Amp = .1 and Amp = .2.

The streamlines and isotherms distributions for $0 \leq Mn \leq 50$ are shown in Fig. 5.13. As observed, when there is no magnetic field, the corrugations' shape is the only factor affecting the flow. By increasing Mn, significant alterations in the thermal and dynamic field structures are detected, which are characterized by substantial distortions of the streamlines leading to the development of several vortices, mainly near the magnetic field's impact zones. It cannot be denied that the configuration of the physical system under investigation is at the origin of some recirculation zones, except that the amplification of the magnetic number is decisive as it leads to different dynamic distributions at Mn = 50. For the thermal field, it is clear that the formed vortices provide mixing to the ferrofluid with a significant drop in

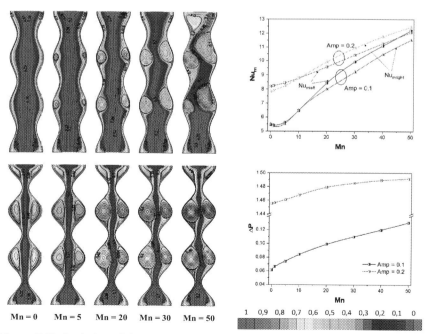

Figure 5.13 Evolution of the streamlines, isotherms, mean Nusselt numbers and pressure drop with Mn for two amplitudes: Ri = 1, R_{int} = 1 and n = 3.

temperatures, which will favorably affect the cooling of the warm walls. The appearance of the ferrofluid vortices can be explained as follows: the ferrofluid is composed of suspended nanoparticles, each of which is a magnetized monodomain with decreasing magnetizing capacity with increasing temperature. The hot ferrofluid is pushed aside as the magnetic number Mn rises, and the cold ferrofluid eventually replaces it until heat exchange causes the heated ferrofluid to lose its ability to magnetize. The ferrofluid is then set in motion and replaced by a renewed, colder layer. In this way, a kind of autonomous circulation of the ferrofluid layers is activated at the magnetic field's impact zones, creating these vortices. It should also be noted that despite the symmetrical distribution of the sources, they generate asymmetrical disturbances due to flow deviations caused at the level of the cavity necks, which act as confined jets whose ejection is subject to dynamic instabilities, especially at large amplitudes, and also because of the reverse flows fed by the buoyancy force near each corrugated wall.

The mean Nusselt numbers curves for Amp $= .1$ reveal an evolution that is initially characterized, between Mn $= 0$ and Mn $= 1$, by a slight diminution in heat exchange efficiency as the magnitude of the magnetic field is insufficient to activate vortices that can produce intense ferrofluid mixing. Above the threshold of Mn $= 1$, the heat transfer rate is enhanced because of the formation of larger vortices. For Amp $= .2$, a considerable increase in the Nusselt numbers with the magnetic number is observed. The difference in values between Nu_{mleft} and Nu_{mright} can be predicted from the streamlines and isotherms, where an asymmetry in the location and volume of the generated recirculation zones is observed. Finally, the consequence of these streamlines and isotherms distributions is significant pressure drops that grow with increasing Mn and Amp.

These outcomes about the impact of the magnetic number on the dynamic and thermal fields are qualitatively similar to those found by Nessab et al. [47] on forced convection flow in a channel with a narrowing at the inlet and by Guerroudj et al. [48] in a vertical channel with porous blocks of diverse forms.

R_{int} is the ratio of the magnetic field intensities produced by the left-hand sources to those produced by the right-hand ones. According to Fig. 5.14, for values of R_{int} less than unity, the disturbances are more pronounced close to the right corrugated wall, where bigger vortices are observed. As the intensity ratio and amplitude increase, the disruptions on the left side become more and more significant and extend into the interspace trapped between the necks created by the corrugations. This situation

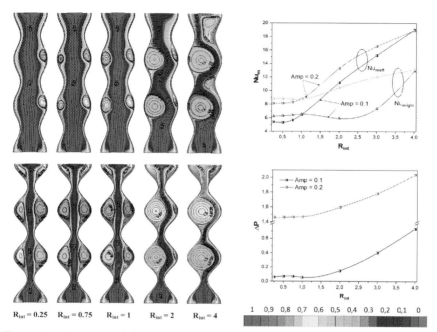

Figure 5.14 Variation of the streamlines, isotherms, mean Nusselt numbers and pressure drop with R_{int} for two amplitudes: $Ri = 1$, $Mn = 10$ and $n = 3$.

results in a continuous ferrofluid flow along the right corrugated wall. However, it is slightly disturbed by small vortices situated at the impact zones of the sources on the right side. From a thermal standpoint, raising the intensity ratio causes the size of the vortices to grow. It thus favors heat transfer, especially where the sources are interacting.

From the evolution of the mean Nusselt numbers, it can be seen that the increase of the intensity ratio at $Amp = .1$ initially leads to a slight decline of the heat exchange rate up to a critical R_{int}, corresponding to a generation of small recirculation cells that act as heat sinks, and inducing poor thermal management on the left wall side at $R_{int} = 0.5$ and on the right wall side at $R_{int} = 2$. Once these critical ratios are exceeded, the impact of the magnetic intensity will be focused on the left side, where the largest Nu_m values are found. For a corrugation amplitude such that $Amp = .2$, the heat transfer is increased with increasing R_{int}.

Regarding the pressure drop, there is a continuous increase in this parameter with the rise of R_{int}. Indeed, the resistances brought on by the amplified disturbances at the sources' impact zones cause more disorder and, as a result, high-pressure reductions.

7. Conclusion

Researchers have become interested in the numerous ways that nanofluids and the magnetic field interact, especially when it comes to methods for improving thermal systems. As a result, this chapter provides an overview of the various interactions between a magnetic field and a nanofluid, specifically MHD and FHD, when there are magnetizable nanoparticles. Also, the diverse developed models to simulate nanofluid flows are provided. These models include the two-phase approach defined by the Buongiorno model and the single-phase technique when the nanofluid is supposed to be a single homogeneous phase. Through some contributions aimed at improving the heat transfer efficiency in thermal systems with smooth or corrugated walls that can be supported by passive heat transfer enhancement techniques such as porous media, the effects of magnetic field intensification on the behavior of nanofluids were studied. The Nusselt number, pressure drop, and entropy generation were calculated after solving the velocity and temperature fields by algorithms driven by either FVM (finite volume method) or FEM (finite element method). The results obtained have shown an appreciable improvement in heat exchange by intensifying the magnetic field for both MHD nanofluids and FHD ferrofluids. However, pressure drops and irreversibilities are inevitably produced, requiring a judicious selection of control parameters to establish a balance between the gain in heat exchange and the damage to dynamic friction.

Nomenclature

(a_i,b_i) Coordinates of ith magnetic source (m)
(u,v) Velocity components (m s^{-1})
(x,y) Spatial coordinates (m)
(Z_1,Z_2) Temperature ratios, $(T_C/\Delta T, T_i/\Delta T)$
Amp Amplitude of corrugation
B, B_0 Magnetic induction (T)
C Inertial coefficient
C_p Specific heat capacity at constant pressure (J kg^{-1} K^{-1})
Da Darcy number, K/H^2
D_B Coefficient of Brownian diffusion (m^2 s)
d_p Nanoparticle diameter (m)
D_T Thermopheritic coefficient (m^2 s)
E Electric field (V m^{-1})
Ec Eckert number, $U_i^2/C_{pbf}\Delta T$
g Gravitational acceleration (m s^{-2})

$\mathbf{H}$ Magnetic field and channel height (A m^{-1}, m)

$\mathbf{Ha}$ Hartman number, $\left(\sigma_{bf} B_0^2 H^2 / \mu\right)^{1/2}$

$\mathbf{I}$ Electric current (A)

$\mathbf{J}$ Electric current density (A m^{-2})

$\mathbf{k}$ Thermal conductivity (W m^{-1} K^{-1})

$\mathbf{K}$ Porous medium permeability (m^2)

$\mathbf{k_B}$ Constant of Boltzmann (J K^{-1})

$\mathbf{Le}$ Lewis number, $k_{bf} / D_B \left(\rho C_p\right)_{bf}$

$\mathbf{M}$ Magnetization (A m^{-1})

$\mathbf{Mn}$ Magnetic number, $\mu_0 K' \Delta T I_l^2 / 4\pi^2 H^2 \rho_{bf} U_i^2$

$\mathbf{n}$ Number of corrugations

$\mathbf{N_B}$ Parameter of Brownian motion, $D_B \left(\rho C_p\right)_p \phi_i / k_{bf}$

$\mathbf{N_T}$ Parameter of thermophoresis effect, $D_T \left(\rho C_p\right)_p \Delta T / k_{bf}$

$\mathbf{Nu}$ Nusselt number

$\mathbf{P}$ Pressure (Pa)

$\mathbf{Pr}$ Prandtl number, $\mu C_{pbf} / k_{bf}$

$\mathbf{q}$ Heat flux density (W m^{-2})

$\mathbf{r}$ Radius of elliptic block (m)

$\mathbf{Re}$ Reynolds number, $\rho_{bf} U_i H / \mu$

$\mathbf{Ri}$ Richardson number, $g\beta_{bf} \Delta T H / U_i^2$

$\mathbf{R_{int}}$ Intensity ratio, I_r / I_l

$\mathbf{R_k}$ Ratio of thermal conductivity, k_{eff} / k_{nf}

$\mathbf{R_m}$ Magnetic Reynolds number

$\mathbf{R_\mu}$ Ratio of dynamic viscosity, μ_{eff} / μ

$\mathbf{S}$ Entropy generation

$\mathbf{T}$ Temperature (K)

$\mathbf{T_c}$ Curie temperature (K)

$\mathbf{T_{fr}}$ Freezing temperature (K)

$\mathbf{\Delta T}$ Temperature difference, $q_h H / k_{bf}$ or $T_h - T_c$

Greek symbols

$\mathbf{\mu}$ Dynamic viscosity (kg m^{-1} s^{-1})

$\mathbf{\sigma}$ Electrical conductivity (Ω^{-1} m^{-1})

$\mathbf{\rho}$ Density (kg m^{-3})

$\mathbf{\varepsilon}$ Porosity

$\mathbf{\theta}$ Dimensionless temperature

$\mathbf{\alpha}$ Orientation angle (°)

$\mathbf{\beta}$ Coefficient of thermal expansion (K^{-1})

$\mathbf{\phi, \varphi}$ Nanoparticles volume fraction

$\mathbf{\mu_0}$ Void magnetic permeability (T m A^{-1})

Subscripts

$\mathbf{bf}$ Base fluid

$\mathbf{c}$ Cold

$\mathbf{eff}$ Effective

$\mathbf{h}$ Hot

i Inlet
l Left
low Lower
m Mean
max Maximum
min Minimum
nf Nanofluid
p Nanoparticle
r Right
up Upper

References

[1] P. Teertstra, M.M. Yovanovich, J.R. Culham, Analytical forced convection modeling of plate fin heat sinks, J. Electron. Manuf. 10 (2000) 253—261.

[2] Rajesh Baby, C. Balaji, Experimental investigations on phase change material based finned heat sinks for electronic equipment cooling, Int. J. Heat Mass Tran. 55 (2012) 1642—1649.

[3] A. Arshad, M. Jabbal, S.P. Talebizadeh, B.M. Anser, H. Faraji, Y. Yan, Transient simulation of finned heat sinks embedded with PCM for electronics cooling, Therm. Sci. Eng. Prog. 18 (2020) 100520.

[4] M. Muneeshwaran, M.-K. Tsai, C.-C. Wang, Heat transfer augmentation of natural convection heat sink through notched fin design, Int. Commun. Heat Mass Tran. 142 (2023) 106676.

[5] S.U.S. Choi, Enhancing thermal conductivity of fluids with nanoparticles, developments and applications of non-Newtonian flows, ASME J. Heat Tran. 66 (1995) 99—105.

[6] J.C. Maxwell, A Treatise on Electricity and Magnetism, second ed., vol. 1, Clarendon Press, Oxford, 1881.

[7] O.A. AbdRabbuh, A.H. Abdelrazek, S.N. Kazi, M.N.M. Zubir, Nanofluids thermal performance in the horizontal annular passages: a recent comprehensive review, J. Therm. Anal. Calorim. 147 (2022) 11633—11660.

[8] M.A. Khattak, A. Mukhtar, S.K. Afaq, Application of nanofluids as a coolant in heat exchangers: a review, Journal of Advanced Research in Materials Science 66 (2020) 8—18.

[9] C.Y. Tsai, H.T. Chien, P.P. Ding, B. Chan, T.Y. Luh, P.H. Chen, Effect of structural character of gold nanoparticules in nanofluid on heat pipe thermal performance, Mater. Lett. 58 (2004) 1461—1465.

[10] H.B. Ma, C. Wilson, B. Borgmeyer, K. Park, Q. Yu, S.U. S Choi, M. Tirumala, Effect of nanofluid on the heat transport capability in an oscillating heat pipe, Appl. Phys. Lett. 88 (2006) 143116.

[11] S.C. Tzeng, C.W. Lin, K.D. Huang, Heat transfer enhancement of nanofluids in rotary blade coup'ing of four-wheel-drive vehicles, Acta Mech. 179 (2005) 11—23.

[12] Z. Zhang, Q. Que, J. Zhang, Synthesis, structure and lubricating properties of dialkyidithiophosphate-modified Mo-S compound nanoclusters, Wear 209 (1997) 8—12.

[13] S.M. You, J.H. Kim, K.H. Kim, Effect of nanoparticles on critical heat flux of water in pool boiling heat transfer, Appl. Phys. Lett. 83 (2003) 3374—3376.

[14] P. Vassallo, R. Kumar, S. D'Amico, Pool boiling heat transfer experiments in silica-water nanofluids, Int. J. Heat Mass Tran. 47 (2004) 407—411.

[15] A. Jordan, R. Scholz, P. Wust, H. Fahling, R. Felix, Magnetic fluid hyperthermia (MFH): cancer treatment with AC magnetic field induced excitation of biocompatible superparamagnetic nanoparticles, J. Magn. Magn. Mater. 201 (1999) 416–419.

[16] Y. Ding, D. Wen, Particle migration in a flow of nanoparticle suspensions, Powder Technol. 149 (2005) 84–92.

[17] S. Mirmasoumi, A. Behzadmehr, Numerical study of laminar mixed convection of a nanofluid in a horizontal tube using two-phase mixture model, Appl. Therm. Eng. 28 (2008) 717–727.

[18] V. Bianco, O. Manca, S. Nardini, Numerical investigation on nanofluids turbulent convection heat transfer inside a circular tube, Int. J. Therm. Sci. 29 (2009) 3632–3642.

[19] A. Akbarinia, R. Laur, Investigating the diameter of solid particles effects on a laminar nanofluid flow in a curved tube using a two phase approach, Int. J. Heat Fluid Flow 30 (2009) 706–714.

[20] J. Buongiorno, Convective transport in nanofluids, ASME J. Heat Mass Trans. 128 (2006) 240–250.

[21] J. Koo, C. Kleinstreuer, A new thermal conductivity model for nanofluids, J. Nanoparticle Res. 6 (2004) 577–588.

[22] R.S. Prasher, P. Bhattacharya, P.E. Phelan, Brownian motion based convective-conductive model for the effective thermal conductivity of nanofluids, ASME J. Heat Mass Trans. 128 (2006) 588–595.

[23] B. Yang, Thermal conductivity equations based on Brownian motion in suspensions of nanoparticles (nanofluids), ASME J. Heat Mass Trans. 130 (2008) 1–5.

[24] J. Tyndall, On Haze and dust, Proc. Roy. Soc. Lond. 6 (1870) 1–6.

[25] P. Berger, N.B. Adelman, K.J. Beckman, D.J. Campbell, A.B. Ellis, G.C. Lisensky, Preparation and properties of an aqueous ferrofluid, J. Chem. Educ. 76 (1999) 943.

[26] L.M. Pop, C.D. Buioca, V. Iusan, M. Zimnicaru, Long-term stability of magnetic fluids in low-gradient magnetic fields, J. Magn. Magn. Mater. 252 (2002) 46–48.

[27] I. Bolshakova, M. Bolshakova, A. Zaichenko, A. Egorov, The investigation of the magnetic fluid stability using the devices with magnetic field microsensors, J. Magn. Magn. Mater. 289 (2002) 108–110.

[28] W. Huang, J. Wu, W. Guo, R. Li, L. Cui, Study on the magnetic stability of iron-nitride magnetic fluid, J. Alloys Compd. 443 (2007) 195–198.

[29] T. Guo, X. Bian, C. Yang, A new method to prepare water based Fe_3O_4 ferrofluid with high stabilization, Physica A 438 (2015) 560–567.

[30] S. Kamali, M. Pouryazdan, M. Ghafari, M. Itou, M. Rahman, P. Stroeve, H. Hahn, Y. Sakurai, Magntization and stability study of cobalt-ferrite-based ferrofluid, J. Magn. Magn Mater. 404 (2016) 143–147.

[31] F. Yu, Y. Chen, X. Liang, J. Xu, C. Lee, Q. Liang, P. Tao, T. Deng, Dispersion stability of thermal nanofluids, Prog. Nat. Sci. Mater. Int. 27 (2017) 531–542.

[32] S.N. Lysenko, S.A. Astafeva, D.E. Yakusheva, M. Balasoiu, Novel parameter predicting stability of magnetic fluids for possible application in nanocomposite preparation, Appl. Surf. Sci. 463 (2019) 217–226.

[33] D. Mavrudieva, Etiquettes magnétiques interrogeables à distance. Application à la mesure de température, Institut National Polytechnique de Grenoble – INPG, 2007.

[34] A. Zebib, Thermal convection in a magnetic fluid, J. Fluid Mech. 321 (1996) 121–136.

[35] V.G. Bashtovoy, B.M. Berkovsky, A.N. Vislovich, Introduction to Thermomechanics of Magnetic Fluids, Hemisphere, Springer-Verlag, Berlin, 1988.

[36] H. Matsuki, K. Yamasawa, K. Murakami, Experimental considerations on a new automatic cooling device using temperature-sensitive magnetic fluid, IEEE Trans. Magn. 13 (1977) 1143–1145.

[37] V.K. Pecharsky, K.A. Geschneidner Jr., Giant magnetocaloric effect in Gd_5 (Si_2Ge_2), Phys. Rev. Lett. 78 (1997) 4494−4497.

[38] E.E. Tzirtzilakis, A mathematical model for blood flow in magnetic field, Phys. Fluids 17 (2005) 077103.

[39] M. Sheikholeslami, D.D. Ganji, Ferrohydrodynamic and magnetohydrodynamic effects on ferrofluid flow and convective heat transfer, Energy 75 (2014) 400−410.

[40] R. Hesham, M. Abdel Aziz, S. Yehia, A.A. Ghani, Magnetothermal properties and magnetocaloric effect in $ErFe_2$ compound, Cryogenics 115 (2021) 103229.

[41] M. Hennel, L. Galdun, R. Varga, Analysis of magnetocaloric effect in Ni_2FeGa-coated microwires, J. Magn. Magn. Mater. 560 (2022) 169646.

[42] G. Salinas, C. Lozon, A. Kuhn, Unconventional applications of the magnetohydrodynamic effect in electrochemical systems, Curr. Opin. Electrochem. 38 (2023) 101220.

[43] H. Alfvén, Existence of electromagnetic-hydrodynamic waves, Nature 150 (1942) 405−406.

[44] R. Moreau, La magnétohydrodynamique ou ces fluides qui conduisent l'électricité, La Houille Blanche 5−6 (1994) 110−117.

[45] B. Fersadou, H. Kahalerras, W. Nessab, D. Hammoudi, Effect of Magnetohydodynamics on heat transfer intensification and entropy generation of nanofluid flow inside two interacting open rectangular cavities, J. Therm. Anal. Calorim. 138 (2019) 3089−3108.

[46] M. Corcione, Empirical correlating equations for predicting the effective thermal conductivity and dynamic viscosity of nanofluids, Energy Convers. Manag. 52 (2011) 789−793.

[47] W. Nessab, H. Kahalerras, B. Fersadou, D. Hammoudi, Numerical investigation of ferrofluid jet flow and convective heat transfer under the influence of magnetic sources, Appl. Therm. Eng. 150 (2019) 271−284.

[48] N. Guerroudj, B. Fersadou, K. Mouaici, H. Kahalerras, Ferrohydrodynamics mixed convection of a ferrofluid in a vertical channel with porous blocks of various shapes, J. Appl. Fluid Mech. 16 (2023) 131−145.

CHAPTER SIX

Impact of Ohmic heating and nonlinear radiation on Darcy—Forchheimer magnetohydrodynamics flow of water-based nanotubes of carbon due to nonuniform heat source

Khilap Singh[1], Padam Singh[2], Manoj Kumar[3]

[1]Department of Mathematics, H. N. B. Government Post Graduate College, Khatima, Uttarakhand, India
[2]Department of Mathematics, Galgotias College of Engineering and Technology, Greater Noida, Uttar Pradesh, India
[3]Department of Mathematics, Statistics and Computer Science, G. B. Pant University of Agriculture and Technology, Pantnagar, Uttarakhand, India

Highlights:

- Hydrothermal performance of magnetohydrodynamic water-based carbon nanotubes from bidirectional stretchable surface.
- Influence of nonuniform heat source, radiation, and Ohmic heating is also analyzed.
- Fourth-fifth order method of Runge—Kutta—Fehlberg and shooting scheme are used to handle relevant equations.

1. Introduction

1.1 Carbon nanotubes

Carbon nanotubes (CNTs) are nanosize cylindrical hollow shapes of concentric graphitic carbon layered by fullerene-like hemispheres. Due to high electric conductivity, slight compactness, and excellent stiffness of CNTs, they have attracted huge interest of investigators in the recent past decades. The first time in 1919, Ijima [1] discovered this new type of materials, that is, CNTs. The CNTs have been used in various engineering and scientific procedures such as biological systems, catalytic membranes, storage devices and electrochemical power conversion, polymer composites, and actuation according to Sadiq et al. [2]. Choi et al. [3] depicted that thermal conductivity of resulting fluid increased if carbon nanotubes are mixed in oil. Xue [4] formed a model to observe conductivity of thermal of a mixture with carbon nanotube. Ding et al. [5] proposed the heat exchange features of watery suspensions of CNT nanofluids (carbon nanotubes with multiple walls) passing through a flat tube. They found that with increasing variations of Reynolds number, the rate of convective heat transport raises. Wang et al. [6] analyzed that experimentally the pressure and heat transport of carbon nanotubes fluid drops from a circular tube which is horizontal. The results depicted that the nanoliquids with small concentration boost the rate of heat exchange by pump power with extra small penalty. Safaei et al. [7] considered the force convection turbulent heat transfer of water based functionalized multi-walled carbon nanotube (FMWCNT) via a forward-facing step. Ellahi et al. [8] observed the natural convection motion of carbon nanotubes with multi and single walls suspended in a water and salt mixture. Subsequently many researchers Hayat et al. [9], Iqbal et al. [10], Khan et al. [11], Hayat et al. [12], Rehman et al. [13], Bilal et al. [14], Kandasamy et al. [15], Turkyilmazoglu [16], Abdulkadhimet al. [17], Sedeh et al. [18], and Sarafraz et al. [19] studied the different aspect of heat transfer and flow features of CNTs via various geometries with different effects.

Generally, CNTs are of two kinds, first carbon nanotube with single–wall (SWCNTs) and second one is carbon nanotube with multi–walls (MWCNTs). These materials, including CNTs of one-dimensional, two-dimensional graphene, and zero-dimension fullerenes according to Mohajeri et al. [20], are shown in Fig. 6.1. Karttunen et al. [21] considered nanostructures based on Ge_9-clusters with zero, one, and two-dimensional. Alshehri et al. [22] studied biomedical utilizations of carbon nanotubes such as remedies of toxicity, mechanisms, and factors. Georgakilas et al. [23] reviewed the noncovalent behavior of graphene and graphene oxide with different materials such as carbon allotropes (carbon nanotubes, nanodiamonds, and fullerenes), graphene analogs (MoS_2, WS_2), magnetic nanostructures, quantum dots, nanoparticles of metal oxide-based, polymers, biomolecules, and drugs. Hong et al. [24] analyzed the nanomedicinal therapy and biological imaging significance of carbon nanomaterials. Lee et al. [25] investigated the biomedical utilizations of biosensors based on graphene oxide. Recently, Pattnaik et al. [26] and Zheng et al. [27] have studied carbon nanomaterials-based

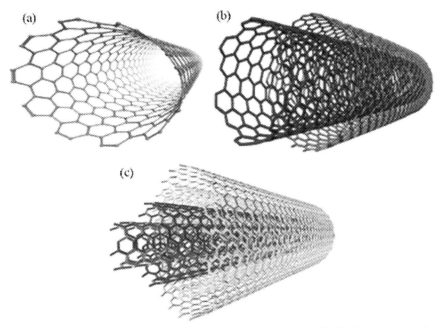

Figure 6.1 Illustrations of **(A)** carbon nanotube with single-wall. **(B)** Carbon nanotube with double-walls. **(C)** Carbon nanotube with multi-walls.

development in nonviral gene transfer, regenerative medicine drug delivery, and tissue engineering. The length of CNTs may be between 0.1 and 10 nm as considered in Bethune et al. [28], Dresselhaus et al. [29], Thomas [30], and Dresselhaus et al. [31]. The thermal, electronic, mechanical, and transport features of both MWCNTs and SWCNTs are better than traditional liquids reported by Refs. [32–44]. The carbon nanotubes are extensively used in various military, spaces, medical and electronic devices due to their unique features by Klaine et al. [45], Mauter and Elimelech [46], and Keller et al. [47]. Also, films formed by CNTs and membrane may be also utilized in filtration of different organic contaminants from the water and a hydro purification system [48–50]. The maximum analysis of CNTs is devoted to restraining their chirality (Zhu et al. [51], and Reich et al. [52]), diameter [53–59], length [60–62], wall number [63], and in all these above investigations, CNTs were of cross-circular section shapes. Autreto et al. [64] discussed carbon nanotubes with square cross-sections. Mizutani and Kohno [65] reported rectangular or square cross-section MWCNTs. Abramyan et al. [66] studied flows and equilibrium characters of nonpolar and polar fluids with square cross-section single-wall carbon nanotubes (SWCNTs) by applying molecular dynamics (MD) simulations, and they found that liquid flows via them and equilibrium liquid formations within such SWCNTs are significantly dissimilar from those circular cross-section SWCNTs. Recently, Zare and Rhee [67], Anuar et al. [68], Freitas et al. [69], Khana et al. [70], Muhammad et al. [71], Anuar et al. [72], Mirantsev and Abramyan [73], and Hosseinzadeh et al. [74] have studied about various aspects of CNTs flow via different geometries influenced by various effects. Electrical conductivity of pure carbon nanotube yarns has been discussed by Miao [75]. Muhammad [76] explored the mixed impacts of heat absorption/generation and nonlinear radiation of thermal magnetohydrodynamics carbon nanotubes via a stretchable sheet. Nadeem et al. [77] evaluated the influence of thermal conductivity on the heat transport of SWCNT and MWCNT. Sun et al. [78] discovered the behavior and utilization of carbon nanotube of different numbers of walls.

1.2 Nonlinear thermal radiation

Due to fast development of human society, various power-related problems occur like storage of global power and environmental pollution. The numerous research communities are working on to developing new resources of sustainable power. The key source of renewable power is solar power which can resolve these problems. The nonlinear thermal radiation is very useful in chemical procedures at high working temperature, combustion chambers, propulsion devices, aircraft, atomic plants, sun-oriented power technology, space technology, furnace design, polymer processing, etc. Cortell [79] first investigated the influence of nonlinear radiation of thermal in fluid motion via a stretching surface. Sheikholeslami et al. [80] discovered the impact of radiation on magnetohydrodynamic motion of nanomaterial. Reddy et al. [81] considered nonlinear radiation effect in magnetohydrodynamics (MHD) ferrofluids motion due to temperature reliant on viscosity. Hayat et al. [82] examined the role of nonlinear radiation on non–Darcy motion of nanotubes of carbon based in water with heat absorption and generation. Mahanthesh et al. [83] deliberated the impact of nonuniform heat generation and radiation on the Marangoni convection MHD flow of MWCNT and SWCNT nanofluids with viscous dissipating. Archana et al. [84] observed the effect of magnetohydrodynamic on three-dimensional flow of a Maxwell nanofluid due to nonlinear thermal radiation. Mahanthesh et al. [85] explained the role of time-dependent velocity and nonlinear radiation of thermal on motion of dusty nanoliquid in an isothermal plate with exponential heat source. Furthermore, investigations about nonlinear thermal radiative heat transfer is discussed in articles Hayat et al. [86], Hayat et al. [87], Khan et al. [88], Gireesha et al. [89], and Ghadikolaei et al. [90].

1.3 Darcy–Forchheimer porous space

The fluid flow via porous space has much utilization, for example, in foams and foamed solids, porous rocks, and aerogels to reinforce polymer blends and soils, microemulsions, cement grouts, alloys, slurries, muds injection, drilling liquids, and so on. The famous law of Darcy's is adequate for fluid motion with low velocity and weak porosity. The law of Darcy's is also applicable in turbulence flows and other high movement impacts in porous medium. Consequently, by use of a nonlinear relationship between pressure gradient and filtration velocity, Darcy's equation improved by Forchheimer [91]. Muskat [92] verified its authority for greater values of Reynolds number. Hommel et al. [93] reviewed relations of permeability-porosity for developing pore medium in view of biogeochemically changed porous space (see Fig. 6.2). Seddeek [94] addressed mixed convection Darcy–Forchheimer

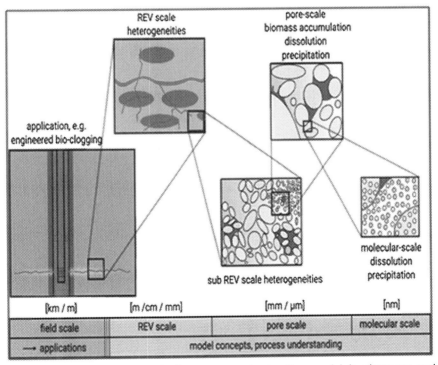

Figure 6.2 Porous medium scales for process understanding, model development, and applications.

flow in a liquid-embedded porous space. Subsequently, many researchers Ahmed [95], Bakar et al. [96], Hayat et al. [97], Umavathi et al. [98], Hayat et al. [99], Alamri et al. [100], and Ellahi et al. [101] have considered the motion in porous space of Darcy–Forchheimer.

Magnetohydrodynamics is the study of dynamics of electronically conducting fluids. Tahir et al. [102] investigated effect of thermal radiation and MHD on nanofluid film of Maxwell fluid through an unstable stretching sheet. In some studies, researchers have taken the non-Newtonian and Newtonian liquids as a property of mass and heat transport from a time-independent and a time-dependent extending surface by using some physical quantities. The Darcy flow is an expression that illustrates liquid motion along a porous space. The study of porous medium and some of its application can be seen in Vafai [103], and Nield and Bejan [104].

Keeping this in mind, the basic objective of current analysis is to study magnetohydrodynamic flow of both multiple and single CNTs based in water through bidirectional stretching surface. By using Darcy–Forchheimer rule, the expression of porous space has been described. Moreover, the influence of nonlinear heat generation and thermal radiation, and Ohmic heating has been also deliberated. The physical significance of governing variables in the Nusselt number and skin-friction coefficients as well as temperature and velocities is carefully examined. It is anticipated that the current findings may provide suitable information for various industrial purpose and academic research.

2. Mathematical modelling of carbon nanotubes flow and heat transfer

An electrically conducting, incompressible steady laminar three-dimensional boundary layer, Darcy–Forchheimer motion of water-based nanotubes of carbon via a nonconducting, stretchable, and bidirectional surface is considered. The multiple (MWCNTs) as well as single (SWCNTs) wall carbon nanotubes are used in the present study. The nonlinear thermal radiation, nonuniform source of heat, and Joule heating effects are combined in energy expression. The graphical representation of model is given together with coordinate system and configuration of flow in Fig. 6.3. We use rectangular Cartesian coordinates (x, y, z); $\vec{V} = (u, v, w)$ is vector of fluid velocity. A strong field of magnetic $\vec{B} = (0, 0, B_0)$ is employed in z-axis with B_0 as the strength of uniform applied magnetic field. Also, it is supposed that velocities of the stretchable surfaces are $V_w = by$ and

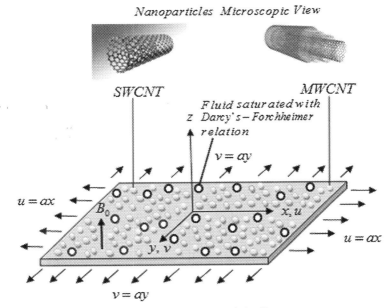

Figure 6.3 Geometry of the flow.

$U_w = ax$, where b and a are the semiradii in y-axis and x-axis, respectively. The ambient and surface temperatures are T_∞ and T_w. The Ion-slip, Hall, and viscous dissipation outcomes are assumed to be insignificant; the pertinent expressions of heat transport and flow in vector form given by Sutton and Sherman [105] are as follows:

Conservation expression of mass:

$$div \vec{V} = 0 \tag{6.1}$$

Conservation expression of momentum:

$$\left(\vec{V}.\nabla\right)\vec{V} = -\frac{1}{\rho_{nf}}\nabla p - \frac{v_{nf}}{k_p}\vec{V} - F_0 \vec{V}.\vec{V} + v_{nf}\nabla^2 \vec{V} + \vec{J} \times \vec{B} \tag{6.2}$$

Conservation expression of energy:

$$\left(\vec{V}.\nabla\right)\vec{T} = \frac{k_{nf}}{\left(\rho C_p\right)_{nf}}\nabla^2 \vec{T} - \frac{1}{\left(\rho C_p\right)_{nf}}\nabla . q_r + \Theta \tag{6.3}$$

Current density vector:

$$\vec{J} = \sigma_{nf}\left(\vec{E} + \vec{V} \times \vec{B}\right) \tag{6.4}$$

Maxwell's equations:

$$div\,\overrightarrow{B}=0 \tag{5a}$$

$$\nabla\times\overrightarrow{H}=\overrightarrow{J} \tag{5b}$$

$$\nabla\times\overrightarrow{E}=0 \tag{5c}$$

where Θ is energy loss function due to Joule heating and nonuniform source of heat, $\overrightarrow{J}$ is the vector of current density, $\overrightarrow{E}$ is intensity of electric force, and $\overrightarrow{H}$ is the intensity of magnetic force. To simplify the model, we suppose that there is no change in heat transfer and flow amounts in z-axis. The expression of the conservation of the electric charge $\Delta.\overrightarrow{J}=0$ indicates that J_z is constant and also the surface is nonconducting; therefore, $J_z=0$ all over in the fluid. Further, $E_x=0$ and $E_y=0$ everywhere in the flow; hence in the present study, we assumed that induced magnetic force is ignored in contrast of used magnetic force, but the Joule heating and nonuniform heat source impacts in fluid are considered.

Under all these above suppositions, by use of Boussinesq approaches and boundary layer into consideration, the above expressions can be simplified into Cartesian coordinates form as follows (Hayat et al. [82]):

$$\frac{\partial u}{\partial x}+\frac{\partial v}{\partial y}+\frac{\partial w}{\partial z}=0 \tag{6.6}$$

$$u\frac{\partial u}{\partial x}+v\frac{\partial u}{\partial y}+w\frac{\partial u}{\partial z}=v_{nf}\frac{\partial^2 u}{\partial z^2}-\frac{v_{nf}}{k_p}u-F_0 u^2-\frac{\sigma_{nf}B_0^{\,2}}{\rho_{nf}}u \tag{6.7}$$

$$u\frac{\partial v}{\partial x}+v\frac{\partial v}{\partial y}+w\frac{\partial v}{\partial z}=v_{nf}\frac{\partial^2 v}{\partial z^2}-\frac{v_{nf}}{k_p}v-F_0 v^2-\frac{\sigma_{nf}B_0^{\,2}}{\rho_{nf}}v \tag{6.8}$$

$$\left(\rho C_p\right)_{nf}\left(u\frac{\partial T}{\partial x}+v\frac{\partial T}{\partial y}+w\frac{\partial T}{\partial z}\right)=k_{nf}\frac{\partial^2 T}{\partial z^2}-\frac{\partial q_r}{\partial z}+\sigma_{nf}B_0^{\,2}\left(u^2+v^2\right)+q''' \tag{6.9}$$

Following boundary conditions are proposed to the above partial differential Eqs. (6.6)−(6.9):

$$u=U_w=ax, \quad v=by, \quad w=0, \quad T=T_w,\, \text{at } z=0,$$
$$u=0, \quad v=0, \quad T\rightarrow T_\infty,\, \text{as } z\rightarrow\infty. \tag{6.10}$$

Table 6.1 Thermophysical properties of nanofluid.

Dynamic viscosity	$\mu_{nf} = \dfrac{\mu_f}{(1-\phi)^{2.5}}$
Kinematic viscosity	$v_{nf} = \dfrac{\mu_{nf}}{\rho_{nf}}$
Density	$\rho_{nf} = \rho_f(1-\phi) + \rho_{CNT}\phi$
Heat capacity	$(\rho C_p)_{nf} = (\rho C_p)_f(1-\phi) + \phi(\rho C_p)_{CNT}$
Thermal conductivity	$\dfrac{k_{nf}}{k_f} = \dfrac{(1-\phi)+2\phi\dfrac{k_{CNT}}{k_{CNT}-k_f}\ln\dfrac{k_{CNT}-k_f}{2k_f}}{(1-\phi)+2\phi\dfrac{k_{CNT}}{k_{CNT}-k_f}\ln\dfrac{k_{CNT}+k_f}{2k_f}}$
Electrical conductivity	$\dfrac{\sigma_{nf}}{\sigma_f} = 1 + \dfrac{3(\sigma-1)\phi}{(\sigma+2)-(\sigma-1)\phi}$, where $\sigma = \dfrac{\sigma_{CNT}}{\sigma_f}$

Thermophysical attributes of nanofluid proposed in Xue [4], Devi and Devi [106], and Saqib et al. [107] are given in Table 6.1 as follows:

Here, ϕ is volume fraction of nanoparticle solid. Physical properties for water-based SWCNTs and MWCNTs are presented in Table 6.2.

2.1 Nonuniform heat source

The temperature and space-related nonuniform heat source are modeled as (Pal and Mondal [108], Sandeep and Sulochana [109]):

$$q''' = \frac{k_{nf}U_w}{x v_f}\left(Q(T_w - T_\infty)f' + Q^*(T - T_\infty)\right) \qquad (6.11)$$

Here Q and Q^* are the internal heat source parameters of space and temperature.

2.2 Rosseland approximation for radiation

The approximation of Rosseland used to optically abundant media and gives the total heat flux of radiation (see Rosseland [110], Pantokratoras and Fang [111]) via equation:

Table 6.2 Physical properties for water-based SWCNTs and MWCNTs.

	Base fluid	Nanoparticles	
Physical properties	Water	SWCNT	MWCNT
$\rho(kg/m^3)$	997	2600	1600
$C_p(J/kgK)$	4179	425	796
$k(W/mK)$	0.613	6600	3000
$\sigma(\Omega m)$	0.05	$10^{-6} - 10^{-7}$	1.9×10^{-4}

$$q_r = -\frac{4}{3k^*} grad(e_b) \tag{6.12}$$

In the above expression, e_b is black body emissive power which is expressed in the form of absolute temperature T using the low of Stefan—Boltzmann of radiation $e_b = \sigma^* T^4$, with constant of Stefan—Boltzmann $\sigma^* = 5.6697 \times 10^{-8} Wm^{-2}K^{-4}$ and k^* is the coefficient of absorption mean.

The boundary layer total radiation heat flux for a plane motion via a hot surface becomes:

$$q_r = -\left(\frac{16\sigma^*}{3k^*}\right) T^3 \frac{\partial T}{\partial z} \tag{6.13}$$

Eq. (6.13) of nonlinear thermal radiation is considered by many researchers Pal and Saha [112], Clouet [113], Krishna et al. [114], Khan et al. [115], Ibrahim and Shankar [116], and Khan et al. [117].

Substituting Eq. (6.12) into Eq. (6.9), the energy expression reduces to:

$$(\rho C_p)_{nf}\left(u\frac{\partial T}{\partial x}+v\frac{\partial T}{\partial y}+w\frac{\partial T}{\partial z}\right)=k_{nf}\frac{\partial^2 T}{\partial z^2}+\left(\frac{16\sigma^*}{3k^*}\right)\frac{\partial}{\partial z}\left(T^3\frac{\partial T}{\partial z}\right)+$$
$$\sigma B_0{}^2\left(u^2+v^2\right)+\frac{k_{nf}U_w}{x\upsilon_f}\left(Q(T_w-T_\infty)f'+Q^*(T-T_\infty)\right) \tag{6.14}$$

2.3 Similarity analysis

The similarity variables anticipated by Hayat et al. [82] are given as:

$$\eta=\sqrt{\frac{a}{\upsilon_f}}z, \qquad \theta=\frac{T-T_\infty}{T_w-T_\infty}, \qquad u=axf'(\eta),$$
$$v=ayg'(\eta), \quad w=-\sqrt{a\upsilon_f}(f(\eta)+g(\eta)) \tag{6.15}$$

By applying the above similarity transformation (15), the expression (6) is satisfied evidently and expressions (7), (8), and (14) are changed by a set of nonlinear ODEs as follows:

$$\frac{A_1}{A_2}(f'''-\Lambda f')+f''(f+g)-(1+F_1)f'^2-\frac{A_3 M}{A_2}f'(\eta)=0 \tag{6.16}$$

$$\frac{A_1}{A_2}(g'''-\Lambda g')+g''(f+g)-(1+F_1)g'^2-\frac{A_3 M}{A_2}g'(\eta)=0 \tag{6.17}$$

$$\left. \begin{array}{l} \dfrac{1}{A_4}\left[\dfrac{k_{nf}}{k_f}\theta'' + R(\theta(\theta_w - 1) + 1)^2 \times \left\{3\theta'^2(\theta_w - 1) + \theta''(\theta(\theta_w - 1) + 1)\right\}\right] + \\[3mm] \Pr\,\theta'(f + g) + \left(Qf' + Q^*\theta\right) + \Pr M\left(Ec_x f'^2 + Ec_y g'^2\right) = 0 \end{array} \right\}$$

$$(6.18)$$

The corresponding conditions of boundary (10) are transformed as follows:

$$
\begin{array}{lllll}
f(0) = 0, & f'(0) = 1, & g(0) = 0, & g'(0) = \alpha, & \theta(0) = 1, \\[2mm]
f'(\infty) \to 0, & g'(\infty) \to 0, & \theta(\infty) \to 0.
\end{array}
$$

$$(6.19)$$

where A_1, A_2, A_3 and A_4 are invariable given by:

$$A_1 = \frac{1}{(1 - \phi)^{2.5}}, \qquad A_2 = 1 - \phi + \frac{\rho_{CNT}}{\rho_f}\phi,$$

$$A_3 = 1 + \left(\frac{3(\sigma - 1)\phi}{2 + \sigma - (\sigma - 1)\phi}\right), \quad A_4 = (1 - \phi) + \frac{(\rho C_p)_f}{(\rho C_p)_{CNT}}\phi$$

$$(6.20)$$

2.4 Declaration of curiosity

2.4.1 Skin friction coefficients

Mathematically, the coefficients of skin-friction in the direction of x and y-axis (C_{fx} and C_{fy}) are described as:

$$C_{fx} = \frac{2\tau_{wx}}{\rho_f U_w^2}, \quad C_{fy} = \frac{2\tau_{wy}}{\rho_f V_w^2}$$

$$(6.21)$$

Here τ_{wy} and τ_{wx} are shear stresses of wall in the direction of y and x-axis, respectively, are specified as follows:

$$\tau_{wy} = \mu_{nf}\left(\frac{\partial v}{\partial z}\right)\bigg|_{z=0}, \quad \tau_{wx} = \mu_{nf}\left(\frac{\partial u}{\partial z}\right)\bigg|_{z=0}$$

$$(6.22)$$

By inserting Eq. (6.22) in Eq. (6.21), one has:

$$C_{fx}\mathrm{Re}_x^{1/2} = \frac{2f''(0)}{(1 - \phi)^{2.5}}, \quad C_{fy}\mathrm{Re}_y^{1/2} = \frac{2g''(0)}{(1 - \phi)^{2.5}}$$

$$(6.23)$$

2.4.2 Nusselt number

Mathematically, the heat flux of wall q_w is defined as:

$$q_w = (q_r)_{z=0} - k_{nf} \left(\frac{\partial T}{\partial z} \right) \Bigg|_{z=0} \tag{6.24}$$

The Nusselt number Nu_x is described as follows:

$$Nu_x = \frac{x q_w}{k_f (T_w - T_\infty)} \tag{6.25}$$

By putting Eq. (6.24) in Eq. (6.25), we have:

$$Nu_x Re_x^{-1/2} = - \left(\frac{k_{nf}}{k_f} + Nr(\theta(0)(\theta_w - 1) + 1)^3 \right) \theta'(0) \tag{6.26}$$

3. Numerical methods of solution

In the present chapter, the shooting scheme and RKF technique are utilized to achieve solution of system of ordinary differential expressions (6.16)−(6.18) together with boundary conditions (6.19).

3.1 Runge−Kutta−Fehlberg Method

The Runge−Kutta−Fehlberg (RKF) scheme is utilized to investigating mathematical outcomes of ODEs formed by Erwin Fehlberg, a German mathematician. The Fehlberg method is related to the Runge−Kutta technique, in which $O(h^4)$ and $O(h^5)$ methods are used, so this method is called as four-fifth RKF technique. Two different approximations are founded and compared at every step of RKF method. The approximations are valid to close harmony and repeated to obtain acceptance for preferred accurateness. The formula of fifth-order RKF scheme is:

$$z_{n+1} = z_n + \left(\frac{16}{135}k_0 + \frac{6656}{12825}k_2 + \frac{28561}{56430}k_3 - \frac{9}{50}k_4 + \frac{2}{55}k_5 \right) h$$

where the coefficients k_0 to k_5 are defined as follows:

$$k_0 = f(x_n, y_n),$$

$$k_1 = f \left(x_n + \frac{1}{4}h, y_n + \frac{1}{4}hk_0 \right),$$

$$k_2 = f \left(x_n + \frac{3}{8}h, y_n + \left(\frac{3}{32}k_0 + \frac{9}{32}k_1 \right) h \right),$$

$$k_3 = f\left(x_n + \frac{12}{13}h, y_n + \left(\frac{1932}{2197}k_0 - \frac{7200}{2197}k_1 + \frac{7296}{2197}k_2\right)h\right)$$

$$k_4 = f\left(x_n + h, y_n + \left(\frac{439}{216}k_0 - 8k_1 + \frac{3680}{513}k_2 - \frac{845}{4104}k_3\right)h\right) \text{ and}$$

$$k_5 = f\left(x_n + \frac{1}{2}h, y_n + \left(-\frac{8}{27}k_0 + 2k_1 - \frac{3544}{2565}k_2 + \frac{1859}{4104}k_3 - \frac{11}{40}k_4\right)h\right).$$

By subtracting fifth-order RK scheme from fourth-order RK scheme, the error is obtained:

$$y_{n+1} = y_n + \left(\frac{25}{216}k_0 + \frac{1408}{2565}k_2 + \frac{2197}{4104}k_3 - \frac{1}{5}k_4\right)h$$

The results may be recomputed by the use of lesser step size, if error goes away from a specified accuracy. The formula to calculating new step size is defined by:

$$h_{new} = h_{old}\left(\frac{\varepsilon h_{old}}{2|z_{n+1} - y_{n+1}|}\right)^{1/4}.$$

For the criterion of convergence, the difference of nondimensional temperature, concentration, and velocity is making sure less than 10^{-6} in any two successive iterations. The RKF-45 scheme confers lesser error in contrast to Runge–Kutta scheme of fourth order or fifth order.

3.2 Shooting technique

The shooting technique converted the solutions of a boundary value model to iterative solutions of an initial value model. The standard methods occupy an error and trial process. The initial points are selected as known boundary points. Any unknown initial points are considered, and the system of initial value is handled by utilizing RK–45 technique. Unless the calculated consent with the well-known boundary equations, the initial equations be adjusted and system of equation is handled once more. The procedure is revised in anticipation of the supposed initial conditions obtained, with respect to particular tolerance, a result that consent through the well-known boundary equations. This technique needs well initial presumptions to absent initial values. The correctness of presumptions is decided related on how closely result satisfies the free stream boundary equations.

In the present chapter, the inbuilt ode-45 function of MATLAB has been employed to run the technique by presumption the absent initial values. The inbuilt ode-45 function is an adaptive step-size algorithm founded by integration method of RKF and utilizes a combination of fourth-order or fifth-order Runge–Kutta technique. This approach has been extensively implemented recently in many heat transfer problems including mass and radiative heat transport flow.

3.3 Implementation of methods

Consider $\theta = y_8$, $\theta = y_7$, $g'' = y_6$, $g' = y_5$, $g = y_4$, $f'' = y_3$, $f' = y_2$, and $f = y_1$. The nonlinear ordinary differential Eqs. (6.16)–(6.18) initially converted into first-order linear ODEs as follows:

$$y_1' = y_2;$$

$$y_2' = y_3;$$

$$y_3' = \Lambda y_2 - \frac{A_2}{A_1}\left(y_3(y_1 + y_4) - (1 + F_1)y_2^2 - \frac{A_3 M}{A_2}y_2\right);$$

$$y_4' = y_5;$$

$$y_5' = y_6;$$

$$y_6' = \Lambda y_5 - \frac{A_2}{A_1}\left(y_6(y_1 + y_4) - (1 + F_1)y_5^2 - \frac{A_3 M}{A_2}y_5\right);$$

$$y_7' = y_8;$$

$$y_8' = \frac{k_f}{k_{nf}}\left(\begin{array}{l}-R(\theta(\theta_w - 1) + 1)^2 \times \{3\theta'^2(\theta_w - 1) + \theta''(\theta(\theta_w - 1) + 1)\} \\ -A_4\left(\mathrm{Pr}\,\theta'(f + g) + (Qf' + Q^*\theta) + \mathrm{Pr}\,M(Ec_x f'^2 + Ec_y g'^2)\right)\end{array}\right)$$

Subject to the related initial conditions:

$$y_8(0) = \beta_3, \quad y_7(0) = 1, \quad y_6(0) = \beta_2, \quad y_5(0) = \alpha, \quad y_4(0) = 0,$$
$$y_3(0) = \beta_1, \quad y_2(0) = 1, \quad y_1(0) = 0.$$

Now, the province of model is discretized and boundary condition is converted as $y_2(\eta_{\max}) = 0$, $y_5(\eta_{\max}) = 0$ and $y_7(\eta_{\max}) = 0$, here $\eta_{\max}$ is maximum estimate of η (related to size of step $\nabla\eta$) at which conditions of boundary (6.19) are satisfied. The PC code type in the MATLAB and ran for various estimates of boundary layer length η and step length $\nabla\eta$, and it is seen that there is only an insignificant change or no change in velocities and temperature for various estimates of $\eta > 5$ and $\nabla \eta > 0.01$. Consequently, in this study, we have put boundary layer length $\eta = 5$ and step size $\nabla \eta = 0.01$ and standard of convergence is five decimals place exactness. In this chapter, five initial conditions are at $\eta = 0$ and three conditions are on boundary $\eta = \infty$. In order to find the result of model, three additional initial values at $\eta = 0$ (i.e., values of β_1, β_2, and β_3) are acquired using shooting scheme. At last, the converted problem of initial value has been handled by the use of RKF-45 method.

4. Result and discussion

In order to understand the behavior of Joule heating and thermal radiation in the occurrence of nonuniform heat source on MHD heat transfer and Darcy–Forchheimer flow field of water-based carbon nanotubes are plotted graphically and numerical results are obtained for the coefficients of heat transfer and skin-friction. To analyze the impacts of various major parameters that have been used in the present problem, some suppositions are taken to account. For nanofluid, multiple (MWCNTs) as well as single (SWCNTs) wall nanotubes of carbon have been utilized as nanomaterial and water as base fluid. Furthermore, assume thermal conductivity around carbon nanotubes is double than base fluid and width of carbon nanotube is 10 nm. A comparison between present outcomes and outcomes of Hayat et al. [82] is done in Table 6.3 for the authentication of the present model. Table 6.3 shows high accuracy of the outcomes founded in the present model. In this chapter, default values of parameters have been selected as $\phi = 0.03, F_1 = 0.2, Ec_x = Ec_y = 1.5, \alpha = 0.5, M = 2, \theta_w = 1.4, \mathrm{Pr} = 6.72, R = 1.4, Q = 1, Q^* = 0.5$ and $\Lambda = 1$ unless otherwise specified.

Table 6.3 Comparison of numerical data of $-C_{fx}\mathrm{Re}_x^{1/2}$ when $Ec_x = Ec_y = M = Q = Q^* = 0, \alpha = 0.4$ and $F_1 = 0.5$.

		Hayat et al. [82]		Present data	
ϕ	Λ	SWCNTs	MWCNTs	SWCNTs	MWCNTs
0.1	0.3	2.90211	2.86991	2.90208	2.86988
0.2	0.3	3.81395	3.71753	3.81391	3.71750
0.3	0.3	5.14122	4.90268	5.141217	4.90266
0.1	0.1	2.80485	2.76345	2.80480	2.76343
0.1	0.5	2.99937	2.97637	2.99933	2.97634

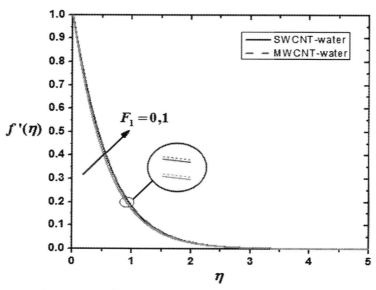

Figure 6.4 Velocity sketch $f'(\eta)$ in respect of various estimates of Forchheimer parameter F_1.

The distributions of temperature $\theta(\eta)$, velocities $f'(\eta)$ and $g'(\eta)$ for $F_1 = 0, 1$ are portrayed in Figs. 6.4–6.6. Figs. 6.4 and 6.5 delineate that the fluid velocity $f'(\eta)$ enhances with a rise in Forchheimer parameter F_1, while velocity profile $g'(\eta)$ declines with same parameter. The velocities of SWCNT-water flow are less than the velocities of MWCNT-water flow. The key reason behind this is that inertial force enhances with greater Forchheimer parameter values which result in decay in fluid velocity. It is clear from Fig. 6.6 that fluid temperature boosts by greater value of Forchheimer parameter F_1 for both SWCNT-water and MWCNT-water cases. Also, this figure depicts that the temperature is almost equal to MWCNT-water and SWCNT-water motions.

In order to understand the role of parameter of magnetic M on velocities $f'(\eta)$ and $g'(\eta)$, temperature $\theta(\eta)$, a graphical demonstration is present in Figs. 6.7–6.9. Figs. 6.7 and 6.8 depict that velocity $f'(\eta)$ and $g'(\eta)$ are decreasing functions of magnetic parameter for both conditions of MWCNT-water and SWCNT-water. Moreover, velocity of MWCNT-water flow is greater than the velocity of SWCNT-water flow. Physically the existence of magnetic field produced Lorentz force due to electric-magnetic field interface through the electro-related dynamic motion in fluid

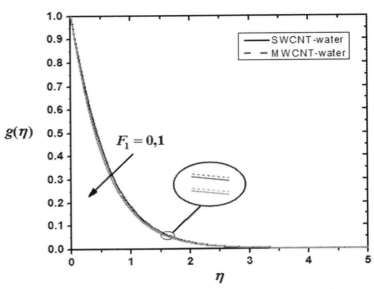

Figure 6.5 Velocity sketch $g'(\eta)$ in respect of various estimates of Forchheimer parameter F_1.

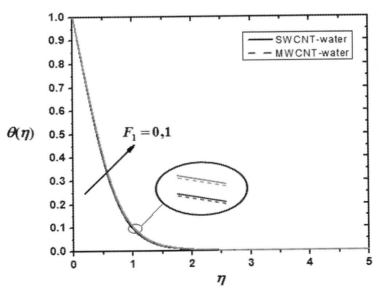

Figure 6.6 Temperature sketch $\theta(\eta)$ in respect of several estimates of Forchheimer parameter F_1.

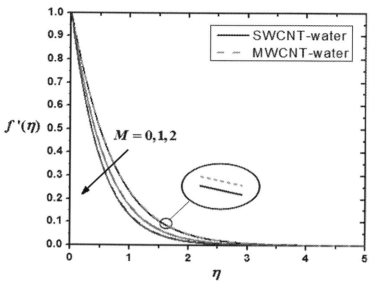

Figure 6.7 Velocity sketch $f'(\eta)$ in respect of various estimates of magnetic parameter M.

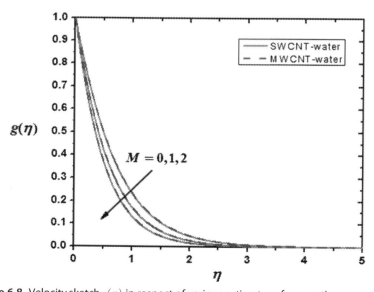

Figure 6.8 Velocity sketch $g(\eta)$ in respect of various estimates of magnetic parameter M.

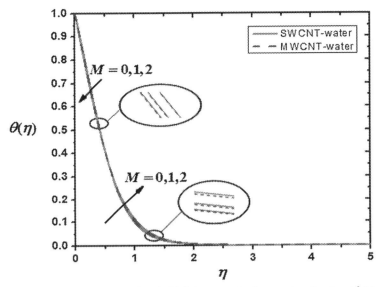

Figure 6.9 Temperature.sketch $\theta(\eta)$ in respect of several estimates of M.

motion which diminishes the liquid velocity. It is seen from Fig. 6.9 that rising values of magnetic parameter cause a decline in temperature near surface, but an opposite trend is seen far from surface. It is also clear from this figure that SWCNT-water and MWCNT-water flows have almost equal temperature.

The role of stretching parameter α on the velocities and temperature curves is illustrated by Figs. 6.10–6.12. Through Figs. 6.10 and 6.11, it is evident that the velocity along x-axis slow down with increase in estimates of α; however, opposite trend is observed for the velocity by the side of y-axis in the existence of SWCNTs as well as MWCNTs. It is obvious as of Fig. 6.12 that fluid temperature lessens with greater estimates of stretching parameter α for both cases of MWCNT and SWCNT.

The decreasing behavior of the temperature curves with incremental variations of Eckert numbers Ec_y and Ec_x in y and x-axis, respectively, and is depicted in Figs. 6.13 and 6.14. These figures depicted that the trend and magnitude of temperature are similar for both MWCNTs and SWCNTs. The main reason behind this is that viscosity of liquid may lose more energy with enhancing variations of Eckert numbers, which reduces internal fluid energy and consequently retards fluid temperature.

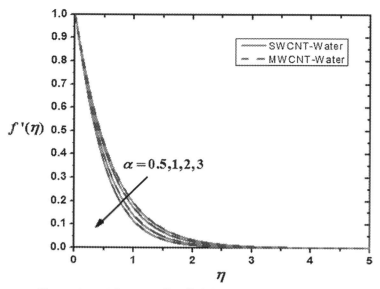

Figure 6.10 Velocity profile $f'(\eta)$ for various estimates of α.

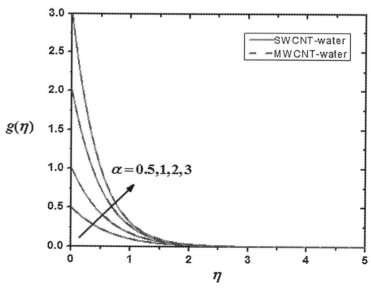

Figure 6.11 Velocity profile $g'(\eta)$ for various estimates of stretching parameter α.

Impact of Ohmic heating and nonlinear radiation on Darcy–Forchheimer 155

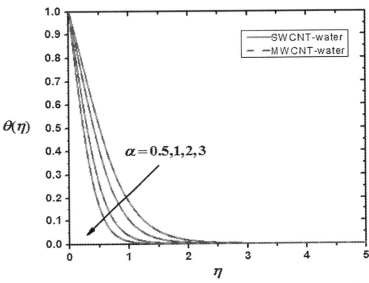

Figure 6.12 Temperature sketch $\theta(\eta)$ in respect of several estimates of α.

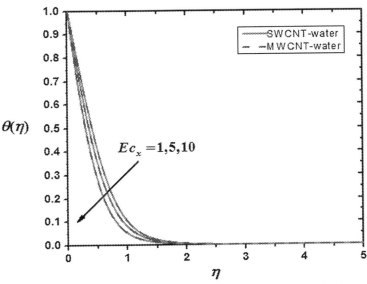

Figure 6.13 Temperature sketch $\theta(\eta)$ in respect of various estimates of Eckert number Ec_x.

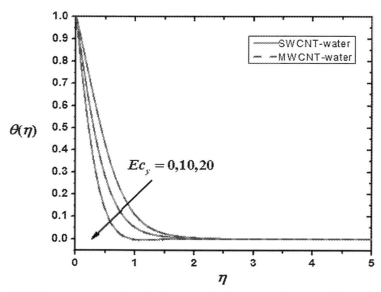

Figure 6.14 Temperature sketch $\theta(\eta)$ in respect of various estimates of Eckert number Ec_y.

Figs. 6.15 and 6.16 reveal that how temperature curves vary by space Q and temperature Q^*-dependent internal heat source parameters. These figures indicate that fluid temperature is a growing function of Q and Q^* for both SWCNT and MWCNT flows. Usually, the rising variations of space Q and temperature Q^*-dependent internal heat source parameters work as heat generator which raises the internal energy of liquid, thus an increment in fluid temperature.

The temperature profiles draw with the variations of thermal radiation parameter $(R = 0, 3, 5)$ as presented in Fig. 6.17. We have seen a rise in temperature profiles with rising estimates of R. Generally, the rate of radiation energy transfer of nanoparticles accelerates with augmenting values of R due to which nanofluid temperature boosts. Further, it is observed from all Figs. 6.4–6.17 that there is negligible variation in the velocities and temperature curves of SWCNTs and MWCNTs.

Table 6.4 reveals the numerical data of the skin-friction coefficients $\mathrm{Re}_x^{1/2} C_{fx}$ and $C_{fy} \mathrm{Re}_y^{1/2}$ in the x and y direction, respectively, for different variations of the nanotubes' solid volume fraction ϕ, Forchheimer parameter F_1, stretching parameter α, and magnetic parameter M. It is quite evident

Impact of Ohmic heating and nonlinear radiation on Darcy–Forchheimer 157

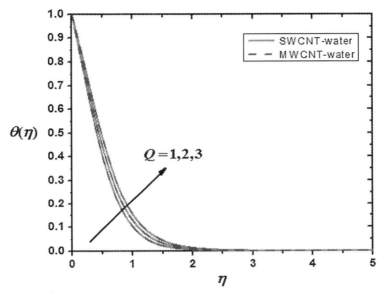

Figure 6.15 Temperature sketch in respect of various estimates of Q.

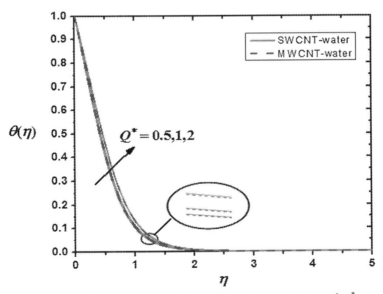

Figure 6.16 Temperature profile $\theta(\eta)$ for various estimates of Q^*.

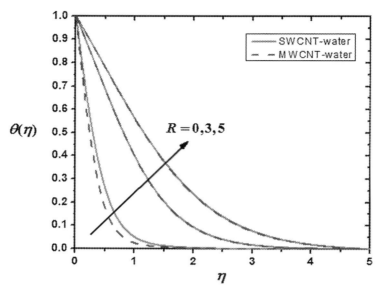

Figure 6.17 Temperature sketch $\theta(\eta)$ in respect of several estimates of R.

Table 6.4 Numerical estimates of $C_{fx}\text{Re}_x^{1/2}$ and $C_{fy}\text{Re}_y^{1/2}$ when $Ec_x = 1.5$, $Ec_y = 1.5$, $Q^* = 0.5$, $\text{Pr} = 6.72$, $R = 1.5$, $\phi = 0.03$, $\alpha = 0.5$, $F_1 = 1.2$, $Q = 1.0$, and $M = 2$.

				$\text{Re}_x^{1/2}C_{fx}$		$\text{Re}_y^{1/2}C_{fy}$	
ϕ	F_1	α	M	SWCNTs	MWCNTs	SWCNTs	MWCNTs
0.01	0.2	0.5	2	−4.2404656	−4.2243512	−2.0196442	−2.0122288
0.02				−4.3333833	−4.3010637	−2.0641374	−2.0492263
0.04				−4.5264496	−4.4610649	−2.1566427	−2.1264880
0.03	0			−4.3610062	−4.3134585	−2.0923469	−2.0701715
	0.8			−4.6259728	−4.5737346	−2.1613528	−2.1379374
	2.5			−5.1452436	−5.0840546	−2.3017451	−2.2758440
	0.2	0.6		−4.4462601	−4.3971874	−2.5682991	−2.5406445
		0.8		−4.4811358	−4.4314310	−3.5197628	−3.4811632
		1		−4.5156345	−4.4653098	−4.5156347	−4.4653109
		0.5	1	−3.8854100	−3.8466209	−1.8226904	−1.8055972
			3	−4.9125542	−4.8553100	−2.3624363	−2.3354435
			5	−5.7599645	−5.6884519	−2.8003579	−2.7659204

from table that both skin-friction coefficients $\text{Re}_x^{1/2}C_{fx}$ and $C_{fy}\text{Re}_y^{1/2}$ are declined with higher variations of volume friction of nanotubes, Forchheimer parameter, stretching parameter, and magnetic parameter for the case of SWCNTs as well as case of MWCNTs. It is also noticed from table that the surface drag force or skin-friction of MWCNT-water flow is more than the skin-friction of SWCNT-water flow.

Table 6.5 is sketched to study the behavior of ϕ, α, F_1, M, and Q on the rate of heat exchange or Nusselt number $Nu_x Re_x^{-1/2}$ for SWCNTs and MWCNTs conditions together. It is examined from Table 6.5 that the rate of heat exchange decelerates with augmenting variations of the nanoparticle volume fraction ϕ, stretching parameter α, space-dependent internal heat source parameter Q, and magnetic parameter M, while this rate accelerated with rising variation of Forchheimer parameter F_1 for SWCNT as well as MWCNT transportation phenomenon.

The variation of Nusselt number $Nu Re_x^{-1/2}$ for various values of Q^*, Ec_x, Ec_y, and R is illustrated in Table 6.6. It is seen from this table that the Nusselt number $Nu Re_x^{-1/2}$ diminishes with higher estimates of the temperature-dependent internal heat source parameter Q^* and Eckert numbers Ec_x and Ec_y. But, this rate boosts with rising variation of thermal radiation parameter R. In addition, Tables 6.5 and 6.6 reveal that heat exchange rate of SWCNTs is higher than MWCNTs.

Table 6.5 Numerical data of $Nu Re_x^{-1/2}$ when $Ec_x = 1.5$, $Ec_y = 1.5$, $Q^* = 0.5$, Pr $= 6.72$, and $R = 1.5$.

					$Nu Re_x^{-1/2}$	
ϕ	α	F_1	Q	M	SWCNTs	MWCNTs
0.01	0.5	1.2	1.0	2.0	−2.93763555	−2.94794475
0.02					−2.97315484	−2.99357214
0.04					−3.04649094	−3.08676022
0.03	**0.6**				−3.25633100	−3.28969413
	0.8				−3.99331628	−4.03586565
	1				−5.03183393	−5.08604145
	0.5	**1**			−3.04114013	−3.07149760
		1.8			−2.92016315	−2.94995154
		3.5			−2.70360447	−2.73215254
		1.2	**0.4**		−2.75642729	−2.78446270
			1.5		−3.22055261	−3.25258035
0.03	0.5	1.2	**3.0**	2.0	−3.85661299	−3.89399567
			1	**1**	−0.34356702	−0.35148475
				3	−5.27298032	−5.32605323
				5	−9.02962251	−9.12638930

Table 6.6 Numerical data of $NuRe_x^{-1/2}$ when $\phi = 0.03, \alpha = 0.5, F_1 = 1.2, Q = 1.0,$ $Pr = 6.72$, and $M = 2$.

				$NuRe_x^{-1/2}$	
Ec_x	Ec_y	Q^*	R	SWCNTs	MWCNTs
0.05	1.5	0.5	1.5	1.27009580	1.27896661
0.6				−0.34933280	−0.35525801
1.2				−2.12255337	−2.14454744
1.5	**0.05**			−1.90480207	−1.92514240
	0.6			−2.32350870	−2.34777681
	1.2			−2.78080684	−2.80897601
	1.5	**1**		−3.62196137	−3.64712699
		1.5		−4.40268077	−4.41670333
		2		−5.50227953	−5.51498142
		0.5		−2.47903540	−2.50148338
				−3.03510125	−3.06587696
				−3.36469880	−3.40049682
			0.0	−3.11489719	−3.14682883
			1.8	−2.98238222	−3.01215446
			5.0	−2.24874655	−2.28529738

5. Conclusions

The three-dimensional magnetohydrodynamic transport of both single and multiple-wall CNTs based in water over bidirectional stretching surface saturated in Darcy–Forchheimer porous medium with the impact of Joule heating, nonlinear thermal radiation, and nonuniform heat source/sink is investigated. Simplified system of nonlinear ODEs is attained by the use of pertinent similarity transformations. The numerical outcomes of simplified system of equations are obtained by utilizing shooting procedure together with RKF fourth-fifth order method (RKF-45). In view of the above discussion, the key outcomes are enlisted as follows:

5.1 Velocities

- The velocity component $f'(\eta)$ along direction of x decreases with increase in magnetic parameter M and stretching parameters α, while same increases with increasing value of Forchheimer parameter F_1.
- The velocity component $g'(\eta)$ along direction of y augments as the value of stretching parameter α enhances; however, a reverse behavior is found for higher values of M and F_1.

5.2 Temperature

- The fluid temperature is an increasing function of Forchheimer parameter F_1, thermal radiation parameter R, space Q, and temperature Q^*-dependent internal heat source parameters.
- Stretching parameters α and Eckert numbers Ec_x and Ec_y decelerate the fluid temperature.

5.3 The surface drag forces

- Both the skin-friction coefficients $Re_x^{1/2}C_{fx}$ and $Re_y^{1/2}C_{fy}$ decelerate with higher variations of nanoparticles, solid volume friction ϕ, Forchheimer parameter F_1, stretching parameter α, and magnetic parameter M.
- In comparison, the surface drag force of MWCNT-water fluid is more than the surface drag force of SWCNT-water fluid.

5.4 The rate of heat exchange

- The rate of heat transfer boosts with increasing estimates of the Forchheimer and thermal radiation parameters.
- The increasing variations of the nanoparticles, solid volume fraction ϕ, the stretching parameter α, space Q and temperature Q^*-dependent internal heat source parameters, magnetic parameter M, and Eckert numbers Ec_x and Ec_y diminished the heat transfer rate.
- The rate of heat transfer is maximum for SWCNTs-water fluid and minimum for MWCNTs-water fluid.

Nomenclatures

(u, v, w) Velocity along (x, y, z)
k_p Porous medium permeability
F_0 Coefficient of Forchheimer
q''' Non-uniform heat source term
q_r Heat flux of radiation
C_p Capacity of specific heat at fixed pressure p
k_{nf} Effective.thermal conductivity
T Nanofluid temperature
Ec_x and Ec_y Eckert numbers along x and y axis
$Pr = (\mu C_p)_f / k_f$ Prandtl number
$M = \sigma B_0^2 / \rho_f a$ Magnetic parameter
$F_1 = x F_0$ Forchhiemer parameter

f' **and** g' The nondimensional components of velocity in x and y-axis

$R = 16\sigma^* T_\infty{}^3/3k^* k_f$ Radiation variable

$\mathbf{Re}_x = xU_w/\upsilon_f$ Local number of Reynold along x-axis

Greek symbols

$\alpha = b/a$ Stretching parameter,

$\theta_w = T_w/T_\infty$ Dimensionless temperature ratio variable

σ_{nf} Electrical conductivity

η Similarity variable

θ Non-dimensional temperature

ρ_{nf} Density of nanofluid $\left[kgm^{-3}\right]$

$\Lambda = \upsilon_f/ak_p$ Porosity variable

υ_{nf} Viscosity of kinematic $\left[m^2 s^{-1}\right]$

Superscripts

$\prime$ derivative with respect to η

References

[1] S. Iijima, Helical microtubules of graphitic carbon, Nature 354 (1991) 56–58.

[2] M.A. Sadiq, F. Haider, T. Hayat, A. Alsaedi, Partial slip in Darcy-Forchheimer carbon nanotubes flow by rotating disk, Int. Commun. Heat Mass Tran. 116 (2020) 104641.

[3] S.U.S. Choi, Z.G. Zhang, W. Yu, F.E. Lockwood, E.A. Grulke, Anomalous thermal conductivity enhancement in nanotube suspensions, Appl. Phys. Lett. 79 (2001) 2252.

[4] Q.Z. Xue, Model for thermal conductivity of carbon nanotube- based composites, Phys. B Condens. Matter 368 (2005) 302–307.

[5] Y. Ding, D. Alisa, R.A. Wen, Heat transfer of aqueous suspensions of carbon nanotubes (CNT nanofluids), Int. J. Heat Mass Tran. 49 (2006) 240–250.

[6] J. Wang, J. Zhu, X. Zhang, Y. Chen, Heat transfer and pressure drop of nanofluids containing carbon nanotubes in laminar flows, Exp. Therm. Fluid Sci. 44 (2013) 716–721.

[7] M.R. Safaei, H. Togun, K. Vafai, S.N. Kazi, A. Badarudin, Investigation of heat transfer enhancement in a forward-facing contracting channel using FMWCNT nanofluids, Numer. Heat Tran. 66 (2014) 1321–1340.

[8] R. Ellahi, M. Hassan, A. Zeeshan, Study of natural convection MHD nanofluid by means of single and multiwalled carbon nanotubes suspended in a salt water solutions, IEEE Trans. Nanotechnol. 14 (2015) 726–734.

[9] T. Hayat, Z. Hussain, T. Muhammad, A. Alsaedi, Carbon nanotubes effects in the stagnation point flow towards a nonlinear stretching sheet with variable thickness, Adv. Powder Technol. 27 (2016) 1677–1688.

[10] Z. Iqbal, E. Azhar, E.N. Maraj, Transport phenomena of carbon nanotubes and bioconvection nanoparticles on stagnation point flow in presence of induced magnetic field, Phys. Exp. 91 (2017) 128–135.

[11] U. Khan, N. Ahmed, S.T. Mohyud-Din, Numerical investigation for three-dimensional squeezing flow of nanofluid in a rotating channel with lower stretching wall suspended by carbon nanotubes, Appl. Therm. Eng. 113 (2017) 1107–1117.

[12] T. Hayat, M.I. Khan, M. Waqas, M. Alsaedi, Numerical simulation for melting heat transfer and radiation effects in stagnation point flow of carbon-water nanofluid, Comput. Methods Appl. Mech. Eng. 315 (2017) 1011–1024.

[13] F.U. Rehman, S. Nadeem, H.U. Rehman, R.U. Haq, Thermophysical analysis for three-dimensional MHD stagnation-point flow of nano-material influenced by an exponential stretching surface, Results Phys. 8 (2018) 316–323.

[14] M. Bilal, M. Sagheer, S. Hussain, Numerical study of magnetohydrodynamics and thermal radiation on Williamson nanofluid flow over a stretching cylinder with variable thermal conductivity, Alex. Eng. J. 57 (2018) 3281–3289.

[15] R. Kandasamy, N.A. Adnan, R. Mohammad, Nanoparticle shape effects on squeezed MHD flow of water-based Cu, Al_2O_3 and SWCNTs over a porous sensor surface, Alex. Eng. J. 57 (2018) 1433–1445.

[16] M. Turkyilmazoglu, Fully developed slip flow in a concentric annulus via single and dual phase nanofluids models, Comput. Methods Progr. Biomed. 179 (2019) 104997.

[17] A. Abdulkadhim, H.K. Hamzah, F.H. Ali, A.M. Abed, I.M. Abed, Natural convection among inner corrugated cylinders inside wavy enclosure filled with nanofluid superposed in porous-nanofluid layers, Int. Commun. Heat Mass Tran. 109 (2019) 104350.

[18] R.N. Sedeh, A. Abdollahi, A. Karimipour, Experimental investigation toward obtaining nanoparticles' surficial interaction with basefluid components based on measuring thermal conductivity of nanofluids, Int. Commun. Heat Mass Tran. 103 (2019) 72–82.

[19] M.M. Sarafraz, O. Pourmehran, B. Yang, M. Arjomandi, R. Ellahi, Pool boiling heat transfer characteristics of Iron-oxide nano-suspension under constant magnetic field, Int. J. Therm. Sci. 147 (2020) 106131.

[20] M. Mohajeri, B. Behnam, A. Sahebkar, Biomedical applications of carbon nanomaterials: drug and gene delivery potentials, J. Cell. Physiol. 234 (1) (2019) 298–319.

[21] A.J. Karttunen, T.F. Fässler, M. Linnolahti, T.A. Pakkanen, Two-, one-, and zero-dimensional elemental nanostructures based on Ge(9)-clusters, ChemPhysChem 11 (9) (2010) 1944–1950.

[22] R. Alshehri, A.M. Ilyas, A. Hasan, A. Arnaout, F. Ahmed, A. Memic, Carbon nanotubes in biomedical applications: factors, mechanisms and remedies of toxicity, J. Med. Chem. 59 (18) (2016) 8149–8167.

[23] V. Georgakilas, J.N. Tiwari, K.C. Kemp, J.A. Perman, A.B. Bourlinos, K.S. Kim, R. Zboril, Noncovalent functionalization of graphene and graphene oxide for energy materials, biosensing, catalytic, and biomedical applications, Chem. Rev. 116 (9) (2016) 5464–5519.

[24] G. Hong, S. Diao, A.L. Antaris, H. Dai, Carbon nanomaterials for biological imaging and nanomedicinal therapy, Chem. Rev. 115 (19) (2015) 10816–10906.

[25] J. Lee, J. Kim, S. Kim, D.H. Min, Biosensors based on graphene oxide and its biomedical application, Adv. Drug Deliv. Rev. 105 (Pt B) (2016) 275–287.

[26] S. Pattnaik, K. Swain, Z. Lin, Graphene and graphene-based nanocomposites: biomedical applications and biosafety, J. Mater. Chem. B 4 (48) (2016) 7813–7831.

[27] X.T. Zheng, A. Ananthanarayanan, K.Q. Luo, P. Chen, Glowing graphene quantum dots and carbon dots: properties, syntheses, and biological applications, Small 11 (14) (2015) 1620–1636.

[28] D.S. Bethune, C.H. Kiang, M.S.D. Vries, G. Gorman, R. Savoy, J. Vazquez, R. Beyers, Cobalt-catalysed growth of carbonnanotubes with single-atomic-layer walls, Nature 363 (1993) 605–607.

[29] M.S. Dresselhaus, G. Dresselhaus, R. Saito, Physics of carbon nanotubes, Carbon 33 (1995) 883–891.

[30] W.E. Thomas, Carbon nanotubes, Phys. Today 49 (1996) 26–32.

[31] M.S. Dresselhaus, G. Dresselhaus, P. Avouris, Carbon Nanotubes, Clarendon Press, Oxford, 1989.

[32] P.G. Collins, P. Avouris, Carbon nanotubes, Sci. Am. 283 (2000) 62–67.

[33] J.K. Holt, H.P. Park, Y. Waang, M. Staderman, A.B. Artyukhin, C.P. Grigopopulos, Fast mass transport through sub-2- nanometer carbon nanotubes, Science 312 (2006) 1034–1037.

[34] M. Majumuder, N. Chopra, R. Andrews, B.J. Hinds, Nanoscale hydrodynamics: enhanced flow in carbon nanotubes, Nature 438 (2005) 930.

[35] A. Majumder, N. Choudhary, S.K. Ghosh, Enhanced flow in smooth single-file channel, J. Chem. Phys. 127 (2007) 054706.

[36] N. Choudhury, B.M. Pettitt, Dynamics of water trapped between hydrophobic solutes, J. Phys. Chem. B 109 (2005) 6422–6429.

[37] B. Mukherjee, P.K. Maiti, C. Dasgupta, A.K. Sood, Strong correlations and Fickian water diffusion in narrow carbon nanotubes, J. Chem. Phys. 126 (2007) 124704.

[38] M. Melillo, F. Zhu, M.A. Snyder, J. Mittal, Water transport through nanotubes with varying interaction strength between tube wall and water, J. Phys. Chem. Lett. 2 (2011) 2978–2983.

[39] N. Chopra, N. Choudhary, Comparison of structure and dynamics of polar and nonpolar fluids through carbon nanotubes, J. Phys. Chem. C 117 (2013) 18398–18405.

[40] S.K. Kannam, D. Todd, J.S. Hansen, P. Davis, How fast does water flow in carbon nanotubes? J. Chem. Phys. 138 (9) (2013) 094701.

[41] J. Su, K. Yang, On the origin of water flow through carbon nanotubes, ChemPhysChem 16 (2015) 3488–3492.

[42] X. Meng, J. Huang, Enhancement of water flow across a carbon nanotube, Mol. Simulat. 42 (2016) 215–219.

[43] A. Sam, S.K. Kannam, R. Hartkamp, S.P. Sathian, Water flow in carbon nanotubes: the effect of tube flexibility and thermostat, J. Chem. Phys. 146 (2017) 234701.

[44] M.E. Suk, N.R. Aluru, Modeling water flow through carbon nanotube membranes with entrance/exit effects, Nano Micro Thermo phys. Eng. 21 (2017) 247–262.

[45] S.J. Klaine, P.J.J. Alvarez, G.E. Batley, T.F. Fernandes, R.D. Handy, D.Y. Lyon, S. Mahendra, M.J. Mclaughlin, J.R. Lead, Nanomaterials in the environment: behavior, fate, bioavailability, and effects, Environ. Phys. Toxicol. Chem. 27 (2008) 1825–1851.

[46] M.S. Mauter, M. Elimelech, Environmental applications of carbon-based nanomaterials, Environ. Sci. Technol. 42 (2008) 5843–5859.

[47] A. Keller, S. McFerran, A. Lazareva, S. Suh, Global life cycle releases of engineered nanomaterials, J. Nanoparticle Res. 15 (2013) 1692.

[48] V.K.K. Upadhyayula, S. Deng, M.C. Mitchell, G.B. Smith, Application of carbon nanotube technology for removal of contaminants in drinking water: a review, Sci. Total Environ. 408 (2009) 1–13.

[49] S. Zhang, T. Shao, S.S.K. Bekaroblu, T. Karanfil, The effects of dissolved natural organic matter on the adsorption of synthetic organic chemicals by activated carbons and carbon nanotubes, Water Res. 45 (2011) 1378–1386.

[50] O.G. Opul, T. Karanfil, Adsorption of synthetic organic contaminants by carbon nanotubes: a critical review, Water Res. 68 (2015) 34–55.

[51] H. Zhu, K. Suenaga, J. Wei, K. Wang, D. Wu, A strategy to control the chirality of single-walled carbon nanotubes, J. Cryst. Growth 310 (2008) 5473–5476.

[52] S. Reich, L. Li, J. Robertson, Control the chirality of carbon nanotubes by epitaxial growth, Chem. Phys. Lett. 421 (2006) 469–472.

[53] T. Hiraoka, S. Bundow, H. Shinohara, S. Iijima, Control on the diameter of single walled carbon nanotubes by changing the pressure in floating catalyst CVD, Carbon 44 (2006) 1853–1859.

[54] M.P. Siegal, D.L. Overmyer, P.P. Provencio, Precise control of multiwall carbon nanotube diameters using thermal chemical vapor deposition, Appl. Phys. Lett. 80 (2002) 2171—2173.

[55] C.L. Cheung, A. Kurtz, H. Park, C.M. Lieber, Diameter-controlled synthesis of carbon nanotubes, J. Phys. Chem. B 106 (2002) 2429—2433.

[56] H. Kataura, Y. Kumazawa, Y. Maniwa, Y. Ohtsuka, R. Sen, S. Suzuki, Y. Achiba, Diameter control of single-walled carbon nanotubes, Carbon 38 (2000) 1691—1697.

[57] S. Bandow, S. Asaka, Y. Saito, A.M. Rao, L. Grigorian, E. Richter, P. Eklund, Effect of the growth temperature on the diameter distribution and chirality of single-wall carbon nanotubes, Phys. Rev. Lett. 80 (1998) 3779—3782.

[58] H. Yasuoka, R. Takahama, M. Kaneda, K. Suga, Confinement effects on liquid-flow characteristics in carbon nanotubes, Phys. Rev. E 92 (2015) 063001.

[59] S.B. Ahmed, Y. Zhao, C. Fang, J. Su, Transport of a simple liquid through carbon nanotubes: role of nanotube size, Phys. Lett. 381 (2017) 3487—3492.

[60] D.B. Geohegan, A.A. Puretzky, N. IvanovI, S. Jesse, G. Eres, J.Y. Hove, In situ growth rate measurements and length control during chemical vapor deposition of vertically aligned multiwall carbon nanotubes, Appl. Phys. Lett. 83 (2003) 1851—1853.

[61] L.C. Venema, J.W.G. Wildoer, H.L.J.T. Tuinstra, C.C. Dekker, A.G. Rinzler, R.E. Smalley, Length control of individual carbon nanotubes by nano structuring-with a scanning tunneling microscope, Appl. Phys. Lett. 71 (1997) 2629—2631.

[62] J. Su, H. Guo, Effect of nanochannel dimension on the transport of water molecules, J. Phys. Chem. B 116 (2012) 5925—5932.

[63] S. Iijima, T. Ichihashi, Single-shell carbon nanotubes of 1-nm diameter, Nature 363 (1993) 603—605.

[64] P.A.S. Autreto, S.B. Legoas, M.Z.S. Flores, D.S. Galvao, Carbon nanotube with square cross-section: an ab initio investigation, J. Chem. Phys. 133 (2010) 124513.

[65] K. Mizutani, H. Kohno, Multi-walled carbon nanotubes with rectangular or square cross-section, Appl. Phys. Lett. 108 (2016) 263112.

[66] A.K. Abramyan, N.M. Bessonov, L.V. Mirantsev, A.A. Chevrychkina, Equilibrium structures and flows of polar and nonpolar liquids in different carbon nanotubes, Eur. Phys. J. B 91 (2018) 48—54.

[67] Y. Zare, K.Y. Rhee, A simulation study for tunneling conductivity of carbon nanotubes (CNT) reinforced nanocomposites by the coefficient of conductivity transferring amongst nanoparticles and polymer medium, Results Phys. 17 (2020) 103091.

[68] N.S. Anuar, N. Bachok, M. Turkyilmazoglu, N.M. Arifin, H. Rosali, Analytical and stability analysis of MHD flow past a nonlinearly deforming vertical surface in carbon nanotubes, Alex. Eng. J. 59 (2020) 497—507.

[69] D.N. De Freitas, B.H.S. Mendonça, M.H. Köhler, M.C. Barbosa, Matheus, J.S. Matos, R.J.C. Batista, A.B. de Oliveira, Water diffusion in carbon nanotubes under directional electric frields: coupling between mobility and hydrogen bonding, Chem. Phys. 537 (2020) 110849.

[70] Z.H. Khan, W.A. Khan, R.U. Haq, M. Usman, M. Hamid, Effects of volume fraction on water-based carbon nanotubes flow in a rightangle trapezoidal cavity FEM based analysis, Int. Commun. Heat Mass Tran. 116 (2020) 104640.

[71] K. Muhammad, T. Hayat, A. Alsaedi, B. Ahmad, Numerical study of entropy production minimization in Bödewadt flow with carbon nanotubes, Phys. Stat. Mech. Appl. 550 (2020) 123966.

[72] N.S. Anuar, N. Bachok, N.M. Arifin, H. Rosali, MHD flow past a nonlinear stretching/shrinking sheet in carbon nanotubes: stability analysis, Chin. J. Phys. 65 (2020) 436—446.

[73] L.V. Mirantsev, A.K. Abramyan, Equilibrium structures and flows of polar and nonpolar liquids and theirmixtures in carbon nanotubes with rectangular cross sections, Comput. Mater. Sci. 172 (2020) 109296.

[74] K. Hosseinzadeh, A. Asadi, A.R. Mogharrebi, J. Khalesi, Entropy generation analysis of $(CH_2OH)_2$ containing CNTsnanofluid flow under effect of MHD and thermal radiation, Case Stud. Therm. Eng. 14 (2019) 100482.

[75] M. Miao, Electrical conductivity of pure carbon nanotube yarns, Carbon 49 (2011) 3755—3761.

[76] S. Muhammad, G. Ali, Z. Shah, S. Islam, S. Hussain, The rotating flow of magneto hydrodynamic carbon nanotubesovera stretching sheet with the impact of non-linear thermal radiation and heat generation/absorption, Appl. Sci. 8 (4) (2018) 482.

[77] S. Nadeem, A.U. Khan, S.T. Hussain, Model based study of SWCNT and MWCNT thermal conductivities effect on the heat transfer due to the oscillating wall conditions, Int. J. Hydrogen Energy 42 (48) (2017) 28945—28957.

[78] C.F. Sun, B. Meany, Y. Wang, Characteristics and applications of carbon nanotubes with different numbers of walls, in: Carbon Nanotubes and Graphene, Elsevier, 2014, pp. 313—339 (Chapter).

[79] R. Cortell, Fluid flow and radiative nonlinear heat transfer over a stretching sheet, J. King Saud Univ. Sci. 26 (2014) 161—167.

[80] M. Sheikholeslami, D.D. Ganji, M.Y. Javed, R. Ellahi, Effect of thermal radiation on magnetohydrodynamics nanofluid flow and heat transfer by means of two -phase model, J. Magn. Magn Mater. 374 (2015) 36—43.

[81] J.V.R. Reddy, V. Surunamma, N. Sandeep, Impact of nonlinear radiation on 3D magnetohydrodynamic flow of methanol and kerosene based ferrofluids with temperature dependent viscosity, J. Mol. Liq. 236 (2017) 39—100.

[82] T. Hayat, S. Ullah, M.I. Khan, A. Alsaedi, Q.Z. Zia, Non- Darcy flow of water-based carbon nanotubes with nonlinear radiation and heat generation/absorption, Results Phys. 8 (2018) 473—480.

[83] B. Mahanthesh, B.J. Gireesha, N.S. Shashikumar, S.A. Shehzad, Marangoni convective MHD flow of SWCNT and MWCNT nanoliquids due to a disk with solar radiation and irregular heat source, Phys. E Low-dimens. Syst. Nanostruct. 94 (2017) 25—30.

[84] M. Archana, B.J. Gireesha, P. Venkatesh, M.G. Reddy, Influence of nonlinear thermal radiation and magnetic field on three- dimensional flow of a Maxwell nanofluid, "J. Nanofluids. 6 (2) (2017) 232—242.

[85] B. Mahanthesh, N.S. Shashikumar, B.J. Gireesha, I.L. Animasaun, Effectiveness of Hall current and exponential heat source on unsteady heat transport of dusty TiO_2-EO nanoliquid with nonlinear radiative heat, J. Computat. Design Eng. 6 (4) (2019) 551—561.

[86] T. Hayat, S. Qayyum, A. Alsaedi, M. Waqas, Simultaneous influences of mixed convection and nonlinear thermal radiation in stagnation point flow of Oldroyd-B fluid towards an unsteady convectively heated stretched surface, J. Mol. Liq. 224 (2016) 811—817.

[87] T. Hayat, S. Qayyum, M. Imtiaz, A. Alsaedi, Comparative study of silver and copper water nanofluids with mixed convection and nonlinear thermal radiation, Int. J. Heat Mass Tran. 102 (2016) 723—732.

[88] M. Khan, M. Irfan, W.A. Khan, Impact of nonlinear thermal radiation and gyrotactic microorganisms on the magneto-burgers nanofluid, Int. J. Mech. Sci. 130 (2017) 375—382.

[89] B.J. Gireesha, M. Umeshaiah, B.C. Prasannakumara, N.S. Shashikumar, M. Archana, Impact of nonlinear thermal radiation on magnetohydrodynamic threedimensional

boundary layer flow of Jeffrey nanofluid over a nonlinearly permeable stretching sheet, Phys. Stat. Mech. Appl. 549 (2020) 124051.

[90] S.S. Ghadikolaei, K. Hosseinzadeh, D.D. Ganji, Numerical study on magnetohydrodynic CNTs-water nanofluids as a micropolar dusty fluid influenced by non-linear thermal radiation and Joule heating effect, Powder Technol. 340 (2018) 389−399.

[91] P.H. Forchheimer, Wasserbewegungdurchboden [movement of water through Soil], Z. Acker- Pflanzenbau 49 (1901) 1736−1749.

[92] M. Muskat, The Flow of Homogeneous Fluids through Porous Media, 1946. Edwards, MI, 1946.

[93] J. Hommel, E. Coltman, H. Class, Porosity-permeability relations for evolving pore space: a review with a focus on (Bio)-geochemically altered porous media, Transport Porous Media 124 (2001) 89−629.

[94] M.A. Seddeek, Influence of viscous dissipation and thermophoresis on Darcy Forchheimer mixed convection in a fluid saturated porous media, J. Colloid Interface Sci. 293 (2006) 137−142.

[95] S.E. Ahmed, Mixed convection in thermally anisotropic non-Darcy porous medium in double lid-driven cavity using Bejan's heat lines, Alex. Eng. J. 55 (2016) 299−309.

[96] S.A. Bakar, N.M. Arifin, R. Nazar, F.M. Ali, Forced convection boundary layer stagnationpoint flow in Darcy-Forchheimer porous medium past a shrinking sheet, Frontiers Heat Mass Transfer 7 (38) (2016) 1−6.

[97] T. Hayat, T. Muhammad, S. Al-Mezal, S.J. Liao, Darcy-Forchheimer flow with variable thermal conductivity and Cattaneo-Christov heat flux, Int. J. Numer. Methods Heat Fluid *Flow*26 (2016) 2355−2369.

[98] J.C. Umavathi, O. Ojjele, K. Vajravelu, Numerical analysis of natural convective flow and heat transfer of nanofluids in a vertical rectangular duct using Darcy Forchheimer-Brinkman model, Int. J. Therm. Sci. 111 (2017) 511−524.

[99] T. Hayat, F. Haider, T. Muhammad, A. Alsaedi, Numerical study for Darcy Forchheimer flow of nanofluid due to an exponentially stretching curved surface, Results Phys. 8 (2018) 764−771.

[100] S.Z. Alamri, R. Ellahi, N. Shehzad, A. Zeeshan, Convective radiative plane Poiseuille flow of nanofluid through porous medium with slip: an application of Stefan blowing, J. Mol. Liq. 273 (2019) 292−304.

[101] R. Ellahi, S.M. Sait, N. Shehzad, Z. Ayaz, A hybrid investigation on numerical and analytical solutions of electro- magnetohydrodynamics flow of nanofluid through porous media with entropy generation, Int. J. Numer. Methods Heat Fluid Flow 30 (2019), 834− 854.

[102] F. Tahir, T. Gul, S. Islam, Z. Shah, A. Khan, W. Khan, L. Ali, Flow of a nanoliquid film of Maxwell fluid with thermal radiation and magneto hydrodynamic properties on an unstable stretching sheet, J. Nanofluids 6 (6) (2017) 1021−1030.

[103] K. Vafai (Ed.), Porous Media: Applications in Biological Systems and Biotechnology, CRC Press, 2010.

[104] D.A. Nield, A. Bejan, Heat transfer through a porous medium, in: Convection in Porous Media, Springer, New York, NY, 2013, pp. 31−46.

[105] G.W. Sutton, A. Sherman, Engineering Magnetohydrodynamics, vol. 70, Macgraw-Hill, NewYork, 1965, p. 548.

[106] S.P.A. Devi, S.S.U. Devi, Numerical investigation of hydromagnetic hybrid Cu-Al_2O_3/water nanofluid flow over a permeable stretching sheet with suction, Int. J. Nonlinear Sci. Numer. Stimul. 17 (5) (2016) 249−257.

[107] M. Saqib, I. Khan, S. Shafie, A. Qushairi, Recent advancement in thermophysical properties of nanofluids and hybrid nanofluids: an overview, City Univ. Int. J. Computat. Anal. 3 (2) (2020) 16−25.

[108] D. Pal, H. Mondal, Soret and Dufour effects on MHD non- Darcian mixed convection heat and mass transfer over a stretching sheet with non-uniform heat source/sink, Physica B 407 (2012) 642−651.

[109] N. Sandeep, C. Sulochana, Momentum and heat transfer behavior of Jeffrey, Maxwell and Oldroyd-B nanofluids past a stretching surface with non-uniform heat source/sink, Ain Shams Eng. J. 9 (2018) 517−524.

[110] S. Rosseland, Astrophysikaud Atom-Theoretischegrundlagen, Springer, Berlin, 1931, pp. 41−44.

[111] A. Pantokratoras, T. Fang, Blasius flow with non- linear Rosseland thermal radiation, Meccanica 49 (6) (2014) 1539−1545.

[112] D. Pal, P. Saha, Influence of nonlinear thermal radiation and variable viscosity on hydromagnetic heat and mass transfer in a thin liquid film over an unsteady stretching surface, Int. J. Mech. Sci. 119 (2016) 208−216.

[113] J.F. Clouet, The Rosseland approximation for radiative transfer problems in heterogeneousmedia, J. Quantit. Spectrosc. Radiat. Tramfer 58 (1997) 33−43.

[114] P.M. Krishna, R.P. Sharma, N. Sandeep, Boundary layer analysis of persistent moving horizontal needle in Blasius and Sakiadis magnetohydrodynamic radiative nanofluid flows, Nucl. Eng. Technol. 49 (8) (2017) 1654−1659.

[115] M.I. Khan, T. Hayat, M.I. Khan, A. Alsaedi, Activation energy impact in nonlinear radiative stagnation point flow of Cross nanofluid, Int. Commun. Heat Mass Tran. 91 (2018) 216−224.

[116] W. Ibrahim, B. Shankar, MHD boundary layer flow and heat transfer of a nanofluid past a permeable stretching sheet with velocity, thermal and solutal slip boundary conditions, Comput. Fluid 75 (2013) 1−10.

[117] M.I. Khan, M. Waqas, T. Hayat, A. Alsaedi, M.I. Khan, Significance of nonlinear radiation in mixed convection flow of magneto Walter-B nanoliquid, Int. J. Hydrogen Energy 42 (2017) 26408−26416.

CHAPTER SEVEN

Thermos-physical properties and heat transfer characteristic of copper oxide—based ethylene glycol/water as a coolant for car radiator

Alhassan Salami Tijani and
Muhammad Yus Azreen Bin Mohd Yusoff
School of Mechanical Engineering, College of Engineering, Universiti Teknologi MARA (UiTM), Shah Alam, Selangor Darul Ehsan, Malaysia

Highlights

- The study investigates thermal performance of nanofluid copper oxide (CuO in car radiators).
- Viscous effect of the nanofluid is mainly due to the formation of clustering.
- Pressure drop penalty of the working fluid is a result of increase in viscous effect.
- Heat transfer characteristics were enhanced with an increase in nanofluid concentration.

1. Introduction

The use of compact car radiators for the thermal management of car engines is very critical to ensure an increase in the efficiency and life span of car engines [1]. Some of the factors that can lead to an increase in the efficiency of car radiators are the geometrical design of the fins and also the type of coolant that is used in order to enhance its performance [2]. The combined effect of these two factors greatly influences the thermal properties of the car radiator and thereby increasing the life span of the car [3]. Traditionally, water is commonly used as a cooling fluid for car radiator [4]; however, due to the poor thermal characteristic of water, a new cooling fluid known as nanofluid which is introduced by Choi in 1995 [5] is now

Advanced Materials-Based Fluids for Thermal Systems
ISBN: 978-0-443-21576-6
https://doi.org/10.1016/B978-0-443-21576-6.00001-7

Copyright © 2024 Elsevier Inc.
All rights are reserved, including those
for text and data mining, AI training,
and similar technologies.

being used as a coolant. Nanofluid is a mixture of small-sized particle suspension in liquid; it has attracted a lot of attention in recent years due to its high thermal characteristic. The increase in thermal characteristics performance is directly related to the size of the particle, the type of nanoparticle being used, and the method of the preparation of the nanofluid [6]. In order to overcome the low thermal conductivity of water that is mixed with ethylene glycol, adding metallic or nonmetallic particles to form nanofluid can enhance the thermal characteristic of the mixture [7].

Radiators consist of vertical tubes welded or brazed at the right and the left side with fins [8]. The thermal management of car radiators depends on the geometrical configuration of the fins and tubes. Hussein et al. [9] research has shown that the flat tube radiator has a significant thermal performance characteristic of up to 40% when using water-SiO_2 as a coolant. Recently, a report by Naraki et al. [10] indicates that using the same flat tube design with water-CuO nanofluid as coolant and different velocities of airflow, there was about 6%–8% of enhancement in heat transfer.

More recent attention has been focused on the numerical study of curved tubes by Akhbarinia and Behzadmehr [11]; their results show that curved tubes lower skin friction by using water-Al_2O_3 as the coolant. Maiga et al. [12] also reported numerical studies involving analyzing circular tubes and parallel disks; they reported that nanocoolant influenced the heat transfer characteristics in radiator, but the skin friction and pressure drop penalty significantly affect the pumping power.

A large and growing body of literature has investigated air-side fins of a car radiator; they play a major role in increasing the thermal performance by lowering the air-side pressure drop and thermal resistance [8]. There have been a few designs that have been implemented on the fin's surface in order to increase the efficiency of the car radiator. One of the studies by Allison and Dally [13] shows that a delta-winglet vortex fin has been designed and resulting in 46% reduction of pumping power compared to louvered fins. On the other hand, a 3D numerical analysis related to louvered fins has been conducted by Hseish et al. [14]; their results show that a fin with a louvered design can increase the heat transfer enhancement up to 118% by using 4° angle, and the air-side friction can be reduced by 119%. Another design such as wavy plate fin channel also has been studied by Malik et al. [15]; it shows that a design of wavy fin will induce and produce a small mini turbulence that will increase the heat transfer along the radiator. Apart from that, Chen et al. [16] also carried out an analysis regarding an oval finned tube with a ratio of heat transfer to pressure drop equal to 1.151

with two staggered winglets and obtained 1.097 with four staggered winglets of oval tube.

An experimental study conducted by Hussein et al. [9] when using different nanofluids which are SiO_2/water and TiO_2/water with concentrations of 0−2.5 vol% shows an increase of Reynold's number of about 25% and 30% for TiO_2/water and SiO_2/water, respectively. The concentration of the nanofluids when obtaining that result is 1vol%. Meanwhile, Devireddy et al. [17] performed an experiment related to TiO_2 nanofluids and is combined with a mixture of 40:60 EG–water as the base fluids resulting in 45.4% enhancement in heat transfer when the volume flow rate is being used at 5 L/min with turbulent flow. Meanwhile, Vajjha et al. [18] stated that through their numerical analysis, Al_2O_3-40:60 EG/W has reached a heat transfer performance of 94%. Meanwhile, another coolant of CuO-60:40 EG/W obtained a lower enhancement than alumina oxide which is 89%. The flow that is being used is laminar with a concentration of 10 vol% Al_2O_3 and 6.0 vol% CuO, respectively. Another numerical analysis by Vajjha et al. [19] also about the same type of nanofluids as above which is Al_2O_3-60:40 EG/W and CuO-60:40 EG/W but with a different flow which is turbulent showing that the same volume; 6.0vol% concentration for each fluid achieved a heat transfer enhancement of 61% and 92.5% for Al_2O_3 and CuO nanofluids, respectively.

In these few decades, researchers have obtained a great interest in managing the heat transfer processes in car radiators by using numerous nanofluids and design of the tube and also fins. The most famous of the nanofluids is the alumina oxide and copper oxide which is being mixed with a combination of water and ethylene glycol as base fluid [19−22]. Meanwhile, for the type of fins, the most famous one is the louvered fins which has great influence in producing good performance of car radiator. Although there is vortex generator that is being design such as rectangular or triangular shape of vortex generator in order to produce small windmill for creating turbulence air, the louvered fins are still higher better than the vortex generator in most cases [23].

The cooling effect on the car radiator has become a great issue which has been discussed this few decades in order to increase the heat performance of the car as well as increase the life span. Research has been carried out regarding different geometrical configurations of car radiators as well as the coolant that is being used to absorb the heat that is being produced and dissipated it to the air. Due to the small space in the engine car, the design of the car radiator and its geometry play a major role in order to increase the thermal characteristic [1]. Louvered fin has been adopted many times in the recent study due to its great contribution in increasing the heat transfer of the car radiator. Vaisi et al. [24] carried out numerical and experimental studies on the compact heat exchanger with louvered fins. Based on their study, the louvered fins can increase the heat transfer of the car radiator to 9.3% and decrease the pressure by 18.2%. Nuntaphan et al. [25] also agreed that the louvered fins do increase the heat transfer as they conducted a numerical study regarding the inclination angle of the louvered fins. The inclination angle that is being used is at the range of 30° to 45°. This study is the first study that indicates the effect of the inclination angle toward the heat performance of the car radiator.

Louvered fins have attracted many researchers to study geometry and its effect toward heat transfer. Having a louvered fin as the designation on top of the surface of the fin radiator will increase the heat transfer coefficient due to increase of air velocity [26]. Besides, increasing the angle and flow depth of the louver will increase the pressure drop and it will only decrease when the pitch between the two fins is increased [26].

Considering all of this evidence, it seems that different nanofluids and radiator geometrical configurations have been reported in the literature with an attempt to enhance heat transfer characteristics of car radiators, nevertheless, novel and advanced radiator configuration needs to be considered and analyzed; this manuscript therefore considers to close the existing gap by investigating effect of concentration of nanofluid on the thermal performance of car radiator and also a novel radiator tube is proposed to enhance the heat transfer characteristics of the radiator.

2. Modelling and simulation

CATIA VR5 was used in designing the tubes and fins in this study. The radiator is made of many fins and tubes; however, due to the limitation in Ansys CFD software, only a small part of the radiator was analyzed. Fig. 7.1 A and B shows the geometrical configuration of the flat tube and fins. Table 7.1 shows the dimensions of the tube.

3. Theoretical background
3.1 Physical properties of nanofluid

In this section, the thermal properties of the working fluid were estimated beforehand so that it can be used in the Ansys software. The CuO

Thermos-physical properties

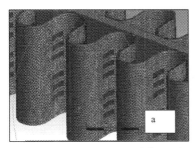
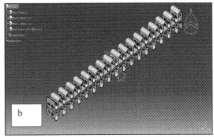

Figure 7.1 Geometry of the radiator part (A) mesh part (B) isometric view.

Table 7.1 Geometry of car radiator.

Parameter of tube	Value (mm)
Length	310.5 mm
Height	3 mm
Thickness	5 mm
Width	20 mm
Pitch	24 mm
Fin length	19 mm
Fin thickness	5 mm
Fin pitch	15 mm
Louver length	2 mm
Louver angle	25°
Hydraulic diameter	0.0054 mm
Type	Aluminum

nanoparticles were combined with the base fluid which is a mixture of water and ethylene glycol and the assumption that is being made throughout the whole system is that the nanofluid is well mixed and dispersed in the tube. The density, viscosity, thermal conductivity, and specific heat capacity of the nanofluid are based on the formula as follows:

$$\rho_{nf} = \varphi \rho_p + (1 - \varphi) \rho_{bf} \tag{7.1}$$

$$(\rho C p)_{nf} = (\rho C p)_p + (1 - \varphi)(\rho C p)_{bf} \tag{7.2}$$

$$k_{nf} = \frac{k_p + (\Phi - 1)k_{bf} - \varphi(\Phi - 1)(k_{bf} - k_p)}{k_p + (\Phi - 1)k_{bf} + \varphi(k_{bf} - k_p)} k_{bf} \tag{7.3}$$

$$\mu_{nf} = \mu_{bf}(1 + 2.5\,\varphi) \tag{7.4}$$

Table 7.2 Thermophysical properties of nanoparticle copper oxide (CuO).

Physical properties	Value
Size of grain (nm)	60
Purity	+98%
Appearance	Brownish-black powder
ρ (kg/m^3)	6500
Cp (J/kg.K)	535.6
k (W/m.K)	69

The Φ is the empirical shape factor which is known as $\Phi = 3/\psi$. Meanwhile, ψ is sphericity of particle which is the ratio of surface area of volume divided by the surface area of the particle and is assumed to be 3 in value. The properties of the nanofluid particles are shown in Table 7.2 below [2].

3.2 Heat transfer coefficient estimation

The heat transfer characteristic is usually obtained by calculating the Prandtl number, heat transfer coefficient, Nusselt number, and lastly, the rate of heat transfer as shown:

$$Pr = \frac{v}{a} = \frac{c_p \mu}{k} \tag{7.5}$$

$$Nu = 0.023 Re^{0.8} \, Pr^{0.3} \tag{7.6}$$

$$h = \frac{Nu.k}{d} \tag{7.7}$$

d is the hydraulic diameter and k is the thermal conductivity.

Lastly the Newton law of cooling is adopted to estimate the heat transfer rate as shown below:

$$\dot{Q} = \dot{m} C_p (T_{inlet} - T_{outlet}) \tag{7.8}$$

3.3 Simulation assumptions

The car radiator model is being numerically simulated using Ansys CFD software; calculation of the data for heat transfer thermos–physical properties was carried out using Excel spreadsheet. In the Ansys software, important assumptions affecting the simulation were set up for the temperature inlet, velocity inlet, and also the wall boundary condition. The condition that has been set up is as follows:

- k-epsilon was adopted for the simulation.

Thermos-physical properties · 175

- A k-epsilon model is being used for the flow in the tube.
- Inlet velocity of 0.007 m/s, 0.014 m/s, 0.021 m/s, and 0.028 m/s, which is equal to volumetric flow rate of 2 L/min, 4 L/min, 6 L/min, and 8 L/min is being used. Meanwhile, the temperature inlet for the coolant is being set at 95°C which is equal to 368K [27].
- The pressure outlet is considered as 0Pa.
- Convection is assumed to occur at the wall of the fins, and flat tube has a constant heat transfer coefficient of 10 $W/m^2.K$ and the temperature of the air is 35°C.

The radiator that is being simulated consists of one flat tube and continuous louvered fins across the tube of the radiator. There are a few assumptions for the simulation in the Ansys Fluent before the simulation took place which are listed as follows:

- The flow condition is steady and compressible flow along the flat tube.
- The thermos-physical properties of the nanofluid and the base fluid are considered constant.

3.4 Grid independent performance

In this study, grid independent was performed to determine the most suitable number of meshing elements for the simulation (refer to Table 7.3). The difference between the temperature outlet and the element size was used to perform sensitivity analysis to select the most accurate and suitable mesh relevance. Based on the simulation that was done on the meshing. A meshing size of 868050 was recorded as the highest using −80 mesh relevance; on the other hand, meshing size of 866218 at −100 mesh relevance was the lowest. The grid independence test was conducted in increments of 5 for the number of mesh relevance. Hence, the mesh relevance of −90 was used in this study as the difference in the number element is smaller while having the same output temperature which is 361.294 K.

Table 7.3 Grid independent.

Mesh relevance	Element size	Outlet temperature (K)	Error
−80	866,834	361.294	–
−85	868,050	361.294	–
−90	866,922	361.294	–
−95	867,128	361.295	0.001
−100	867,218	361.295	–

4. Discussion of findings

4.1 Base fluid and thermophysical properties of nanofluid

Before conducting the simulation in the Ansys software, the important physical properties of the working fluid were first calculated. Tables 7.4 and 7.5 b show the properties of the nanofluid and base fluid. It can be observed from the tables that, the nanofluid has a better thermal conductivity than the base fluid and also the highest concentration of 0.06 recorded a thermal conductivity of 0.399 W/m.K.

4.2 Heat transfer performance characteristic

Tables 7.6 and 7.7 show the heat transfer performance of the nanofluid and base fluid at four different velocities of 0.007, 0.014, 0.021, and 0.021 m/s which are equivalent to 2, 4, 6, and 8 L/min of volume flow rate,

Table 7.4 Base fluid thermal properties.

Material	Density (kg/^{3})	Specific heat capacity, (J/kg.K)	Thermal conductivity (W/m.K)	Viscosity (kg/m.s)
Base fluid	1080.697	2803.754	0.3357	0.0004353

Table 7.5 Thermal properties of CuO nanofluid.

Materials		Properties		
Water + EG + CuO	Density (kg/m3)	Heat capacity, (J/kg.K)	Thermal conductivity (W/m.K)	Viscosity (kg/m.s)
0.01	1134.89	2673.8472	0.3457	0.0004462
0.03	1243.28	2448.0084	0.3663	0.000468
0.06	1405.85	2174.5426	0.3990	0.0005006

Table 7.6 Thermal characteristics of base fluid.

Material	Prandtl number	Velocity (m/s)	Reynolds number	Nusselt number	Heat transfer coefficient (W/m^2.K)
Water + EG	3.636	0.007	930.397	8.032	49.934
	3.636	0.014	1863.476	14.001	87.040
	3.636	0.021	2792.533	19.351	120.298
	3.636	0.028	3724.271	24.363	151.458

Table 7.7 Thermal characteristics of CuO nanofluid.
Material

Water + EG + CuO	Velocity (m/s)	Prandtl number	Reynolds number	Nusselt number	Heat transfer coefficient (W/m^2.K)
0.010	0.007	3.451	953.223	8.793	56.298
0.030		3.126	995.689	9.105	61.785
0.060		2.727	1052.464	9.518	70.348
0.010	0.0014	3.451	1909.193	15.327	98.133
0.030		3.126	1994.247	15.871	107.699
0.060		2.727	2107.961	16.591	122.625
0.010	0.0021	3.451	2861.042	21.184	135.630
0.030		3.126	2988.502	21.935	148.850
0.060		2.727	3158.909	22.930	169.479
0.010	0.0028	3.451	3815.639	26.671	170.762
0.030		3.126	3985.625	27.617	187.406
0.060		2.727	4212.890	28.870	213.379

respectively. The range of Reynold numbers recorded for both base fluid and nanofluid is 930 until 4212 which is clearly showing that the flow for the fluids is laminar flow. It can be seen that increasing the velocity for both fluids increases the thermal performance of the car radiator. It was observed that at a velocity of 0.028 m/s, the recorded Nusselt number is 24.363 and for the highest nanofluid concentration, the Nusselt number was 28.87. The heat transfer coefficient for both fluids also shows a big difference where the highest heat transfer coefficient of both base fluid and nanofluid are 151.458 W/m^2.K and 213.379 W/m^2.K, respectively (refer to Tables 7.6 and 7.7).

One such interesting observation from this study is that, increasing nanofluid concentration from 0.01 to 0.06 shows a considerable increase in nanofluid thermal conductivity (refer to Fig. 7.2). The highest recorded thermal conductivity for 0.06 concentration is 0.399 W/m.K. Another finding that stands out from the results reported earlier is that there is enhancement in overall heat transfer coefficient with volume flowrate of 2 L/min until 8 L/min; this is due to increase in thermal conductivity of the nanofluid. It can be observed in Fig. 7.3 that, at volume flowrate of 8 L/min and nanofluid concentration of 0.06, the heat transfer coefficient for the CuO nanofluid which is 213.379 W/m^2.K. It is interesting to note that there is tremendous enhancement in Nusselt number which is a very important parameters in determining the thermal characteristics of the working fluid. For example, as shown in Fig. 7.4, at nanofluid concentration of 0.06, the results of the Nusselt number show 28.87.

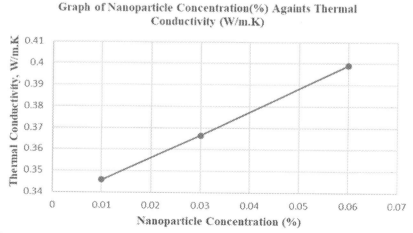

Figure 7.2 Graph of nanoparticle concentration (%) against thermal conductivity (W/m.K).

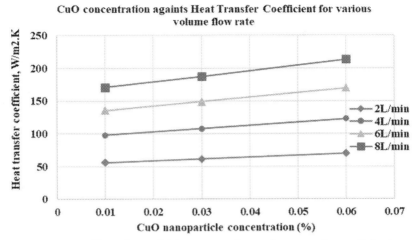

Figure 7.3 Effect of nanofluid concentration on heat transfer coefficient.

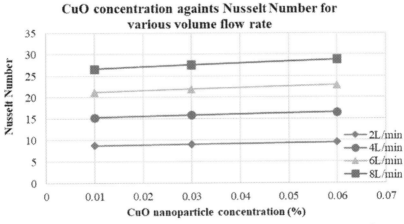

Figure 7.4 CuO concentration against Nusselt number for various volume flow rate.

4.3 Validation of data

The comparison of the data from the study and literature is being done in order to validate the study that has been done on the car radiator. The data have been compared with the Nusselt number that has been taken from the literature which is from the study. An experiment has been done on the CuO nanofluid in the car radiator by Heris et al. Based on Table 7.8, the average error of the study that has been conducted is 12.65%. The error is due to the difference in geometry of the car radiator as well as some differences in value for the thermophysical properties.

Table 7.8 Validation of data.

Experimental data				My work				Percentage error (average) %			
Volume flow rate	Nanoparticle concentration			Volume flow rate	Nanoparticle concentration			Nanoparticle concentration			
L/min	0	0.01	0.03	0.06	L/min	0	0.01	0.03	0.06	0	0.01	0.03	0.3
2	9.37	9.91	10.52	11.21	2	8.03	8.79	9.10	9.51	16.68	12.74	15.60	17.87
4	16.31	17.67	18.01	19.20	4	14.00	15.33	15.87	16.59	16.50	15.25	13.48	15.73
6	22.58	23.63	25.23	27.12	6	19.35	21.18	21.94	22.93	16.70	11.56	14.50	18.27
8	28.62	30.54	32.97	34.14	8	24.36	26.67	27.61	28.87	17.48	14.51	19.41	18.25
					Average error					16.84	13.52	15.74	17.53
					Total average error					12.65			

Thermos-physical properties

Table 7.9 Heat transfer rate for different working fluids at 8 L/min.

Working fluids	Exit fluid temperature (K)	Temperature difference	Heat transfer rate (W)
Base fluid	365.436	2.564	23.30
Base fluid + 0.01% CuO	360.747	7.253	74.31
Base fluid + 0.03% CuO	360.687	7.313	82.23
Base fluid + 0.06% CuO	360.610	7.390	94.61

Table 7.9 and also Fig. 7.5 illustrate the heat transfer rate of the base fluid and also the CuO nanofluid. The most obvious finding to emerge from the analysis is that the nanofluid has a better cooling effect than the base fluid because the exit temperature for base fluid is lower than that of the nanofluid. Another finding that stands out from the results reported earlier is that increasing nanofluid concentration enhances the rate of heat transfer. The highest rate of heat transfer recorded is 94.61W when using 0.06 concentration. Meanwhile, base fluid only exhibits 23.30W of heat transfer.

Table 7.10 shows the temperature distribution across the radiator tube and fins. Table 7.10 reveals very interesting insights, thus it shows that the temperature decreases gradually along the length of the flat tube which shows that at a volume flowrate of 8 L/min and different concentrations of 0.01, 0.03, and 0.06, there is convective heat transfer from the radiator to the working fluid. The most striking observation to emerge from the data comparison was that as the nanofluid concentration increases the temperature profile across the whole length of the flat tube and fins increases as well. This confirms that a large portion of the fluid is dissipated from the working fluid to the surface of the flat tube. Table 7.11 illustrates the

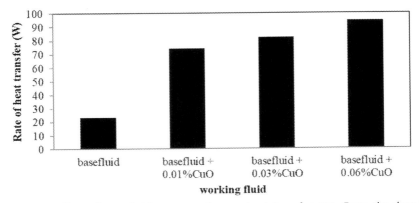

Figure 7.5 Effect of nanofluid concentration on heat transfer rate. *From the data in Table 7.10, it is apparent that the length of time left between.*

Table 7.10 Temperature contour at 8 L/min (a) base fluid, (b) 0.01% CuO nanofluid, (c) 0.03% CuO nanofluid, (d) 0.06% CuO nanofluid.

Base fluid

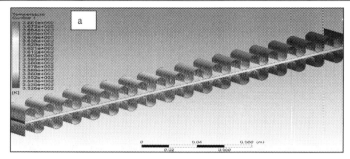

0.01 CuO nanofluid

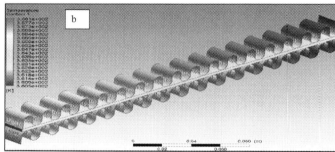

0.03 CuO nanofluid

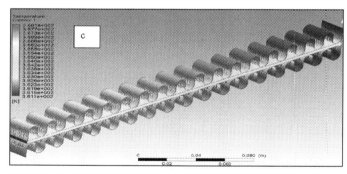

0.06 CuO nanofluid

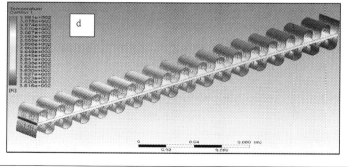

Thermos-physical properties 183

Table 7.11 Streamline profile at 8 L/min (a) base fluid (b) 0.01% CuO nanofluid (c) 0.03% CuO nanofluid (d) 0.06% CuO nanofluid.

Base fluid

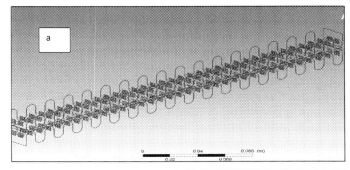

0.01 CuO nanofluid

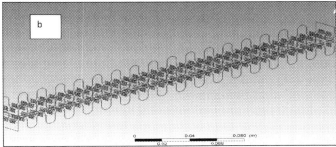

0.03 CuO nanofluid

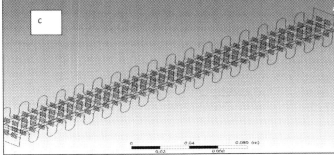

0.06 CuO nanofluid

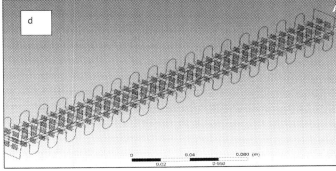

velocity profile of the working fluid, it shows that the streamlines across each channel are under turbulent conditions, this situation enhances the heat transfer characteristics of the working fluid.

5. Conclusion and recommendation

The present manuscript presents computational modeling of thermal characteristics and performance enhancement of car radiator with nanofluid as heat transfer fluid. The heat transfer characteristics at different nanofluid concentrations were established. The flow situation in the flat tube was laminar, and the k-epsilon model was used for the flow in the tube. Some significant findings to emerge from this study are that:

- The nanofluid has a better thermal conductivity than the base fluid, and also the highest concentration of 0.06 recorded a thermal conductivity of 0.399 W/m.K.
- Based on the data obtained, the highest Nusselt number enhancement obtained when using copper oxide nanoparticles is 67% when using 8 L/min flow rate which shows a great cooling effect for the car radiator. Heat transfer improvement is important to increase the lifespan of the car radiator itself as well as prolong the lifespan of the usage of a car to avoid overheating and breakdown.

Abbreviation

Cp Specific heat capacity
d Hydraulic diameter
h Heat transfer coefficient
k Thermal conductivity
$\dot{m}$ Mass flow rate
Nu Nusselt number
Pr Prandtl number
$\dot{Q}$ Rate of heat transfer
Re Reynold's number
T Temperature
ρ Density
φ Particle sphericity
Φ Empirical shape factor
μ Viscosity

Subscript
bf Base Fluid
EG Ethylene glycol
f Nanofluid
p Nanoparticle

References

[1] G. Huminic, A. Huminic, The heat transfer performances and entropy generation analysis of hybrid nanofluids in a flattened tube, Int. J. Heat Mass Tran. 119 (2018) 813−827, https://doi.org/10.1016/j.ijheatmasstransfer.2017.11.155.

[2] S.Z. Heris, M. Shokrgozar, S. Poorpharhang, M. Shanbedi, S.H. Noie, Experimental study of heat transfer of a car radiator with CuO/ethylene glycol-water as a coolant, J. Dispersion Sci. Technol. 35 (5) (2014) 677−684, https://doi.org/10.1080/01932691.2013.805301.

[3] A. Vaisi, M. Esmaeilpour, H. Taherian, Experimental investigation of geometry effects on the performance of a compact louvered heat exchanger, Appl. Ther. Eng. 31 (2011), https://doi.org/10.1016/j.applthermaleng.2011.06.014.

[4] M. Abdollahi-Moghaddam, K. Motahari, A. Rezaei, Performance characteristics of low concentrations of CuO/water nanofluids flowing through horizontal tube for energy efficiency purposes; an experimental study and ANN modeling, J. Mol. Liq. 271 (2018) 342−352, https://doi.org/10.1016/j.molliq.2018.08.149.

[5] S.U.S. Choi, J.A. Eastman, Enhancing thermal conductivity of fluids with nanoparticles, ASME Int. Mech. Eng. Congr. Expo. 66 (March) (1995) 99−105, https://doi.org/10.1115/1.1532008.

[6] Y.C. Chiu, K.M. Yu, Experimental investigation of heat transfer potential of Al2O3/water-mono ethylene glycol nanofluids as a car radiator coolant, J. Dispersion Sci. Technol. 13 (3) (1992) 293−314, https://doi.org/10.1016/j.csite.2017.11.009.

[7] S.M. Peyghambarzadeh, S.H. Hashemabadi, S.M. Hoseini, M. Seifi Jamnani, Experimental study of heat transfer enhancement using water/ethylene glycol based nanofluids as a new coolant for car radiators, Int. Commun. Heat Mass Tran. 38 (9) (2011) 1283−1290, https://doi.org/10.1016/j.icheatmasstransfer.2011.07.001.

[8] Y. Mukkamala, Contemporary trends in thermo-hydraulic testing and modeling of automotive radiators deploying nano-coolants and aerodynamically e ffi cient airside fi ns, Renew. Sust. Energy Rew. 76 (2017) 1208−1229. January 2016.

[9] A.M. Hussein, R.A. Bakar, K. Kadirgama, Study of forced convection nanofluid heat transfer in the automotive cooling system, Case Stud. Therm. Eng. 2 (2014) 50−61, https://doi.org/10.1016/j.csite.2013.12.001.

[10] M. Naraki, S.M. Peyghambarzadeh, S.H. Hashemabadi, Y. Vermahmoudi, Parametric study of overall heat transfer coefficient of CuO/water nanofluids in a car radiator, Int. J. Therm. Sci. 66 (2013) 82−90, https://doi.org/10.1016/j.ijthermalsci.2012.11.013.

[11] N. Sohrabi, N. Massoumi, A. Behzadmehr, S.M.H. Sarvari, Numerical study of laminar mixed convection of a nanofluid in a horizontal tube using two phase mixture model with variables physical properties, Int. Conf. Fluid Mech. Aerodyn. (2008) 50−58.

[12] S.J. Palm, G. Roy, C.T. Nguyen, Heat transfer enhancement with the use of nanofluids in radial flow cooling systems considering temperature-dependent properties, Appl. Therm. Eng. 26 (2006) 2209−2218, https://doi.org/10.1016/j.applthermaleng.2006.03.014.

[13] A. Cb, D. Bb, Effect of a Delta-Winglet vortex pair on the performance of a tube-fin heat exchanger, Int. J. Heat Mass Trans 50 (25-26) (2007) 5065−5072.

[14] B. Ameel, J. Degroote, H. Huisseune, P. De Jaeger, J. Vierendeels, M. De Paepe, Numerical optimization of louvered fin heat exchanger with variable louver angles, J. Phys. Conf. Ser. 395 (1) (2012) 1—8, https://doi.org/10.1088/1742-6596/395/1/012054.

[15] R.M. Manglik, Low Reynolds number forced convection in three-dimensional wavy-plate-fin compact channels : fin density effects, Int. J. Heat Mass Trans 48 (2005) 1439—1449, https://doi.org/10.1016/j.ijheatmasstransfer.2004.10.022.

[16] J.E.O. Brien, M.S. Sohal, T.D. Foust, P.C. Wallstedt, Heat Transfer Enhancement for Finned-Tube Heat Exchangers with Vortex Generators : Experimental and Numerical Results 12 Th International Heat Transfer Conference Heat Transfer Enhancement for Finned-Tube Heat Exchangers with Vortex Generators : Experimen, 2002.

[17] V.V.R. Devireddy Sandhya, M. Chandra Sekhara Reddy, Improving the cooling performance of automobile radiator with ethylene glycol water based TiO2 nanofluid 36 (2009) 674—679.

[18] R.S. Vajjha, D.K. Das, D.R. Ray, Development of new correlations for the Nusselt number and the friction factor under turbulent flow of nanofluids in flat tubes, Int. J. Heat Mass Tran. 80 (2015) 353—367, https://doi.org/10.1016/j.ijheatmasstransfer.2014.09.018.

[19] R.S. Vajjha, D.K. Das, P.K. Namburu, Numerical study of fluid dynamic and heat transfer performance of Al2O3and CuO nanofluids in the flat tubes of a radiator, Int. J. Heat Fluid Flow 31 (4) (2010) 613—621, https://doi.org/10.1016/j.ijheatfluidflow.2010.02.016.

[20] P. Gunnasegaran, N.H. Shuaib, M.F. Abdul Jalal, E. Sandhita, Application of nanofluids in heat transfer enhancement of compact heat exchanger, AIP Conf. Proc. 1502 (1) (2012) 408—425, https://doi.org/10.1063/1.4769160.

[21] M.M.P. Jadhav, D.B. Jadhav, M.E. Nimgade, Title : Heat Transfer Enhancement Using Nanofluids in Automotive Cooling System, 2017, pp. 1035—1042. June.

[22] M.S. Yadav, S.A. Giri, V.C. Momale, Algorithm Sizing Analysis of Louvered Fin Flat Tube Compact Heat Exchanger by Genetic Algorithm, 2017.

[23] C. Cuevas, D. Makaire, L. Dardenne, P. Ngendakumana, Thermo-hydraulic characterization of a louvered fin and flat tube heat exchanger, Exp. Therm. Fluid Sci. 35 (2011) 154—164, https://doi.org/10.1016/j.expthermflusci.2010.08.015.

[24] K. Javaherdeh, A. Vaisi, R. Moosavi, M. Esmaeilpour, Experimental and numerical investigations on louvered fin-and-tube heat exchanger with variable geometrical parameters, J. Therm. Sci. Eng. Appl. 9 (2) (2017) 024501, https://doi.org/10.1115/1.4035449.

[25] A. Nuntaphan, S. Vithayasai, T. Kiatsiriroat, C.C. Wang, Effect of inclination angle on free convection thermal performance of louver finned heat exchanger, Int. J. Heat Mass Tran. 50 (1—2) (2007) 361—366, https://doi.org/10.1016/j.ijheatmasstransfer.2006.06.008.

[26] P. Gunnasegaran, N.H. Shuaib, M.F. Abdul Jalal, The effect of geometrical parameters on heat transfer characteristics of compact heat exchanger with louvered fins. ISRN Thermodyn. 2012 (2012) 1—10, https://doi.org/10.5402/2012/832708.

[27] A.S. Tijani, A.S. bin Sudirman, Thermos-physical properties and heat transfer characteristics of water/anti-freezing and Al2O3/CuO based nanofluid as a coolant for car radiator, Int. J. Heat Mass Tran. 118 (2018) 48—57, https://doi.org/10.1016/j.ijheatmasstransfer.2017.10.083.

CHAPTER EIGHT

Discussion on the stability of nanofluids for optimal thermal applications

Taoufik Brahim[1], Abdelmajid Jemni[2]

[1]University of Sousse, Higher Institute of Applied Sciences and Technology of Sousse (ISSAT-Sousse-Tunisia), Sousse, Tunisia
[2]University of Monastir, National Engineering School of Monastir, Laboratory Studies of Thermal and Energy Systems- LESTE, Monastir, Tunisia

Highlights

- The nanofluids' stability and dispersion are explicitly investigated.
- Characterization of nanofluid stability is undertaken.
- Relevant findings and recommendations for optimized methods are summarized.

1. Nanofluids discussion

1.1 Techniques for the preparation of nanofluids

Nanofluids are defined as diluted suspensions of solids smaller than 100 nm, where the high thermal conductivity of the solid nanoparticles produces a high-conductivity thermal fluid. Nanofluids must satisfy the same technical specifications as the following [1] in order to be used as alternatives for conventional thermal fluids (homogeneous fluids like water or ethylene glycol):

- Resistance to degradation at the specified temperatures, to provide a longer lifetime without residue sedimentation.
- Stability and homogeneity over time.
- Good heat transfer characteristics, high thermal conductivity.
- Be able to operate in systems at low pressures by having low steam pressures at working temperatures.
- Have low viscosity to minimize pumping losses (losses from friction), which will increase electrical energy consumption.

Advanced Materials-Based Fluids for Thermal Systems
ISBN: 978-0-443-21576-6
https://doi.org/10.1016/B978-0-443-21576-6.00008-X

Copyright © 2024 Elsevier Inc.
All rights reserved, including those
for text and data mining, AI training,
and similar technologies.

187

- Avoid corrosion of materials in contact with thermal fluids.
- Provide affordable cost.
- Low fouling potential or little tendency for "fouling" on the exchanger walls.
- Low freezing point.
- That they are biodegradable.

In this chapter, a bibliographic review has been carried out on the methodology utilized by several research teams to produce nanofluids and their subsequent thermal characterization, with the aim of defining the key factors in the methodology to use. In order to develop nanofluids with eventual commercial use, it was first required to dive into the principles of heat transfer at the micro and nanoscale and obtain the required information on dispersion of nanoparticles in aqueous environments and colloidal stability.

Theoretical improvements in conductivity for particle suspensions were first put forth by Maxwell [2] in 1873, and Maxwell's expression predicted that these improvements would depend on both the volumetric fraction of solid nanoparticles in the fluid and the thermal conductivities of the constituent elements. Recent technological developments have produced nanometric-sized particles known as "nanoparticles" and novel suspension techniques are now possible. Because surface phenomena in solid particles predominate in nanofluids, Brownian motion can maintain nanoparticles stably suspended in the nanofluid and prevent them from sedimenting [3]. Given that the nanofluid does not sediment over time and that the size of the particles it contains prevents clogging of the channels through which it flows, nanofluids are the best candidates for use as thermal fluids, as long as agglomeration phenomena do not occur. With respect to the base fluid and different types of nanoparticles at low volume fractions (0.5% −4%), results obtained by different research groups show that nanofluids have very different thermal properties than conventional fluids, and the thermal conductivity obtained can increase by up to 25% [3].

Fig. 8.1 shows the different solid materials used for the preparation of nanofluids as well as their comparative histories. It can be seen that the solid oxides $Al2O_3$, TiO_2, and CuO have been the most used in nanofluid preparation studies made to date.

Table .8.1 gives the summary obtained from the bibliographic review (in historical order) for the preparation of nanofluids in different base fluids (water, ethylene glycol, and water/EG) and at different volumetric fractions (greater than 1% for oxides and less than 1% for multiwall carbon nanotubes,

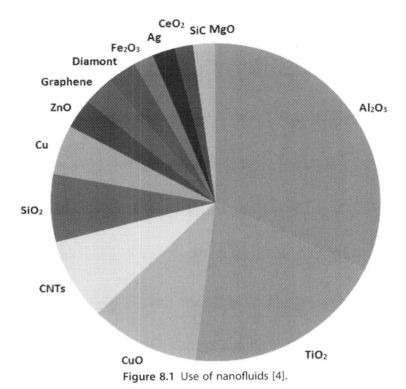

Figure 8.1 Use of nanofluids [4].

MWCNTs). In most studies, the relative increase in thermal conductivity is achieved at temperatures close to ambient [15−17].

There are several different methods for making nanofluids that may be divided primarily into two main categories: "two-step" and "single-step."

In two-step technique, powdered nanoparticles (obtained by synthesis methods in an independent first stage of the process) are dispersed in a second stage or step in the basis fluid chosen [6]. Due to their high surface charge, they tend to agglomerate, for which it is necessary to apply considerable mechanical agitation or ultrasound in this second stage to achieve a high degree of dispersion.

This technique is very attractive for its subsequent industrial application, as it is simple since the already synthesized nanoparticles can be acquired and just proceed to their dispersion. However, it should be noted that complete dispersal until the nanoparticles are totally individualized in the base fluid is very difficult because the nanoparticles initially present a high state of aggregation. For this reason (the nanoparticles fail to disperse completely), it is

Table 8.1 Studies carried out with different types of nanoparticles and base fluids.

Nanofluid type	Volume fraction (VF)	Thermal conductivity relative increase (%)	References
Al_2O_3-water	4.3	30	[5]
Al_2O_3-water	5.0	21	[1]
Al_2O_3-water	0.4	13	[6]
SiO-water	2.0	1.01	[5]
TiO_2-water	1–5	20–30	[7]
CuO-water	0.4	17	[8]
CuO-water	3.4	12	[9]
Cu-water	2.0	1.24	[4]
Cu-water	0.1	1.006	[10]
Cu-EG[a]	0.56	1.10	[11]
Ag-water	0.4	1.11	[12]
Ag-EG[a]	0.12	1.07	[13]
MWCNTs-water[a]	0.05	1.07	[14]
MWCNTs-water[a]	1.5	1.17	[10]
MWCNTs-EG[a]	1.0	1.12	[10]
Graphene-water	0.2	1.17	[13]
Graphene-oil	1.3	1.36	[13]

[a]*MWCNTs*, multiwalled carbon nanotubes; *EG*, ethylene glycol.

necessary to increase volumetric fractions to obtain a significant increase in thermal conductivity.

This two-step approach may be more successful if the nanoparticles are modified superficially by chemical methods such as:

- Surfactants: by adsorbing the surfactant on the surface of the particle, which changes its surface charge. Once the nanoparticles are loaded superficially, repulsion phenomena occur if the charges are of the same sign.
- Control of the pH of the suspension is necessary in order to move away from the isoelectric point, where the zeta potential is zero and the attraction between particles is maximum, and be at high repulsion potentials.
- Functionalization of nanoparticles: the addition of atoms or functional groups to the surface of particles to change their structure.

The nanoparticles and nanofluid are produced simultaneously in the one-step technique (either physically, via "Physical Vapor Deposition" (PVD), or chemically). The aforementioned method minimizes the drying, storage, transport, and subsequent dispersion of the nanoparticles in the base fluid while reducing agglomeration processes. The disadvantage of this method is the difficulty of its industrial application. Physical methods of nanoparticle synthesis, such as the PVD technique, do not allow for large-scale production, while chemical methods have a number of disadvantages, such as the residues of the reagents that can be left in the nanofluid due to the incomplete reaction. It is difficult to know the influence of these kinds of impurities on the thermal conductivity. The transport properties (conductivity, temperature, and final viscosity) of a nanofluid depend on the state of aggregation of the nanoparticles and their stability. The process utilized for producing the nanofluids will thus have a significant impact on the final characteristics of the nanofluid.

The two-stage approach, which enables the nanofluid to be generated in the chosen base fluid with the desired additives, is the most frequently employed method for creating nanofluids, according to the reviewed bibliography. Regardless of the synthesis method used to obtain the nanoparticles, it has been observed in some cases that the reagents used in the synthesis remain as impurities in the prepared nanofluids, affecting their end behavior. Furthermore, when using the one-step technique, the volumetric fractions of nanofluids present are lower than when using the two-step method, according to the literature. It can be stated that there are two critical phenomena in the preparation of nanofluids: the dispersion of nanoparticles and the phenomena of aggregation and subsequent sedimentation that affect their stability. The stability of prepared nanofluidics [6,15] is influenced by the following factors:

- Nanofluids are thermodynamically unstable as they are multiphase systems with high surface loads.
- Dispersed nanoparticles have a high Brownian motion, which can cause their sedimentation by coalescence and gravity.
- Over time, Van der Waals forces can lead to the aggregation of the nanoparticles.
- There should be no reaction between the particles and the base fluid, which could cause the solubilization.

1.2 Dispersion of nanoparticles in the base fluid

As previously stated, the way in which particles are dispersed in the liquid has a significant impact on the final properties of nanofluids because the increase in thermal conductivity attained as a result of the solid particles' high conductivity with respect to the base fluid is a function of the aforementioned fluid's Brownian motion. Therefore, Brownian motion which depends on the degree of dispersion reached at suspending the nanoparticles in the fluid and their subsequent stability, avoiding the formation of new particles. The intensity of the energy used for dispersion and the length of time it is applied determine how quickly nanoparticles disperse and deagglomerate in liquids. According to the bibliography, the energy applied in the dispersion can be carried out in baths of ultrasonic [18,19], ultrasonic probes [10,20,21], magnetic stirrers, "stirrers of high shear", ball mills, and homogenizers [22,23]. It is observed that the treatment with an ultrasound probe is the most effective [24,25], since long ultrasound application times (important in terms of time period) imply an increase in the temperature of the nanofluid due to the applied energy, so there will be an increase in Brownian motion and therefore the collision between the nanoparticles will increase.

The time required to reach a certain degree of dispersion also depends on the type of treatment used and the intensity of the energy applied by each of them [25,26]. Thus, in the case of magnetic stirrers, the size of the agglomerate is modified gradually, ceasing to produce significant changes in the agglomeration size when the energy is not enough to break the agglomerations. However, in the case of ultrasonic probes, which are highly energetic, the time necessary for the dispersion to be complete and reproducible is much lower, making the size of the aggregates practically invariable from the first minute of treatment.

A bibliographical review has been carried out to see which mechanical dispersion techniques are used in the dispersion of nanoparticles in an aqueous base. The results obtained after this review are summarized in Table 8.2. Among the different methods of mechanical dispersion, the following have been chosen for their application in the research:
- Ultrasonic bath
- Magnetic stirring
- Homogenizer
- Ultrasonic probe

Among all these mentioned systems, the ultrasonic probe is the equipment most used by researchers due to its effectiveness in dispersing the nanoparticles in the base fluid [29−32].

Aggregates can be broken by wear mechanisms when applying a shear force or by rupture of the entire mass due to impacts. The rupture mechanism depends on the size of the aggregates and the intensity of energy applied. As the aggregates become smaller as the deagglomeration process progresses, the superficial forces become more important. As a result, the breakdown of large aggregates is relatively simple, whereas the breakdown of aggregates smaller than 1 μm can be very complicated, to the point where

Table 8.2 Nanofluid examples prepared with different dispersion mechanisms.

Nanofluid	Volumetric fraction (%)	Method of preparation	Time of dispersion	Observations	References
Al_2O_3	0.8–5	Ultrasonic bath	6–15 h	Few minutes sediment	[25]
Al_2O_3	0.08	Ultrasonic probe + surfactant + adjustment pH	15 min	Do not sediment	[25]
Al_2O_3	1	Ultrasonic probe	30 min	Do not sediment	[25]
SiO2	1	Ultrasonic probe	2 h	Do not sediment	[6]
CuO	0.003	Ultrasonic probe	2 h	Do not sediment	[25]
CuO	0.1	Ultrasonic probe +surfactant	1 h	Do not sediment	[25]
MWCNTs	0.2	Ultrasonic probe +surfactant			[26]
CNTs	0.2	Ultrasonic bath, Ultrasonic probe, Homogenizer	120 min 60 min 3 passes 18,000 min	Sediment Reduces the size and not sediment	[23]
Graphene	0.5	Ultrasonic probe	–	Stable for more two years	[27]
Graphene	0.2	Ultrasonic probe +surfactant	30 min	do not sediment	[28]

particles as small as 10 and 100 nm cannot be separated mechanically. Depending on the size of the agglomerate and the shape of the nanoparticle, it is necessary to apply a specific amount of energy and time determined for each case. The application time of the mechanical energy is also a critical variable because some studies have also found that once the optimum stirring time has elapsed, they can even observe problems of reagglomeration of the nanoparticles [25]. If the Van de Waals forces are strong enough, the nanoparticles will agglomerate for an excessively long period and form aggregates, which may be the cause of this unfavorable effect. In addition, long ultrasonic application times imply an increase in the temperature of the nanofluid due to the applied energy, which will cause an increase in Brownian motion, and therefore the collision between the nanoparticles will increase. Some researchers have attempted to break up nanoparticle agglomerates using the high cavitation energy generated by equipment called homogenizers, as is, for example, the case with organic-based carbon nanotubes [23]. The cavitation energy is achieved by passing the suspension through microchannels. This step reduction generates a high dispersion of the particles in the fluid due to the microbubbles formed by cavitation.

After the dispersion of the nanoparticles is attained with the application of mechanical energy, it is possible that the particles will reagglomerate once the energy dispersant is stopped since the Van der Waals forces between nanoscale particles are starting to become substantial. Therefore, it is essential to stabilize the suspensions of nanofluids once the dispersion process has finished if the particles are to remain apart for a long period of time [18].

1.3 Stability of the nanoparticles dispersed in the base fluid

The stability of a colloidal dispersion is obtained when the ratio of particles per unit volume remains constant. Colloidal particles are small enough to be unaffected by gravitational forces (1 μm). Within the colloidal state, they are called nanoparticles for materials with a size less than 100 nm. Colloidally stable implies that the solid particles do not sediment or aggregate at a significant rate [33−35].

The stability of nanofluids is one of the most important requirements, as it directly affects their thermophysical properties and heat transfer properties over time. Stability studies of these systems can be classified into four types:

- Thermodynamic stability
- Kinetic stability
- Chemical stability
- Dispersion degree stability

For the application of nanofluids, the degree of dispersion and kinetic stability must be guaranteed. The motion of the dispersed nanoparticles is governed by Brownian motion, which leads to an increase in collisions, favoring the interaction between them. The suspensions lose their stability due to the aggregation of the colloidal particles over time and their subsequent sedimentation due to gravitational forces. The frequency of collisions and the probability of cohesion during the collision both affect the nanoparticles' rate of aggregation [36]. Different factors might cause nanoparticles to bind or aggregate:

- Gelation (the formation of a coherent network of particles that occupy the entire volume, as well as capillarity mechanisms that keep the liquid in place)
- Coagulation (formation of compact particle groups with a higher fraction than in the original suspension)
- Flocculation (union of particles by means of bridges formed by a flocculent agent, forming an open and voluminous structure).

According to the DVLO theory (Derjaguin, Verway, Landau, and Overbeek), the combined effect of the attractive Van der Waals forces and the repulsive forces that exist between particles as they move toward one another owing to Brownian motion determines how stable a particle is in a solution. The two particles will collide and aggregate into a suspension that won't be stable over time if the attracting force is higher than the repulsive force. It is vital to keep the particles apart in order to prevent agglomerates from developing. This is accomplished by adding repulsive forces that are stronger and more pervasive than the attractive ones into the system, keeping the particles as far apart as possible [36]. Repulsive forces come from steric or electrostatic interactions, as shown in Fig. 8.2. In general, electrostatic interactions are important in aqueous systems, while steric interactions are also effective in nonaqueous systems.

The final stability of nanofluid suspensions will depend on these interactions and the result of the balance between attractive and repulsive forces. The efficiency of heat transfer at the surface is improved when the specific surface area of the nanoparticles increases with decreasing size, but the interactions between the particles which affect the stability of nanofluids also increase in the reverse direction [37]. This phenomenon will be critical in obtaining nanofluids since the agglomeration of nanoparticles in a nanofluid leads to a decrease in the thermal conductivity of nanofluids in addition to problems of sedimentation and clogging of the microchannels in the exchangers when the increase in size occurs. Because of this, research on

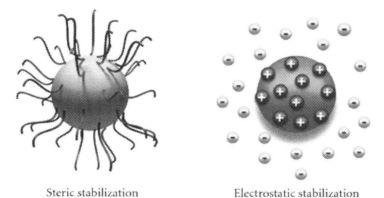

Steric stabilization   Electrostatic stabilization
Figure 8.2 Types of repulsive forces in colloidal stability.

nanofluid stability is a crucial topic that directly affects the characteristics of nanofluids for their following uses. Therefore, in order to use solutions that enhance stability, it is required to investigate and evaluate the elements that affect the dispersion and stability of nanofluids made from powdered nanoparticles, or "two-way technique" steps [36].

2. Mechanisms to increase the stability of nanofluids

The preparation of homogeneous and stable nanofluids is challenging since the nanoparticles always have a tendency to form aggregates due to the strong forces of Van der Waals interaction [38]. As the volume fraction increases, it improves the thermal conductivity, but the hydrodynamic interactions as well as the probability of collision between particles become more important, favoring aggregation processes in nanofluids. The state of agglomeration and therefore the stability of nanofluids can be modified by controlling the factors that affect the surface properties of particles and the forces of interaction between them. These factors include the pH of the suspension, the presence of electrolytes and surfactants, the viscosity of the base fluid, the particle diameter, density, and volume fraction [20,36,39]. Mechanisms that can improve the stability of prepared fluids are classified into three types:
- Addition of surfactants
- Techniques for modifying the surface of nanoparticles without surfactants, through functionalization
- Modification of the zeta potential.

2.1 Modification of the surface of the nanoparticles; surfactants

Dispersing agents or surfactants have usually been used to disperse water-based nanoparticles [16]. The molecule of every surfactant has both a polar (or hydrophilic) group and a nonpolar (hydrophobic or lipophilic) group. The nonpolar group is often an alkylaromatic or paraffinic hydrocarbon, whereas the polar group is typically a functional group including heteroatoms (O, S, N, and P). The surfactants in aqueous media migrate toward the surfaces of the particles so that their water-soluble component remains in the aqueous phase and the hydrophobe remains out of that phase on the surface of the nanoparticle (Fig. 8.3), thus modifying its behavior against aggregation.

Surfactants are added to the base fluid to maintain the nanoparticles' dispersion, avoiding steric hindrance or electrostatic attraction from causing them to aggregate. Polymers are frequently utilized to accomplish steric stabilization; these polymers will adsorb on the surface of the particles and produce an extra steric repulsive force [36]. For example, zinc oxide nanoparticles modified with polymethyl methacrylate (PMMA) have favorable compatibility with polar solvents. The protective role of the addition polyvinylpyrrolidone (PVP) makes the silver nanofluids exceedingly stable. PVP has also been found to be an effective agent for improving graphite suspension stability. The nanoparticles are superficially charged using one or more of the following ways to induce electrostatic stabilization: the physical adsorption of charged species into the surface, isomorphic substitution of ions, dissociation of charged species at the surface, accumulation or depletion of electrons, and preferential adsorption of ions. Table 8.3 summarizes the several surfactant types most frequently utilized by researchers for making nanofluids.

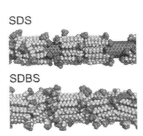

Figure 8.3 Incorporating the surfactant sodium dodecyl sulfate, SDS, on a carbon nanotube.

Table 8.3 Nanofluids prepared with different types of surfactants.

Nanofluid material	Surfactant type	References
Al_2O_3	PVA	[40]
Al_2O_3	SDS	[40]
Al_2O_3	Triton	[40]
Al_2O_3	SDBS	[41]
$Mg(OH)_2$	CTAB,SDS	[29]
TiO_2	CTAB	[42]
SiO_2	CTAB	[40]
CuO	CTAB	[40]
CuO	SDBS	[41]
Cu	SDS	[40]
CNTs[a]	SDS	[43]
CNTs[a]	Triton	[40]
CNTs[a]	Arabic gum	[40]
CNTs[a]	Arabic gum, SDS	[26]
CNTs[a]	SDS	[44]
MWCNTs[a]	Triton	[45]
MWCNTs[a]	SDS	[21]
MWCNTs[a]	PEG	[40]
Graphene	SDS, CTAB, CO890	[28]

[a]*MWCNTs*, multiwalled carbon nanotubes; *CNTs*, carbon nanotubes.

Although the method of dispersion and stabilization of the nanoparticles with the use of surfactants is widespread, it has been found experimentally that sometimes the binding bond between the surfactant and the surface of the nanoparticles is affected by the temperature, decreasing its effect [46].

2.2 Modification of the surface of the nanoparticles; functionalization

The modification of the surface of nanoparticles through their functionalization is another method used to improve the dispersion and stability of nanofluids. The objective of functionalization is to introduce hydrophilic functional groups through chemical and physical reactions. For example, the functionalization of carbon nanotubes is a method widely used to increase the stability of suspensions. The functionalization methods used with nanotubes include an acid treatment to introduce functional groups followed by a washing operation. Abbasi et al. [45] used nitric acid treatment for 4 h and a treatment with a mixture of sulfuric acid and nitric acid (3:1). With both treatments, he managed to prepare stable suspensions of hybrid nanofluids (multiwalled carbon nanotubes and Al_2O_3) in a water-based fluid. Despite this, he confirmed that the treatment affects the morphology of nanotubes, and if this is altered, it can also affect thermal conductivity. Talaei et al. [46] investigated the effect of the number of functional groups introduced on the nanotube surface on the stability of the dispersion in the base fluid (water), demonstrating that the more functional groups there are, the more stable the dispersion. Aravind et al. [47] studied the effect of the functionalization time (reflux from 1 to 4 h in a mixed 3:1 sulfuric acid and nitric acid solution of the nanotubes) on their stability by dispersing them in water and an ethylene glycol base. After analysis of the UV absorption spectrum, it was observed that the maximum absorbance (best dispersion of the nanotubes) was obtained with the functionalized nanotubes for 2.5 h. This technique is one of the most promising since it does not imply the incorporation of any additive (such as a surfactant) in the preparation of nanofluids and has been widely used in the consulted bibliography, obtaining good results in the stability of nanofluids [44,46,48].

2.3 Stabilization of nanofluids by modification of the zeta potential

Zeta potential measurements provide an immediate analysis of the degree of stability and dispersion of suspended particles. Nanoparticles suspended in the nanofluid are electrically charged on their surfaces, with this charge being compensated with a charge of the opposite sign and forming an electrical double layer. The zeta potential is the electric potential that exists in the shear plane, which delimits the fixed part of the ionic double layer of the mobile part. It is a measure directly proportional to the charge of the particles, so that the higher its value, the more charged they will be and the greater the electrostatic potential and the consequent repulsion that will prevent the particles from being attracted and sticking together to form aggregates. A zeta potential close to zero (mV) indicates the maximum agglomeration of nanoparticles and their precipitation. A high absolute value (from a magnitude of 100 mV) indicates excellent stability. The isoelectric point of a suspension is the pH value at which the zeta potential is zero. This point can be determined by measuring the zeta potential as a function of its pH using the assay of valuation. Once the isoelectric point is established, high values of zeta potential can be reached with a pH far from the pH of the isoelectric point. Modifying the zeta potential of a nanofluid suspension by regulating its pH is a technique widely used by researchers to achieve good stability. Haddad et al. [40] experimentally verified that nanofluids with a potential zeta of 30 mV are already physically stable, while below 20 mV, their stability is limited and below 5 mV they already agglomerate. Pastoriza-Gallego et al. [49] and Mondragon et al. [50] demonstrated that in silica nanofluids, the zeta potential depends on the nanoparticles' volumetric fraction. Increases in mass content from 0.10 to 0.20 decrease the zeta potential. In nanofluids prepared from carbon nanotubes and CuO, the maximum potential measured was 20 mV, which indicates that the repulsion forces between particles are not enough to overcome the attraction between them. In nanotube nanofluids, the addition of a surfactant at pH 7 increases the zeta potential to 50 mV.

3. Characterization of nanofluid stability

As already mentioned in the previous point, the methodology used to obtain nanofluids is that of two stages: the acquisition of nanoparticles in powder and subsequent dispersion in an aqueous medium by stirring, a methodology in which the stability of the nanofluids is lower due to the tendency of nanoparticles to agglomerate. The effectiveness of a nanofluid depends on the sedimentation of the nanoparticles; therefore, the evolution of their size over time should be analyzed, determining the tendency to agglomeration and subsequent sedimentation. In the bibliography, various techniques are used to measure the stability of fluids [22,30−32,43,51,52], ranging from visual techniques, in which the sedimentation of the prepared suspensions is detected using images, to centrifugation methods [30], which thus accelerate the sedimentation processes, to the measurement of the variation of the volumetric fraction by measuring the density of the nanofluid [22], to the zeta potential in the double layer, to the measurement of the zeta potential in the double layer, up to the measurement techniques that are based on the interference generated by the particles in suspension against the passage of a beam of light [40]. Particle or agglomerate size measurement techniques using dynamic dispersion light scattering (DLS) are used by several authors [25,53] to determine the size of the agglomerates in the suspension and their evolution. It should be noted that although these measurements do not allow a direct measurement of the size of the nanoparticles dispersed in the nanofluid (the sample is diluted in order to carry out the extent, thereby greatly modifying the conditions that affect the particle agglomeration, such as pH, dispersant, and temperature), these measurements are sufficiently representative of the state of agglomeration of the nanoparticles in the base fluid and allow comparisons of the degree of dispersion achieved in nanofluids. Finally, the stability of suspensions containing prepared nanoparticles was assessed using the two methods described below (measurement of light backscattered in Turbiscan equipment and absorbance measurement in a UV spectrophotometer), each depending on the nature of the nanoparticles. The stability or sedimentation of the different samples of nanofluids prepared has also been characterized by visual analysis.

3.1 Characterization of nanofluids' stability by backscattered light

One of the methods for measuring the stability of nanofluids is the dispersion technique of light in the Turbiscán equipment [6,23,32,43,51,52]. The Turbiscan technology consists of measuring the light intensities in transmission (defined as T) and reflection (defined as BS), more commonly referred to as backscatter, in function of the height of the sample to detect the change of particle size in processes of coalescence and flocculation and the separation of phases in processes of sedimentation or flotation. For this, the equipment has an optical head with an infrared light source (880 nm) and two detectors (one for transmission, T, and another that detects reflection, BS) that cover the entire height of the sample that is inside a glass cell.

With the data collected on light intensity, absorbance profiles are obtained with respect to time that allow characterizing the stability of the suspension and detecting processes such as sedimentation, flocculation, coalescence, etc.

When analyzing the stability measures in the Turbiscan, three scenarios can arise [20,39]:

- Stability: This situation occurs when the amount of light backscattered by the sample is stable at the bottom and in the center of the cell without presenting variations in the time of the backscatter or transmission values.
- Sedimentation: The percentage backscatter (% BS) at the bottom of the cell increases with time with an increasing particle fraction and instead decreases at the top of the cell. In this case, when the nanoparticles settle, they are placed at the bottom of the cell, allowing less light to pass and measuring the largest detector backscatter. In contrast, there are fewer particles, a smaller fraction, and thus a lower BS percentage at the top of the cell.
- Agglomeration or flocculation: The amount of backscattered light decreases with time when the size of the particles is greater than the size of the incident beam (600 nm). If the particle size when agglomerated is smaller than the incident beam (600 nm), we would have an increase in the BS backscatter signal (%).

3.2 Characterization of nanofluids' stability by UV spectrophotometry

Characterization of the stability of the nanofluid using the spectrophotometry technique is possible as long as the nanoparticles have the characteristic absorption bands within the range 200–1100 nm. This technique is an easy method to determine the fraction of nanoparticles in the upper area of the container containing the nanofluid and evaluate their variation with sedimentation time [6,54]. The evolution of the amount of sedimented particles, which can be measured over time if a sample is pipetted from the top, can be used to determine the stability of nanofluids. The absorbance values (A) measured in the sample are proportional to the amount of particles present; the lower this value, the fewer particles in the sample, indicating that it is more clarified as a result of the sedimentation process. The fraction of particles on the surface is proportional to the absorbance since the absorbance is related to the fraction of the substance, (c), by the Lambert–Beer law, which is summarized with the equation:

$$A = \varepsilon.b.c$$

where:

c is the fraction of the substance (mol/l)

b is the optical path length (width of the cell containing the solution of the substance) (cm)

ε is the molar absorptivity, a property of every substance, proportional to the amount of radiation absorbed at a given wavelength as measured per unit fraction (l/(mol.cm))

In order to apply the Lambert–Beer law, it is necessary to previously select a length since both the absorbance (A) and the absorptivity (ε) vary with it. It had previously obtained the substance's absorption spectrum, which is a representation of absorbance values against wavelength expressed in nanometers (nm). The wavelength value for which the absorbance is maximum can be chosen from the absorption spectrum, and assays can be performed over time at that fixed wavelength. To carry out a test to characterize the stability of nanofluids by spectrophotometry, the T60 spectrophotometer from PG Instruments was used. The methodology followed to obtain reproducible absorbance values was:

- Sample preparation: 10 pL are pipetted from the surface of the sample to be tested and then diluted in distilled water until the measuring device's cuvette is leveled.

- The absorbance of the sample is measured at the selected wavelength.
- Several readings are made in order to obtain an average of the measurement and assign the absorbance for the time in which the sample was taken.
- The process of collecting the sample, preparing it, and measuring it is repeated at a fixed interval.
- Representation of the evolution of relative absorbance values over the sampling and measurement period (>72 h).

3.3 Characterization of nanofluids by image analysis

3.3.1 Transmission electron microscopy

Transmission electron microscopy (TEM) is a widely used technique for the structural and chemical characterization of materials in the fields of inorganic chemistry, organic chemistry, materials engineering, biology, and biomedicine. In the TEM, a sample is irradiated with a beam of electrons with an energy between 100 and 200 keV. Among the different interactions that are produced when electrons impinge on matter, transmission and scattering are used for image formation, electron diffraction is used to obtain crystallographic information, and the emission of characteristic X-rays is used to determine the elemental composition of the sample. The TEM is used for the tests. TEM has been found not to be suitable for measuring the degree of dispersion of nanoparticle agglomerates in the base fluid after mechanical dispersion, since with this technique, it is not possible to distinguish whether the image obtained is that of an agglomerated particle or a superposition of particles that in their original state were individualized, since the electron beam passing through the thin layer passes through all these particles, observing in the image only a global projection.

3.3.2 Scanning electron microscopy

The samples were observed and photographed with the signal of backscattered electrons and secondary electrons from the field emission scanning electron microscope (SEM).

The backscattered electron signal provides information about the topography and composition; it is more intense the greater the average atomic number of the sample, so that the lighter areas contain heavier elements (contrast of composition). The secondary electron signal is shallower, so it provides information on the morphology of the sample, highlighting the irregularities of the surface, such as cracks, pores, grain boundaries, or crystals.

The observation has been made with electrons accelerated at two different potentials, 10 and 20 keV, so the lower the accelerating potential of the electrons, the more superficial the sign.

4. Conclusion

This chapter summarizes the conclusions derived from the dispersion study carried out in the preparation of aqueous-based nanofluids from nanoparticles of different natures and its subsequent thermophysical characterization. First, the general conclusions obtained about the methodology are summarized.

- It has been found that for the preparation of nanofluids by the two-stage method is essential to carry out, together with mechanical dispersion, a modification of the surface of the particles, either by stabilization of the suspension by means of pH modification or by functionalization of the surface of the nanoparticles by the covalent or noncovalent method, in order to avoid agglomeration of the dispersed nanoparticles.
- It has been verified that among all the dispersion techniques studied, the application of the ultrasound probe is the one that leads to the best results after setting the application time.
- It has been found that high energy, such as that used in probe ultrasound, is required to achieve the dispersion of nanoparticle aggregates in water, with the application time being a critical variable in the dependent process and, consequently, the nature of the material to be dispersed.
- TEM has been found not to be suitable for measuring the degree of dispersion of nanoparticle agglomerates in the base fluid after mechanical dispersion. With the technique of SEM, collects information from the surface of the particle, being the most representative size obtained. It should be noted that none of the two techniques can measure liquid samples, so it is necessary conditioning the sample with a drying process that can change the state of particle aggregation.

As future lines that can give continuity to the research work on dispersion of nanoparticles developed in this chapter, we can cite:

- Study the dispersion of the same materials with nanometric size in a fluid base other than water, where the increases in thermal conductivity obtained may be higher than those observed in the bibliographical review carried out. In addition, these base fluids would facilitate the dispersion of carbon nanoparticles, characterized by their hydrophobic nature.

Furthermore, the stability of hybrid nanofluids (mixtures of different materials) will be another challenge.
- Apply knowledge about the preparation and characterization of suspensions of materials of nanometric size at a high fraction for their application in other areas, such as conductive inks based on carbon nanotubes or graphene, or inkjet inks to obtain ceramic coatings with new functionalities.

Nomenclature

Symbols
- A Absorbance (dimensionless)
- b Optical path length (cm)
- c Substance fraction $\left(\frac{mol}{l}\right)$
- ε Molar absorptivity (l/(mol.cm)

Abbreviations
- **CNT** Carbon nanotube
- **CTAB** Cetyltrimethylammonium bromide
- **DLS** Dynamic dispersion light scattering
- **DTAB** Dodecyl trimethyl ammonium bromide
- **EG** Ethylene glycol
- **MWCNT** Multiwalled carbon nanotube
- **PMMA** Polymethyl methacrylate
- **PVA** Polyvinyl alcohol
- **PVP** Polyvinylpyrrolidone
- **SDBS** Sodium dodecyl benzenesulfonate
- **SDS** Sodium dodecyl sulfate
- **SEM** Scanning electron microscope
- **TEM** Transmission electron microscopy
- **UV** Ultraviolet light

References

[1] R. Saidur, K.Y. Leong, H.A. Mohammad, A review on applications and challenges of nanofluids, Malaysia: Renew. Sustain. Energy Rev. 15 (2010) 1646–1668, https://doi.org/10.1016/j.rser.2010.11.035.

[2] S.K. Das, S.U.S. Choi, W. Yu, Nanofluids: Science and Technology, John Wiley & Sons, 2007.

[3] W. YU, D.M. France, S.U.S. Choi, J.L. Routbort, Review and Assessment of Nanofluid Technology for Transportation and Other Applications, Energy systems Division, Argone National Laboratory, 2007.

[4] Y. Xuan, Q. Li, W. Hu, Aggregation structure and thermal conductivity of nanofluids, AIChE J. 49 (4) (2003) 1038–1043, https://doi.org/10.1002/aic.690490420.

[5] H. Masuda, A. Ebata, K. Teramae, N. Hishinuma, Alteration of thermal conductivity and viscosity of liquid by dispersing ultra-fine particles (Dispersion of Al2O3, SiO2 and TiO2 ultra-fine particles), Netsu Bussei 7 (1993) 227233.

[6] A. Ghadimi, R. Saidur, H.S.C. Metselaar, A review of nanofluid stability properties and characterization in stationary conditions, Int. J. Heat Mass Tran. 54 (2011) 4051−4068.

[7] S.M.S. Murshed, K.C. Leong, C. Yang, Investigations of thermal conductivity and viscosity of nanofluids, Int. J. Therm. Sci. 47 (5) (2008) 560−568.

[8] Y. Li, Jing'en Zhou, S. Tung, E. Schneider, S. Xi, Review on development of nanofluid preparation and characterization, Powder Technol. (2009) 196 89−10. Journal ISSN : 0032-5910.

[9] J.H. Lee, S.H. Lee, C. Choi, S. Jang, S. Choi, A review of thermal conductivity data, mechanisms and models for nanofluids, Int. J. Micro-Nano Scale Trans. 1 (4) (2010) 269−322.

[10] J. LIU, et al., Fullerene pipes, Science 280 (5367) (1998) 1253−1256.

[11] S.U. S Choi, J.A. Eastman, Enhancing thermal conductivity of fluids with nanoparticles, in: Internatiomal Mechanical Engineering Congress and Exhibition, vols. 12−17, 1995, pp. 99−105.

[12] Y. Lu, W. Kang, J. Jiang, J. Chen, D. Xu, P. Zhang, H. Wu, Study on the stabilization mechanism of crude oil emulsion with an amphiphilic polymer using the β-cyclodextrin inclusion method, RSC Adv. 7 (14) (2017) 8156−8166.

[13] Y. Rao, Nanofluids: stability, phase diagram, rheology and applications, Particuology 8 (2010) 549−555 (Chinese Society of Particuology and Institute of Process Engineering, Chinese Academy of Sciences. Published by Elsevier B.V).

[14] H. Jiang, Q. Zhang, L. Shi, Effective thermal conductivity of carbon nanotubebased nanofluid, J. Taiwan Inst. Chem. Eng. 55 (2015) 76−81, https://doi.org/10.1016/j.jtice.2015.03.037.

[15] S. Ozerinc, S.K. Kakac, A. Guvenc, Enhanced thermal conductivity of nanofluids; a state -of -the-art review. 145-170, Ankara, Turkey, Microfluid. Nanofluidics 8 (2009), https://doi.org/10.1007/s10404-009-0524-4.

[16] R.S. Vajjha, D.K. Das, A review and analysis on influence of temperature and concentration of nanofluids on thermophysical properties, heat transfer and pumping power, Int. J. Heat Mass Tran. 55 (2012) 4063−4078.

[17] T. Brahim, A. Jemni, Numerical case study of packed sphere wicked heat pipe using Al_2O_3 and CuO based water nanoffuid, Case Stud. Therm. Eng. 8 (2016) 311−321, https://doi.org/10.1016/j.csite.2016.09.002.

[18] S. Lazzari a, L. Nicoud b, B. Jaquet b, M. Lattuada c, M. Morbidelli, Fractal-like structures in colloid science, Adv. Colloid Interface Sci. 235 (2016) 113, https://doi.org/10.1016/j.cis.2016.05.002.

[19] H. Tyagi, P. Phelan, R.S. Prasher, Predicted efficiency of a low-temperaturenanofluidbased direct absorption solar collector, J. Sol. Energy Eng. 131 (4) (2009) 041004.

[20] Y. Li, J. Zhou, S. Tung, E. Scheneider, S. Xi, A review on development of nanofluid preparation and characterization, Powder Thecnol. 196 (2009) 89−101, https://doi.org/10.1016/j.powtec.2009.07.025.

[21] M.J. Assael, C.-F. Chen, I. Metaxa, W.A. Wakeham, Thermal conductivity of suspensions of carbon nanotubes in water, Int. J. Thermophys. 25 (4) (July 2004).

[22] B. Munkhbayar, M. Bat-Erdene, B. Ochirkhuyag, D. Sarangerel, B. Battsengel, H. Chung, H. Jeong, An experimental study of the planetary ball milling effect on dispersibility and thermal conductivity of MWCNTs-based aqueous nanofluids, Mater. Res. Bull. 47 (2012) 4187−4196.

[23] Y. Hwang, J.-K. Lee, J.-K. Lee, Y.-M. Jeong, S.-ir Cheong, Y.-C. Ahn, S.H. Kim, Production and dispersion stability of nanoparticles in nanofluids, Powder Technol. 186 (2008), https://doi.org/10.1016/j.powtec.2007.11.020, 145-15.

[24] R. Mondragon, C. Segarra, J.C. Jarque, Experimental characterization and modeling of thermophysical properties of nanofluids at high temperature conditions for heat transfer applications, Powder Technol. 249 (2013) 516–529.

[25] S. Umer Ilyas, R. Pendyala, N. Marneni, Preparation, sedimentation, and agglomeration of nanofluids, Chem. Eng. Technol. 37 (12) (2014) 2011–2021, https://doi.org/10.1002/ceat.201400268.

[26] R. Sadri, G. Ahmadi, H. Togun, M. Dahari, S. Newaz, E. sadeghinezhad, n. Zubir, An experimental study on thermal conductivity and viscosity of nanofluids containing carbon nanotubes, Nanoscale Res. Lett. 9 (2014) 151.

[27] Y. Wanga, A.I. Hussein, Al-Saaidi, M. Kong, J. Alvarado, Thermophysical performance of graphene based aqueous nanofluids, Int. J. Heat Mass Tran. 119 (2018) 408–417.

[28] N.W. Pu, C.A. Wang, Y.M. Liu, Y. Sung, D.S. Wang, M.D. Ger, Dispersion of graphene in aqueous solutions with different types of surfactants and the production of graphene films by spray or drop coating, J. Taiwan Inst. Chem. Eng. 43 (1) (2012) 140–146.

[29] A. Amin, M. Asadi, M. Siahmargoi, T. Asadi, M.G. Andarati, The effect of surfactant and sonication time on the stability and thermal conductivity of water-based nanofluid containing Mg(OH)2 nanoparticles: an experimental investigation, Int. J. Heat Mass Tran. 108 (2017) 191–198.

[30] I.M. Mahbubul, I.M. Shahrul, S.S. Khaleduzzaman, R. Saidur, M.A. Amalina, A. Turgut, Experimental investigation on effect of ultrasonication duration on colloidal dispersion and thermophysical properties of alumina–water nanofluid, Int. J. Heat Mass Tran. 88 (2015) 73–81.

[31] I.M. Mahbubul, R. Saidur, M.A. Amalina, E.B. Elcioglu, T. Okutucu-Ozyurt, Okutucu-Ozyurt Effective ultrasonication process for better colloidal dispersion of nanofluid, Ultrason. Sonochem. 26 (2015) 361–369.

[32] A. Ghadimi, I.H. Metselaar, The influence of surfactant and ultrasonic processing on improvement of stability, thermal conductivity and viscosity of titania nanofluid, Exp. Therm. Fluid Sci. 51 (2013) 1–9.

[33] M.U. Sajid, H.M. Ali, Thermal conductivity of hybrid nanofluids: a critical review, Int. J. Heat Mass Tran. 126 (2018) 211–234, https://doi.org/10.1016/j.ijheatmasstransfer.2018.05.021.

[34] H. Babar, H. Ali, Towards hybrid nanofluids: preparation, thermophysical properties, applications, and challenges, J. Mol. Liq. 281 (2019), https://doi.org/10.1016/j.molliq.2019.02.102.

[35] T. Shah, H. Ali, Applications of hybrid nanofluids in solar energy, practical limitations and challenges: a critical review, Sol. Energy 183 (2019) 173–203, https://doi.org/10.1016/j.solener.2019.03.012.

[36] W. Yu, H. Xie, A review on nanofluids: preparation, stability mechanisms, and applications, J. Nanomater. 2012 (2012), https://doi.org/10.1155/2012/435873. Article ID 435873.

[37] V. Bianco, O. manca, S. Mardini, K. Vafai, Heat Transfer Enhacement with Nanofluids, Editorial: CRR Press, 2015 (Taylor and Francis Group, LLC).

[38] J. Buongiorno, D.C. Venerus, N.P. Thomas, A benchmark study on the thermal conductivity of nanofluids, J. Appl. Phys. 106 (2009) 9.

[39] Formulaction, Stability of Pigment Inkjet Inks. Application Paper, 2009, pp. 1–5.

[40] Z. Haddad, C. Abid, H.F. Oztop, A. Mataoui, A review on how the researches prepare their nanofluids, Int. J. Therm. Sci. 76 (2013) 168–189 (Turkey).

[41] M.A. Khairul, K. Shah, E. Doroodchi, R. Azizian, B. Moghtaderi, Effects of surfactant on stability and thermo-physical properties of metal oxide nanofluids, Int. J. Heat Mass Tran. 98 (2016) 778–787, https://doi.org/10.1016/j.ijheatmasstransfer.2016.03.079.

[42] S.M.S. Murshed, K.C. Leong, C. Yang, Enhanced thermal conductivity of TiO2-water based nanofluids .367-373, s.l, Int. J. Therm. Sci. 44 (2005), https://doi.org/10.1016/j.ijthermalsci.2004.12.005.

[43] A. Ghozatloo, A.M. Rashidi, M. Shariaty-Niasar, Effects of surface modification on the dispersion and thermal conductivity of CNT/water nanofluids, Int. Commun. Heat Mass Tran. 54 (2014) 1—7, https://doi.org/10.1016/j.icheatmasstransfer.2014.02.013.

[44] A. Nasiri, M. Shariaty-Niasar, A. Rashidi, A. Amrollahi, R. Khodafarin, Effect of dispersion method on thermal conductivity and stability of nanofluid, Exp. Therm. Fluid Sci., Theran, Iran 35 (4) (2011) 717—723, https://doi.org/10.1016/j.expthermflusci.2011.01.006.

[45] S.M. Abbasi, A. Rashidi, A. Nemati, K. Arzani, The effect of functionalisation method on the stability and the thermal conductivity of nanofluid hybrids of carbon nanotubes/gamma alumina, Iran. Ceramics International, Rev. Heat Trans. Nanofluids: Conduct. Convect. Radiat. Exp. Results 39 (4) (2012) 3885—3891.

[46] Z. Talaei, A. Reza, A. Rashidi, A. Amrollahi, M. Emami, The effect of functionalized group concentration on the stability and thermal conductivity of carbon nanotube fluid as heat transfer media. Theran, Iran, Int. Commun. Heat Mass Tran. 38 (2011) 513—517.

[47] S.J. Aravind, P. Baskar, T.T. Baby, R.K. Sabareesh, S. Das, S. Ramaprabhu, Investigation of structural stability, dispersion, viscosity, and conductive heat transfer properties of functionalized carbon nanotube based nanofluids, J. Phys. Chem. C 115 (34) (2011) 16737—16744.

[48] M. Abbasi, et al., The effect of functionalisation method on the stability and the thermal conductivity of nanofluid hybrids of carbon nanotubes/gamma alumina, Ceram. Int. 39 (4) (2013) 3885—3891.

[49] M.J. Pastoriza-Gallego, C. Casanova, R. Paramo, B. Baroes, J.L. Legido, M.M. Pineiro, Study on stability and thermo physical properties (density and viscosity) of Al2O3 in water nanofluids, J. Appl. Phys. 106 (2009) 06430.

[50] R. Mondragon, J. Enrique Julia, A. Barba, J.C. Jarque, Characterization of silica water nanofluids dispersed with an ultrasound probe: a study of their physical properties and stability, Powder Technol. 224 (2012) 138e146.

[51] W. Yu, D.M. France, J.L. Routbort, S.U.S. Choi, Review and comparison of nanofluid thermal conductivity and heat transfer enhancements, Heat Tran. Eng. 29 (5) (2008) 432—460, https://doi.org/10.1080/01457630701850851.

[52] C.H. Chon, K.D. Kihm, S.U.S. Choi, S.P. Lee, Empirical correlation finding the role of temperature and particle size for nanofluid Al2O3 thermal conductivity enhancement, Appl. Physics Lett. 87 (2005) 153107.

[53] N.A.C. Sidik, H.A. Mohammed, O.A. Alawi, S. Samion, A review on preparation methods and challenges of nanofluids, Int. Commun. Heat Mass Tran. 54 (2014) 115—125.

[54] A. Nasiri, M. Shariaty-Niasar, A.M. Rashidi, R. Khodafarin, Effect of CNT structures on thermal conductivity and stability of nanofluid, Int. J. Heat Mass Tran. 55 (2012) 1529—1535.

CHAPTER NINE

Entropy optimization of magnetic nanofluid flow over a wedge under the influence of magnetophoresis

Kalidas Das[1], Md Tausif Sk[2]

[1]Department of Mathematics, Krishnagar Government College, Krishnanagar, West Bengal, India
[2]Department of Mathematics, A. B. N. Seal College, Cooch Behar, West Bengal, India

Highlights

- Unsteady Falkner—Skan nanofluid flow is studied flowing over an expanding wedge along with the influence of magnetophoresis and induced magnetic force.
- The mathematical model of the flow system is presented by using suitable partial differential equations along with no mass flux boundary conditions.
- The entropy generated in the system and Bejan number of the flow system is analyzed via suitable numerical procedure.
- The influence of the magnetophoresis and induced magnetic force field on the entropy generation and Bejan number is observed to regulate the decay in the system of the flow.
- After utilizing the numerical technique, it is established that magnetophoresis is a significant factor to study the entropy produced in the flow.

1. Literature review

Magnetophoresis is a phenomenon in which microscopic particles move in light of an applied magnetic field. This method is generally utilized in the field of biotechnology [1] and has numerous applications in areas such as drug delivery, separation of biological cells, and analysis of biological molecules. In this essay, we will examine the basic principles of magnetophoresis, the different types of magnetophoresis, and some of its applications. The principle of magnetophoresis depends on the connection between a magnetic force and nanoparticles [2]. The magnetic particles, also known as

Advanced Materials-Based Fluids for Thermal Systems
ISBN: 978-0-443-21576-6
https://doi.org/10.1016/B978-0-443-21576-6.00011-X

Copyright © 2024 Elsevier Inc.
All rights reserved, including those
for text and data mining, AI training,
and similar technologies.

magnetic beads or magnetic nanoparticles, are typically made of iron oxide or iron–cobalt amalgams also, are covered with a surfactant to avoid aggregation. At the point when a magnetic field is applied, these magnetic particles experience a power proportion to the magnetic field strength and the magnetic susceptibility of the particle. This force can be directed toward or away from the magnetic field, depending on the magnetic susceptibility of the particle. There are several types of magnetophoresis, each with its own unique application. One of the most widely used types of magnetophoresis is magnetic bead separation [3]. This strategy includes the utilization of magnetic dots that are formed with a particular ligand that ties to an objective particle. The magnetic dabs and the objective particles are suspended in a fluid and exposed to a magnetic field. The magnetic power makes the magnetic dabs to move toward the magnetic field, accordingly separating the target molecules from the solution. One more kind of magnetophoresis is magnetic cell detachment [4]. This procedure is utilized to isolate natural cells in light of their magnetic susceptibility. For instance, cells can be marked with magnetic nanoparticles and afterward exposed to magnetic field. The magnetic power makes the marked cells move toward the magnetic field, while the unlabeled cells remain in the solution. This technique has been widely used in the isolation of specific cell populations for research and clinical applications. Magnetophoresis has additionally been utilized for the investigation of organic atoms, like proteins and nucleic acids [5]. In this application, magnetic beads are conjugated with specific antibodies or oligonucleotides that bind to the target molecule. The magnetic beads and target molecules are subjected to a magnetic field, causing the magnetic beads to move toward the magnetic field. The separated target molecules can then be analyzed using various techniques, such as mass spectrometry or fluorescent imaging. Finally, magnetophoresis has also been used for drug delivery [6]. In this application, magnetic nanoparticles are used to carry therapeutic drugs to specific target cells or tissues. The magnetic nanoparticles are subjected to a magnetic field, causing them to move toward the target cells or tissues, where they release the therapeutic drugs. The procedure can possibly work on the proficiency and explicitness of drug delivery, thereby reducing the side effects of traditional drug delivery methods. In conclusion, magnetophoresis is a powerful technique that has numerous applications in the field of biotechnology. Whether it is used for the separation of magnetic particles, cells, or biological molecules, or for drug delivery, magnetophoresis has the potential to revolutionize the way we approach biological and medical problems.

Magnetohydrodynamic (MHD) nanofluid stream over a wedge is a peculiarity that has acquired critical consideration as of late because of its likely applications in different fields, including warm administration, biomedical designing, and energy change [7,8]. MHD nanofluid stream alludes to the progression of a nanofluid, which is a combination of a base liquid and nanoscale particles, within the sight of a magnetic field. The magnetic field creates a Lorentz force that connects with the liquid, prompting changes in the liquid's speed and strain conveyance. This connection is depicted by the Navier—Stokes equations, which are combined with the energy condition and the condition of the magnetic field [9].The stream of a nanofluid over a wedge is an intricate issue that has been concentrated on utilizing mathematical and insightful strategies [10]. The wedge geometry makes a shift in the stream course, prompting a boundary layer development and an expansion in heat move rate. The presence of a magnetic field further convolutes the stream, as the magnetic field influences the speed and temperature circulations of the nanofluid. The utilizations of MHD nanofluid stream over a wedge are various and different. For example, it has been utilized to concentrate on heat move in microelectronics cooling [11], where the nanofluid is used to cool down the high-heat-generating components. It has also been used in biomedical engineering [12], where the flow of nanofluids in blood vessels can be modeled to understand the transport of nanoparticles in the human body. In spite of the many advantages of MHD nanofluid stream over a wedge, there are as yet a few difficulties that should be survived. One of the greatest difficulties is to precisely demonstrate the way of behaving of nanofluids in a magnetic field, as the collaborations between the nanofluid and the magnetic field are exceptionally complicated and not well understood. Around here, a few researchers have contributed their works like Shah et al. [13] researched the impact of higher-order chemical reaction on the MHD nanofluid stream over a wedge. Krishna and Chamkha [14] examined the Hall effect and ion slip on MHD nanofluid stream. Sujatha et al. [15] looked at the nanofluid stream over a cone and a wedge. They examined the impact of chemical reaction on the MHD nanofluid stream over two mathematically various surfaces thinking about thermal radiation, viscous dispersal, and Joule heating alongside heat source/sink. Ahmad et al. [16] explored MHD Casson nanofluid stream over a wedge and concentrated on the impacts of wedge angle and the magnetic field on the qualities of the stream. Unsteady Carreau nanofluid stream is concentrated by Khan et al. [17]. Ali et al. [18] concentrated on the significance of slip conditions on the MHD nanofluid stream over a nonstraightly extending wedge.

Entropy generation is an essential idea in thermodynamics and is a significant proportion of the presentation of energy frameworks. With regards to MHD nanofluid stream, entropy generation [19] alludes to how much entropy produced due to the progression of a nanofluid, which is a combination of a base liquid and nanoscale particles, within the sight of a magnetic field. The entropy generation in MHD nanofluid stream is brought about by different sources, including fluid friction, heat transfer, and magnetic field interactions [20]. The liquid friction produces entropy because of the dispersal of mechanical energy, while the heat move produces entropy because of the exchange of thermal power from the hot to the chilly areas. The entropy generation in MHD nanofluid stream can be broken down utilizing the entropy generation condition, which is gotten from the first and second laws of thermodynamics [21]. The entropy generation condition gives data about the pace of entropy generation and its appropriation all through the stream field. These data are essential for figuring out the presentation of energy frameworks and for advancing their design. The uses of entropy generation examination in MHD nanofluid stream are various and different. For instance, it has been utilized to concentrate on heat move in microelectronics cooling [22], where the entropy generation can be utilized to assess the presentation of cooling frameworks and to distinguish regions for development. It has additionally been utilized in biomedical designing, where entropy generation can be utilized to concentrate on the vehicle of nanoparticles in the human body.

The investigation of MHD nanofluid stream over a wedge is a quickly developing field with various applications and the possibility to make huge commitments to a large number of logical and innovative fields. The proceeded with investigation of this peculiarity will give significant experiences into the principal conduct of liquids in a magnetic field and will prompt new and creative answers for certifiable issues. In spite of the many advantages of entropy generation examination in MHD nanofluid stream, there are as yet a few difficulties that should be survived. One of the greatest difficulties is to precisely demonstrate the way of behaving of nanofluids in a magnetic field, as the cooperations between the nanofluid and the magnetic field are exceptionally complicated and not surely knew. Essentially, the investigation of entropy generation in MHD nanofluid stream is a rapidly emergent domain with various applications and the possibility to make huge commitments to many logical and mechanical fields. The continued with examination of this peculiarities will give critical encounters into the basic way of behaving of liquids in a magnetic field and will prompt new and creative answers for true issues. The fate of this field looks brilliant, and almost certainly, previously unheard-of improvements will arise as more examination is directed around here. Taking motivation from past researchers, authors have attempted to knowledge into the impact of magnetophoresis occasions on the unsteady nanofluid stream over a nonlinearly extending wedge. Likewise to quantify the productivity of the stream framework, authors have researched the entropy generation and Bejan number of the stream.

2. Mathematical formations

Allow us to consider the unsteady 2D Falkner–Skan stream of an incompressible nanofluid past an extending wedge within the sight of magnetophoresis and an induced magnetic field. It is expected that liquid stream is initiated by an extending wedge with the speed along with the free stream speed $u_e = \frac{ax^m}{1-ct}$ where $a, b, c,$ and m are positive constants with $0 \leq m \leq 1$. It should be noted that $u_w > 0$ corresponds to an extending wedge surface speed and $u_w < 0$ corresponds to a contracting wedge surface velocity (see Fig. 9.1). The wedge angle is assumed to be $\Omega = \chi\pi$, $\chi = \frac{2m}{m+1}$. The joined impacts of magnetophoresis, thermophoresis, and Brownian movement are considered due to the nanoparticles.

Under the aforementioned suspicions and in the wake of applying the boundary layer examination, the fundamental boundary layer equations overseeing the preservations of mass, momentum, energy, and nanoparticle

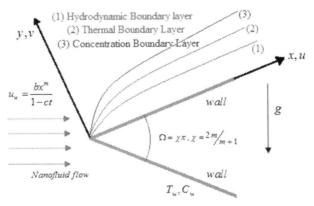

Figure 9.1 Physical model of the flow.

density for the nanofluid within the sight of time-subordinate magnetic field can be communicated as [17,23]:

$$\frac{\partial u}{\partial x} + \frac{\partial v}{\partial y} = 0$$

$$\frac{\partial H_1}{\partial x} + \frac{\partial H_2}{\partial y} = 0$$

$$\frac{\partial u}{\partial t} + u\frac{\partial u}{\partial x} + v\frac{\partial u}{\partial y} = \frac{\partial u_e}{\partial t} + u_e\frac{\partial u_e}{\partial x} + \frac{\mu}{\rho}\frac{\partial^2 u}{\partial y^2} + \frac{\mu_m}{4\pi\rho}\left[H_1\frac{\partial H_1}{\partial x} + H_2\frac{\partial H_1}{\partial y} - H_e\frac{\partial H_e}{\partial x}\right]$$

$$\frac{\partial H_1}{\partial t} + u\frac{\partial H_1}{\partial x} + v\frac{\partial H_2}{\partial y} = \alpha^*\frac{\partial^2 H_1}{\partial y^2} + H_1\frac{\partial u}{\partial x} + H_2\frac{\partial u}{\partial y}$$

$$\frac{\partial T}{\partial t} + u\frac{\partial T}{\partial x} + v\frac{\partial T}{\partial y} = \alpha\frac{\partial^2 T}{\partial y^2} + \tau\left[D_B\frac{\partial T}{\partial y}\frac{\partial C}{\partial y} + \frac{D_t}{T_\infty}\left(\frac{\partial T}{\partial y}\right)^2 - D_m\frac{\partial H_1}{\partial y}\frac{\partial T}{\partial y}\right]$$

$$\frac{\partial C}{\partial t} + u\frac{\partial C}{\partial x} + v\frac{\partial C}{\partial y} = D_B\frac{\partial^2 C}{\partial y^2} + \frac{D_t}{T_\infty}\frac{\partial^2 T}{\partial y^2} - D_m\frac{\partial^2 H_1}{\partial y^2}$$

$$(9.1)$$

with the boundary conditions:

$$u = u_w, v = 0, \frac{\partial H_1}{\partial y} = 0, H_2 = 0, T = T_w, D_B\frac{\partial C}{\partial y} + \frac{D_t}{T_\infty}\frac{\partial T}{\partial y} = 0 \; at \; y = 0$$

$$u \to u_e, H_1 \to H_e, T \to T_\infty, C \to C_\infty \; as \; y \to \infty$$

$$(9.2)$$

Here T_w = temperature of the wedge wall, ν = kinematic viscosity of the flow, μ_m = magnetic permeability, D_B = Brownian diffusion, D_T = thermophoresis diffusion, D_m = magnetophoresis constant, and $()_\infty$ fluid properties far away from the wedge.

To solve system (1) with (2), we need to introduce similarity transformations to transform the set of partial differential equations (PDEs) to ordinary differential equations (ODEs) and solve it numerically:

$$\eta = y\sqrt{\frac{(m+1)u_e}{2\nu x}}, \psi = f(\eta)\sqrt{\frac{2\nu x u_e}{(m+1)}}, H_e = \frac{dx^m}{1-ct}, H_1 = H_e h'(\eta),$$

$$H_2 = -\left(\frac{m+1}{2x}\right)\sqrt{\frac{2\nu x d H_e}{a(m+1)}}h(\eta) - \left(\frac{m+1}{2x}\right)y H_e h'(\eta), \theta = \frac{T - T_\infty}{T_w - T_\infty}, \phi = \frac{C - C_\infty}{C_\infty}$$

$$(9.3)$$

So applying the transformations (3) in (1−2), we obtained the transformed set of ODEs as:

$$f''' + ff'' - \frac{2m}{m+1}\left[(f')^2 - 1\right] - \beta\left[hh'' - \frac{2m}{m+1}\left\{(h')^2 - 1\right\}\right] - K(2f' + \eta f'' - 2) = 0$$

$$\frac{1}{Pr_m}h''' + \frac{2m}{m+1}f'h' + hf'' + \frac{d}{a}\left\{hh'' - \frac{2m}{m+1}(h')^2\right\} - K(2h' + \eta h'') = 0$$

$$\frac{1}{Pr}\theta'' + f\theta' + (N_{BM}\phi' + N_{TP}\theta' - N_{MG}h' - K\eta)\theta' = 0$$

$$\phi'' + LePr(f - K\eta)\phi' + \frac{N_{TP}}{N_{BM}}\theta'' - \frac{N_{MG}}{N_{BM}}h''' = 0$$

$$(9.4)$$

and

$$\left. \begin{array}{l} f = 0, f' = b/_a, h''(0) = 0, h(0) = 0, \theta(0) = 1, N_{BM}\phi'(0) + N_{TP}\theta'(0) = 0 \\ f \to 1, h' \to 1, \theta \to 0, \phi \to 0 \ as \ \eta \to \infty \end{array} \right\}$$

$$(9.5)$$

Here, $\beta = \frac{d^2\mu_m}{4\pi a^2\rho}$ induced magnetic parameter, $K = \frac{c}{a(m+1)x^{m-1}}$ is unsteadiness parameter, $Pr_m = \frac{\nu}{\alpha^*}$ magnetic Prandtl number, $Pr = \frac{\nu}{\alpha}$ Prandtl number, $N_{BM} = \frac{\tau C_\infty D_B}{\nu}$ Brownian motion parameter, $N_{TP} = \frac{\tau D_B(T_w - T_\infty)}{T_\infty \nu}$ thermophoresis parameter, $N_{MG} = \frac{\tau D_m H_e}{\nu}$ magnetophoresis parameter (note: it is local parameter), and $Le = \frac{\alpha}{D_B}$ Lewis number.

For engineering benefits, we need to calculate the shear stress and heat transmission at the surface of the sphere, we can be represented as:

Skin friction: $Cf = \frac{\tau_s}{\rho u_e^2}, \tau_s = \mu\left(\frac{\partial u}{\partial y}\right)_{y=0}$ which transforms to reduced

skin friction $Cfr = CfRe^{\frac{1}{2}} = \sqrt{\frac{m+1}{2}}f''(0)$

Nusselt number: $Nu = \frac{xq_s}{\kappa(T_w - T_\infty)}, q_s = -\kappa\left(\frac{\partial T}{\partial y}\right)_{y=0}$ which becomes

$Nur = NuRe^{-\frac{1}{2}} = -\sqrt{\frac{m+1}{2}}\theta'(0)$

Sherwood number: $Sh = \frac{xq_m}{D_B C_\infty}, q_m = -D_B\left(\frac{\partial C}{\partial y}\right)_{y=0}$ which becomes

$Shr = ShRe^{-\frac{1}{2}} = -\sqrt{\frac{m+1}{2}}\phi'(0)$

where $Re = \frac{u_e x}{\nu}$ is local Reynolds number.

3. Entropy generation

The entropy in nondimensional form for the nanofluid is characterized as:

$$S_{gen}''' = \frac{\kappa}{T_\infty^2}\left(\frac{\partial T}{\partial \gamma}\right)^2 + \frac{\mu}{T_\infty}\left(\frac{\partial u}{\partial \gamma}\right)^2 + \frac{RD}{C_\infty}\left(\frac{\partial C}{\partial \gamma}\right)^2 + \frac{RD}{T_\infty}\left(\frac{\partial C}{\partial \gamma}\frac{\partial T}{\partial \gamma}\right)$$

After using similarity transformation (3), we obtain the dimensionless entropy generation as:

$$Ng = \frac{S_{gen}'''}{S_0} = \left(\frac{m+1}{2}\right)Re\left[(\theta')^2 + \frac{Br}{\omega}(f'')^2 + \frac{\lambda}{\omega^2}(\phi')^2 + \frac{\lambda}{\omega}\theta'\phi'\right],$$

where $S_0 = \frac{\kappa(T_w - T_\infty)^2}{x^2 T_\infty^2}$, $Br = \frac{\mu u_e^2}{\kappa(T_w - T_\infty)}$ Brinkman number, $\omega = \frac{T_w - T_\infty}{T_\infty}$ temperature difference ratio, and $\lambda = \frac{RDC_\infty}{\kappa}$ diffusion characteristic.

Bejan number is defined as $Be = \dfrac{(\theta')^2}{(\theta')^2 + \dfrac{Br}{\omega}(f'')^2 + \dfrac{\lambda}{\omega^2}(\phi')^2 + \dfrac{\lambda}{\omega}\theta'\phi'}$.

4. Numerical solution methodology

Numerically, the arrangement of nonlinear differential Eq. (9.4) with the fitting boundary conditions (5) has been addressed by utilizing two efficient numerical methodologies: explicitly, the shooting strategy with Fehlberg equation, Newton's systems, and the bvp4c functions in Matlab. A careful numerical estimation is performed for different potential outcomes of the pertinent boundary conditions, specifically the induced magnetic parameter β, magnetic Prandtl number Pr_m, and magnetophoresis parameter N_{MG}. To exhibit the believability of the cultivated numerical results, a relationship with the ongoing composing is similarly coordinated in confined cases. The estimations of Cfr for different upsides of "m" are determined mathematically from the first equation of (4) considering $\beta = K = 0$, $b = a = 1$ and contrasted with the outcomes derived by Rajagopal et al. [7], Kuo [8], Ishaq et al. [9] (Table 9.1). Here we notice an elevated degree of concurrence with before referenced.

5. Result and discussion

The nonlinear arrangement of differential Eq. (9.4) alongside the boundary conditions (5) is settled mathematically. Also, the local entropy

Table 9.1 values of skin friction for different values of "m."

m	Rajagopal et al. [7]	Kuo [8]	Ishaq et al. [9]	Present result
0	–	0.469600	0.4696	0.469584
1/19	0.587035	0.587880	0.5870	0.587055
3/17	0.774755	0.775524	0.7748	0.774759
1/3	0.927680	0.927905	0.9277	0.927681
1	1.232585	1.231289	1.2326	1.232588

generated (Ng) in the system and dimensionless Bejan number (Be) are also being computed numerically. The outcomes of the arduous calculations are staged here through proper graphs and charts considering $m = 0.25$, $K = 0.2$, $d = 0.1$, $a = 0.1$, $N_{BM} = 0.2$, $N_{TP} = .2$, $Pr = 1$, and $Le = 1$ unless otherwise specified. Our main purpose is to observe the influence of induced magnetic factor (β), magnetic Prandtl number (Pr_m), and magneto-phoresis parameter (N_{MG}) on the fluid velocity ($\frac{df}{d\eta}$), temperature ($\theta(\eta)$), nanoparticle concentration ($\phi(\eta)$), local entropy generated (Ng), and Bejan number (Be) and summarize the overall exhibition of the stream system to optimize the decay simultaneously.

5.1 Induced magnetic parameter (β)

The induced magnetic parameter can have significant effects on nanofluid stream. At the point when magnetic field is applied to a nanofluid, it can induce a magnetic dipole moment in the nanoparticles suspended in the fluid, which can alter the fluid flow behavior in various ways. The presence of a magnetic field can increase the viscosity of the nanofluid due to the formation of chains or clusters of nanoparticles. These structures can obstruct the fluid flow and increase the fluid's resistance to deformation, prompting an expansion in consistency. Accordingly, the stream speed gradually lessens with more powerful β as shown in Fig. 9.2. Also, it reduces the skin friction of the flow near the surface as seen in Table 9.2. The induced magnetic field to a nanofluid can improve its warm conductivity because of the arrangement of the nanoparticles in the direction of the magnetic field. This can lead to improved heat transfer performance in various applications. That's why as staged in Fig. 9.3, we witness that higher β elevates the temperature of the flow. But it lessens the heat move rate (Nur) close to the surface. The induced magnetic parameter can also affect the nanoparticle concentration in nanofluid flow. At the point when a magnetic field is applied to a nanofluid containing

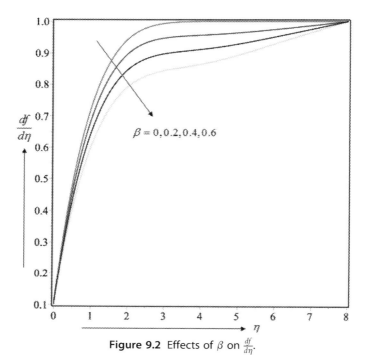

Figure 9.2 Effects of β on $\frac{df}{d\eta}$.

Table 9.2 values of physical quantities for various parameters.

β	Pr_m	N_{MG}	Cfr	Nur	Shr
0.0	0.5	0.3	0.736547	0.308668	−0.30867
0.1			0.707996	0.303501	−0.3035
0.2			0.678489	0.297861	−0.29786
0.3			0.648042	0.291701	−0.2917
0.4			0.616669	0.284972	−0.28497
0.2	0.1	0.3	0.695589	0.276108	−0.27611
	0.4		0.680344	0.294791	−0.29479
	0.7		0.67597	0.302535	−0.30254
	1.0		0.673625	0.307603	−0.3076
	1.3		0.672087	0.311461	−0.31146
0.2	0.5	0.0	0.678489	0.357133	−0.35713
		0.2	0.678489	0.317045	−0.31704
		0.4	0.678489	0.27927	−0.27927
		0.6	0.678489	0.243924	−0.24392
		0.8	0.678489	0.211109	−0.21111

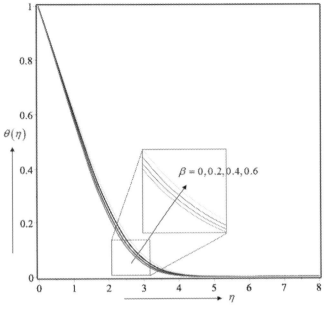

Figure 9.3 Effects of β on $\theta(\eta)$.

nanoparticles, the magnetic force acting on the nanoparticles can alter their distribution and concentration in the fluid. It can cause the nanoparticles to aggregate and form clusters. These clusters can increase the local concentration of nanoparticles near the surface of the fluid flow as shown in Fig. 9.4. Also, in Fig. 9.5, we observe that the Ng of the flow is initially decreasing with higher β. Then after some distance from the surface, Ng starts increasing with more β. The purpose for this is that the magnetic field can induce fluid flow instabilities, which lead to increased energy dissipation and heat transfer within the fluid. The resulting increase in entropy production is a consequence of this enhanced dissipation. Then again, the induced magnetic parameter can also reduce the entropy production of the system by promoting the arrangement of the nanoparticles in the course of the field. This can lead to a more ordered and structured flow behavior, which can reduce the amount of heat energy dissipated as waste. Again in Fig. 9.6, we observe that Be initially diminishes with elevated β. Then after some distance from the wedge wall, its worth gradually decreases with elevated β.

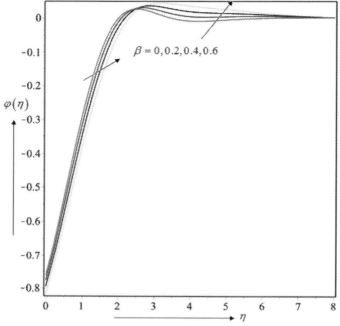

Figure 9.4 Effects of β on $\phi(\eta)$.

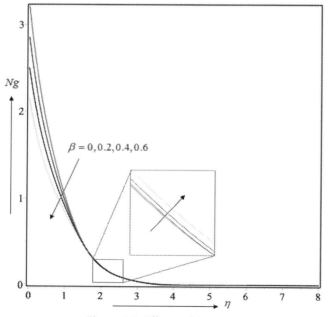

Figure 9.5 Effects of β on Ng.

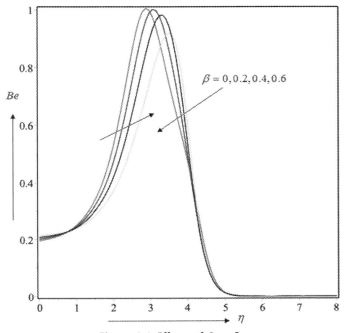

Figure 9.6 Effects of β on Be.

5.2 Magnetic Prandtl number (Pr_m)

The magnetic Prandtl number is a significant factor in the investigation of MHD flows, as it characterizes the relative significance of magnetic and viscous powers in the liquid. If the magnetic Prandtl number is much greater than 1, then the magnetic forces are much stronger than the viscous forces, also, the stream is supposed to be magnetically dominated. Conversely, if the magnetic Prandtl number is much less than 1, then the viscous forces are much stronger than the magnetic forces, and the flow is said to be viscously dominated. The magnetic Prandtl number is a key parameter in many astrophysical and geophysical contexts, such as in the study of the Earth's core or the dynamics of accretion disks around black holes. When the magnetic Prandtl number is large, the magnetic forces rule over the viscous forces, also, the stream conduct can be strongly inclined by the magnetic field. In this case, the nanoparticles can be magnetically aligned, leading to anisotropic flow properties and enhanced heat transfer. Additionally, the magnetic field can induce eddy currents in the fluid, leading to enhanced mixing and turbulence. Then again, when the magnetic Prandtl number is little, the viscous forces dominate, and the flow behavior is less affected by the magnetic field. In this case, the nanoparticles are more

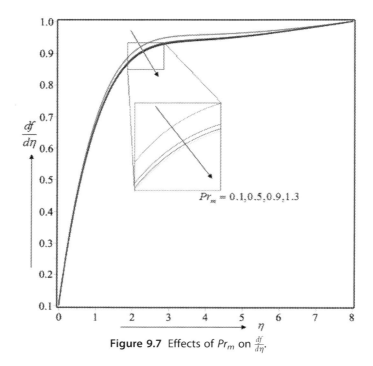

Figure 9.7 Effects of Pr_m on $\frac{df}{d\eta}$.

likely to remain randomly oriented, and the flow properties may be more isotropic. That's why in Fig. 9.7 and in Table 9.2 due to the dominance of magnetic force, the velocity of the flow as well as skin friction decreases with higher Pr_m. The magnetic Prandtl number can also have a significant effect on the temperature of nanofluid flows, which are suspensions of nanoparticles in a fluid. The presence of magnetic fields can induce various thermomagnetic effects, such as the magnetocaloric effect and the thermo-magnetoconvective flow, which can influence the temperature dispersion in the liquid. The extent of these impacts relies upon the magnetic Prandtl number, as well as other factors such as the particle size and concentration, the potency of the magnetic field, and the fluid velocity. In Fig. 9.8, we monitor that temperature of the flow decreases with higher Pr_m, but the heat transfer rate increases along with the mass transfer rate as we witness in Table 9.2. Because when the magnetic Prandtl number is large, the magnetic forces rule over the viscous forces, and the heat transfer properties can be strongly influenced by the magnetic field. In this case, the magnetocaloric effect can occur, where the magnetic field can induce a change in

Entropy optimization of magnetic nanofluid flow

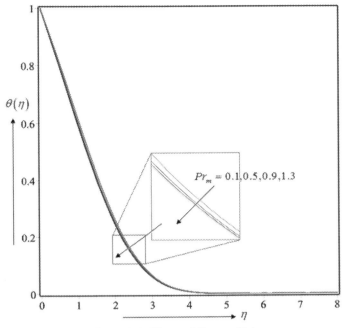

Figure 9.8 Effects of Pr_m on $\theta(\eta)$.

temperature in the fluid due to the magnetic properties of the nanoparticles. Additionally, the thermo-magnetoconvective flow can occur, where the magnetic field can induce convective motion in the fluid, leading to enhanced heat transfer. Also, higher Pr_m leads to anisotropic flow properties and a higher concentration of nanoparticles in certain regions of the flow. Additionally, the magnetic field can induce particle migration, leading to concentration gradients and nonuniform particle distributions as it has been observed in Fig. 9.9. In Fig. 9.10, the effect of Pr_m on Ng is staged. The magnetocaloric effect, where the magnetic field can induce a change in temperature of the liquid caused by the magnetic properties of the nanoparticles, can prompt changes in the entropy of the system. Alternatively, when the magnetic Prandtl number is tiny, the viscous forces dominate, and the magnetic field has less of an impact on the entropy of the fluid. In this case, the entropy of the framework might be more affected by different factors like particle concentration, flow rate, and temperature. In Fig. 9.11, the effect of Pr_m on Be is staged. The magnetic field can induce convective motion in the fluid, leading to enhanced heat transfer and a

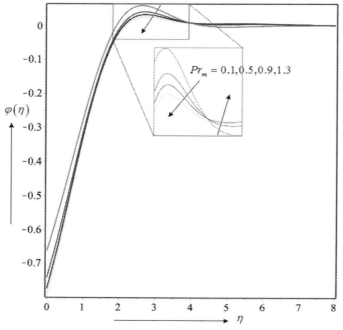

Figure 9.9 Effects of Pr_m on $\phi(\eta)$.

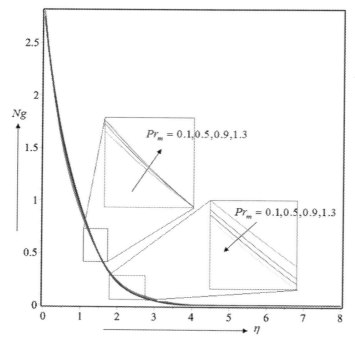

Figure 9.10 Effects of Pr_m on Ng.

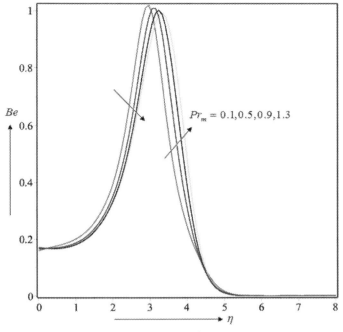

Figure 9.11 Effects of Pr_m on Be.

higher Bejan number. The thermo-magnetoconvective flow can also contribute to the Bejan number by inducing convective motion in the fluid due to the magnetic field.

5.3 Magnetophoresis parameter (N_{MG})

The magnetophoresis parameter is a dimensionless boundary that portrays the overall strength of the magnetic powers following up on nanoparticles in a nanofluid stream. It can have important effects on the behavior of nanofluid flows. In general, a higher magnetophoresis parameter indicates that the magnetic forces are relatively stronger than the viscous forces, which can lead to more pronounced effects on the flow behavior. When the magnetophoresis parameter is high, the magnetic forces can induce convective motion in the nanofluid, leading to enhanced heat transfer as we notice in Fig. 9.12 and a higher convective heat transfer coefficient as staged in Table 9.2. Similarly, when the magnetophoresis parameter is high, the magnetic forces can rule over different powers such as diffusion and gravity, leading to the accumulation of nanoparticles in specific regions of the flow as shown in Fig. 9.13. This can have important implications for various applications such as magnetic drug delivery and magnetic separation. Overall, the effect of the magnetophoresis parameter on nanofluid flows is complex and

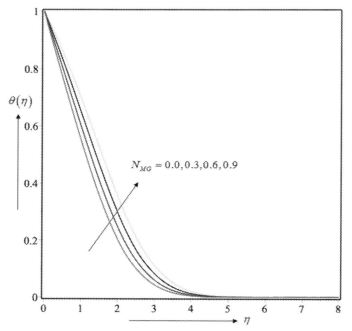

Figure 9.12 Effects of N_{MG} on $\theta(\eta)$.

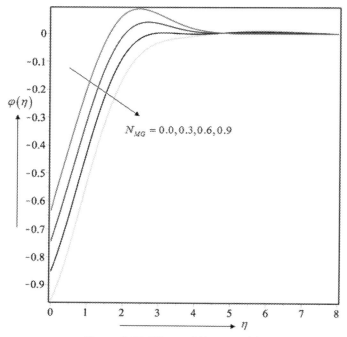

Figure 9.13 Effects of N_{MG} on $\phi(\eta)$.

depends on several factors. However, understanding this parameter is important for predicting and controlling the behavior of nanofluids in magnetic fields, which has applications in fields, for example, energy, gadgets cooling, and materials science. To observe the influence of N_{MG} on the entropy generated in the flow and the Bejan number, we need to focus on Figs. 9.14 and 9.15. We observe that N_{MG} consecutively elevates and diminishes the entropy generation rate of the flow and Be. This is happening because first, the magnetophoresis parameter can affect the convective heat transfer properties of the nanofluid, which in turn affects the entropy generation in the flow. When the magnetophoresis parameter is high, the magnetic forces can induce convective motion in the nanofluid, leading to enhanced heat move and a higher convective heat move coefficient. This results in a reduction in the entropy generation in the nanofluid flow, as the more efficient heat transfer prompts a decrease in the thermal gradients and subsequently a lower pace of entropy generation. Second, the magnetophoresis parameter can affect the Bejan number of the nanofluid flow by influencing the convective heat transfer properties of the nanofluid. The Bejan number is a dimensionless number that signifies the ratio of convective heat transfer rate to the heat transfer rate due to conduction.

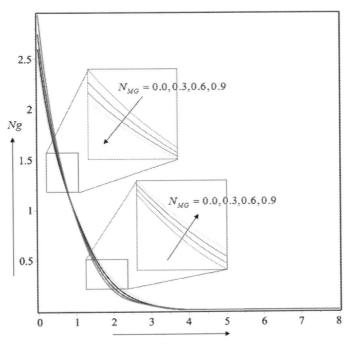

Figure 9.14 Effects of N_{MG} on Ng.

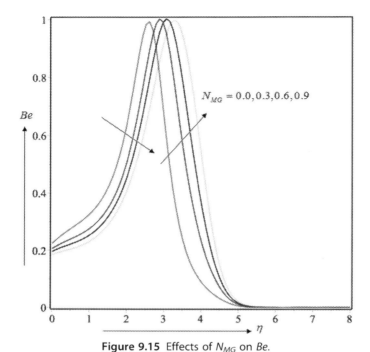

Figure 9.15 Effects of N_{MG} on Be.

When the magnetophoresis parameter is high, the magnetic forces can induce convective motion in the nanofluid, leading to a higher Bejan number. This is because the convective motion induced by the magnetic field can enhance the rate of heat transfer in the nanofluid, leading to a higher rate of heat transfer due to convection and hence a higher Bejan number. However, it is important to note that the effect of the magnetophoresis parameter on the entropy and Bejan number of nanofluid flows depends on several factors, such as the strength and course of the magnetic field, the size and magnetic properties of the nanoparticles, and the flow properties of the nanofluid. Thus, it is crucial to carefully consider these factors when analyzing the effect of the magnetophoresis parameter on the entropy and Bejan number of nanofluid flows.

6. Conclusions

In this article, the locally tantamount numerical plans have been figured for the unsteady 2D Falkner–Skan stream of MHD nanofluid past a moving wedge affected by the magnetophoresis, thermophoresis, and

Brownian developments inside seeing the no mass movement at the surface. Moreover, the entropy created in the structure along with the Bejan number was calculated for the stream assessment. The time-subjected nonlinear partial differential equations were changed to a bunch of semicoupled nonlinear ordinary differential equations and settled numerically by utilizing two particular numerical methodologies to be explicitly the shooting system with Fehlberg recipe and Newton's Raphson strategy as well as bvp4c capacity in Matlab. The critical discoveries of this article are recorded under:

❖ The induced magnetic parameter (β) is significant for diminishing the velocity of the flow system as it helps to form nanoparticle clusters to which hinders the flow.

❖ The entropy generated in the system is slower when the value of the induced magnetic parameter (β) is higher. It helps us to conclude that to regulate the entropy in the system, we should induce the magnet of the system.

❖ The magnetic Prandtl number (Pr_m) and the magnetophoresis parameter (N_{MG}) can affect the heat and mass transfer of the system more efficiently.

❖ In case of the entropy generation, the magnetophoresis parameter (N_{MG}) initially slows the entropy generation in the system, but then higher N_{MG} elevates the generation of the entropy of the system.

Nomenclature

u_e Free stream velocity of the flow
u_w Stretching speed of the wedge
Ω Wedge angle
(u, v) Velocity component of the flow
(H_1, H_2) Induced magnetic field
a, b, c, d, m Constants
μ Viscosity of the fluid
μ_m Magnetic permeability
ρ Density of the nanofluid
D_B Brownian motion diffusion
D_t Thermophoresis diffusion
D_m Magnetophoresis diffusion
T Temperature of the fluid
C Concentration of the nanofluid
S_{gen}''' Entropy generation in the flow
Ng Dimentionless entropy generated
ω Temperature difference ratio
Be Bejan number

ψ Stream function

θ Nondimensional temperature of the nanofluid

ϕ Nondimensional concentration of the nanofluid

β Induced magnetic parameter

K Unsteadiness parameter

Pr_m Magnetic Prandtl number

Pr Prandtl number

N_{BM} Brownian motion parameter

N_{TP} Thermophoresis parameter

N_{MG} Magnetophoresis parameter

Le Lewis number

Cfr Reduced skin friction

Nur Reduced Nusselt number

Shr Reduced Sherwood number

Re Reynolds number

Br Brinkman number

λ Diffusion characteristic

References

[1] M. Suwa, H. Watarai, Magnetoanalysis of micro/nanoparticles: a review, Anal. Chim. Acta 690 (2) (2011) 137–147.

[2] S.S. Leong, Z. Ahmad, S.C. Low, J. Camacho, J. Faraudo, J. Lim, Unified view of magnetic nanoparticle separation under magnetophoresis, Langmuir 36 (28) (2020) 8033–8055.

[3] J. Ruan, W. Zhang, C. Zhang, N. Li, J. Jiang, H. Su, A magnetophoretic microdevice for multi-magnetic particles separation based on size: a numerical simulation study, Eng. Appl. Computat. Fluid Mech. 16 (1) (2022) 1781–1795.

[4] H.N. Thu, H.B. Thu, H.T.T. Thuy, T.T. Bui, L. Do Quang, A combination of dielectrophoresis and magnetophoresis microfluidic chip for cancer cells separation, in: 2022 2nd International Conference on Intelligent Cybernetics Technology & Applications (ICICyTA), IEEE, December 2022, pp. 128–133.

[5] D. Kirby, J. Siegrist, G. Kijanka, L. Zavattoni, O. Sheils, J. O'Leary, R. Burger, J. Ducrée, Centrifugo-magnetophoretic particle separation, Microfluid. Nanofluidics 13 (2012) 899–908.

[6] S.N. Murthy, S.M. Sammeta, C. Bowers, Magnetophoresis for enhancing transdermal drug delivery: mechanistic studies and patch design, J. Contr. Release 148 (2) (2010) 197–203.

[7] K.R. Rajagopal, A.S. Gupta, T.Y. Na, A note on the Falkner-Skan flows of a non-Newtonian fluid, Int. J. Non Lin. Mech. 18 (4) (1982) 313–320.

[8] B.L. Kuo, Application of the differential transformation method to the solutions of Falkner-Skan wedge flow, Acta Mech. 164 (2003) 161–174.

[9] A. Ishaq, R. Nazar, I. Pop, Moving wedge and flat plate in a micropolar fluid, Int. J. Eng. Sci. 44 (2006) 1225–1236.

[10] A.K. Pandey, M. Kumar, Effect of viscous dissipation and suction/injection on MHD nanofluid flow over a wedge with porous medium and slip, Alex. Eng. J. 55 (4) (2016) 3115–3123.

[11] A. Mishra, M. Kumar, Numerical analysis of MHD nanofluid flow over a wedge, including effects of viscous dissipation and heat generation/absorption, using Buongiorno model, Heat Transfer 50 (8) (2021) 8453–8474.

[12] S.M. Vahedi, A.H. Pordanjani, A. Raisi, A.J. Chamkha, Sensitivity analysis and optimization of MHD forced convection of a Cu-water nanofluid flow past a wedge, European Phys. J. Plus 134 (2019) 1–21.

[13] S.Z. Shah, H.A. Wahab, A. Ayub, Z. Sabir, A. haider, S.L. Shah, Higher order chemical process with heat transport of magnetized cross nanofluid over wedge geometry, Heat Transfer 50 (4) (2021) 3196–3219.

[14] M.V. Krishna, A.J. Chamkha, Hall and ion slip effects on unsteady MHD convective rotating flow of nanofluids—application in biomedical engineering, J. Egypt. Mathematical Soc. 28 (1) (2020) 1.

[15] T. Sujatha, K.J. Reddy, J.G. Kumar, Chemical reaction effect on nonlinear radiative MHD nanofluid flow over cone and wedge, Defect Diff. Forum 393 (2019) 83–102 (Trans Tech Publications Ltd).

[16] K. Ahmad, Z. Hanouf, A. Ishak, MHD Casson nanofluid flow past a wedge with Newtonian heating, European Phys. J. Plus 132 (2) (2017) 87.

[17] M. Khan, M. Azam, A. Munir, On unsteady Falkner-Skan flow of MHD Carreau nanofluid past a static/moving wedge with convective surface condition, J. Mol. Liq. 230 (2017) 48–58.

[18] M. Ali, M.A. Alim, Influence of slip parameter, viscous dissipation and joule heating effect on boundary layer flow and heat transfer over a power-law stretching wedge-shaped surface with the correlation coefficient and multiple regressions, Int. J. Appl. Mech. Eng. 27 (2) (2022) 1–21.

[19] G.S. Seth, A. Bhattacharyya, R. Kumar, A.J. Chamkha, Entropy generation in hydromagnetic nanofluid flow over a non-linear stretching sheet with Navier's velocity slip and convective heat transfer, Phys. Fluids 30 (12) (2018) 122003.

[20] M.M. Rashidi, S. Abelman, N.F. Mehr, Entropy generation in steady MHD flow due to a rotating porous disk in a nanofluid, Int. J. Heat Mass Tran. 62 (2013) 515–525.

[21] R. Rehman, H.A. Wahab, U. Khan, Heat transfer analysis and entropy generation in the nanofluids composed by aluminum and $\gamma-$ aluminum oxides nanoparticles, Case Stud. Therm. Eng. 31 (2022) 101812.

[22] Y.L. Zhai, G.D. Xia, X.F. Liu, J. Wang, Characteristics of entropy generation and heat transfer in double-layered micro heat sinks with complex structure, Energy Convers. Manag. 103 (2015) 477–486.

[23] M. Kumari, G. Nath, Steady mixed convection stagnation-point flow of upper convected Maxwell fluids with magnetic field, Int. J. Non Lin. Mech. 44 (10) (2009) 1048–1055.

CHAPTER TEN

Nonaxisymmetric homann stagnation-point flow of nanofluid toward a flat surface in the presence of nanoparticle diameter and solid—liquid interfacial layer

Kalidas Das[1], Shib Sankar Giri[2], Nilangshu Acharya[3]

[1]Department of Mathematics, Krishnagar Government College, Krishnanagar, West Bengal, India
[2]Department of Mathematics, Bidhannagar College, Kolkata, West Bengal, India
[3]Department of Mathematics, P.R. Thakur Govt. College, Ganti, West Bengal, India

Highlights

- Nonaxisymmetric Homann stagnation-point flow configured by a flat plate is considered here.
- The mode of heat transmission in flow is stimulated through melting heat conditions and nonlinear radiation.
- Consequence diameter of nanoparticles and nanolayer is comprised in the flow.
- Temperature circulation in the flow declines when the diameter of the nanoparticles enhances.

1. Introduction

Flows occurring at stagnation points are tracing the flow in proximity to stagnation zone at ventral associate with a brusque-shaped object, prevail on every solid entity while the fluid in motion. Stagnation zone/region meets maximum measured magnitudes of pressure, heat transference, and maximum mass flow rates of deposition as well. Stationary or moveable object in proximity to flow at the point of stagnation has charmed multiple analysts on account of its countless manufacturing and commercial implementations like drag lightening in the fast-paced aircraft. Hiemenz

Advanced Materials-Based Fluids for Thermal Systems
ISBN: 978-0-443-21576-6
https://doi.org/10.1016/B978-0-443-21576-6.00004-2

Copyright © 2024 Elsevier Inc.
All rights are reserved, including those
for text and data mining, AI training,
and similar technologies.

[1] came to be leading and foremost to inquire about 2D planner stagnation-point viscous flow employing a combination of self-similarity conversions to transform Navier—Stokes equations into a design of nonlinear ordinary differential equations (ODEs). Behind the investigations involving motion behavior, nearby stagnation points were amplified in numerous manners to incorporate various physical consequences. Further axisymmetric fluid motion nearby stagnation points were looked into by the researcher Homann [2]. Another prime upshot acknowledged by Howarth [3] with introducing a pair of normal Hiemenz-type flows characterizes the latest collection of nonaxisymmetric stagnation flows. Though Davey [4] remained the supreme researcher to recognize the fallacy in Howarth's article. The exploration of 3D stagnation flow has already been talked about by Davey and Schofield [5]. Bachok and associates [6] judged stagnation-point nanofluids flow moving across a stretching/shrinking plate. The same batch of analysts also inspected a 2D stagnation flow of nanoliquids past the exponential slab [7]. Features of convection and Joule heating for stagnation-point 2D flow of nanoliquid passing over a porous arched shrinking/stretching top were explored by Zhang et al. [8]. Rajput et al. [9] have derived multiple outcomes along with stability in a stagnation flow. Magnetohydrodynamic (MHD) stagnation-point 2D flow past a planar top layer was explored by Dawar et al. [10].

An exploration on nonaxisymmetric Homann flow at stagnation point [2] with variation of periodic radial plus azimuthal velocities was originated by Weidman [11]. Later, Lok et al. [12] expanded by the enclosure of heat conduction plus induced buoyancy phenomena. They contemplated nonaxisymmetric viscous flow at stagnation-point through a vertical regular panel. Mahapatra and Sidui [13] depicted heat conveyance throughout a nonaxisymmetric viscous flow at a stagnation point over a linearly stretched plate. Kudenattia and Kirsur [14] analyzed nonaxisymmetric 3D stagnation-point boundary-layer flow inclusive of an applied magnetic field. Recently, nonaxisymmetric hybrid nanofluid ($Cu-Al_2O_3$/water) flow through a lamina was evaluated by Khashi'ie et al. [15]. Mahapatra and Sidui [16] exposed a 3D nonaxisymmetric flow of viscoelastic liquid over a fixed plate. Khan et al. [17] scrutinized numerous solutions of an unsteady nonaxisymmetric flow. Khan et al. [18] disk explored non-axisymmetric Walter's B nanoliquid flow through cylindrical shaped. Intensification of heat transport for a nonaxisymmetric flow of Maxwell liquid analyzed explored nonaxisymmetric Walter's B nanoliquid flow through cylindrical shape by Jagwal et al. [19].

Stimulus of interfacial layer (nanolayer) at particle/fluid interface thinks about a vital implement for intensifying nanoliquid's thermal conductivity. The solid-like coating appears as a thermal link connecting base fluid plus solid nanoparticle. Actually, fluid molecules form a layered structure at solid exteriors whose microscopic formations are more organized compared to the structure of bulk fluid. An impactful thermal conductivity model of nanofluid taking into account the interface influence between solid particles and base fluid was documented by Xue [20]. Yu and Choi [21] made a noteworthy amending on Maxwell equation for competent thermal conductivity of solid and liquid suspensions to incorporate the influence of this systematic nanolayer. Impression of terminal nanolayer on the functional thermal conductivity of nanofluid was explored by Xie et al. [22]. They deduced an execution in connection with nanolayer for computing magnified thermal conductivity of nanofluid. Leong et al. [23] anticipated an enriched model for stimulating the effectual thermal conductivity of nanofluids because of interfacial layer at solid—liquid terminal. A joint theoretical and experimental analysis on effective thermal conductivity plus viscosity of nanofluids was carried out by Murshed et al. [24]. Alongside the nanolayer, the diameter of nano-sized particles also plays a prime responsibility in thermal integrity concerning the flow. Rana and Beg [25] inspected 2D incompressible Al2O3—water nanofluid flow to display the impact of nanoparticles' size along with nanolayer. Consequence of nanolayer as well as nanoparticle diameter in flow of Buongiorno's nonhomogenous nanofluid model was analyzed by Rana et al. [26]. The resultant of nanoparticle diameter and nanolayer on an engine oil-based nanoliquid flow through three unlike slandering fine pointers of nonidentical outlines was conveyed by Sk et al. [27]. Giri et al. [28] have accomplished an analysis of nanofluid flow inside rotating channel in company of thermal conductivity co-opt with the consequences of nanoparticles diameter as well as nanolayer. A captivating exploration connecting with encouragement of nanolayer and nanoparticle diameter over slippery surface was revealed by Acharya [29]. Acharya et al. [30] have disclosed nanofluid flow through porous plate incorporating with consequences of nanolayer and nanoparticles' diameter.

The melting procedures have numerous utilizations, like heat engines, chemical treating plants, gas production, etc. Gorlaa et al. [31] explored melting heat transmission of nanoliquid flow over unceasing moving surface. Radiative flow in incidence of melting heat was reviewed by Das [32]. Hayat et al. [33] demonstrated melting of the plate for a fluid motion at stagnation point. A numerical analysis has been executed by Gireesha et al. [34] for MHD stagnation-point, 2D flow by considering melting impact. Consequences of melting heat on a steady, laminar nanofluid flow taking account of the Buongiorno model were exemplified by Sheikholeslami and Rokni [35]. Imtiaz et al. [36] examined upshot of melting heat transference within two distinct water-based nanofluids. Phenomenon of melting heat transmission in bioconvective nanofluid flow was achieved by Liu et al. [37]. Analysis of melting heat transmission in nanofluid flow past absorbent cylinder was scrutinized by Singh et al. [38]. A study on hydromagnetic water-based nanoliquid flow through an enlarging surface in incidence of melting heat was conveyed by Das [39].

Encouraged by preceding exploration, the focus of contemporary analysis is to carry out a comparative exploration of flow features between a pair of dissimilar water-based nanofluids, nonaxisymmetric flow at stagnation-point past a stretched flat sheet. The novelty of ongoing research is to inspect the impact of nanolayer accompanying nanoparticles' diameter in the viscous flow of incompressible nanofluids in incidence of melting heat and nonlinear radiation. The governing flow equations are translated into self-similar ODE by the benefit of the appropriate similarity alterations and then have been decoded numerically by employing a standard technique namely Runge—Kutta—Fehlberg fourth-fifth order scheme with shooting strategy for velocity and temperature distributions.

2. Mathematical formulation

Consider a 3D nonaxisymmetric Homann stagnation-point viscous flow over a stretched flat plate. Water-based nanofluid consisting of SWCNTs (single-walled carbon nanotubes) and MWCNTs (multi-walled carbon nanotubes) as nanoparticles (Table 10.1) is considered here to execute flow analysis. Accept that admitted flat plate is stretched along $x - $ axis and $y - $ axis linearly with velocities $uw\ (x) = cx$ and $vw\ (y) = cy$,

Table 10.1 Physical properties of the base fluid and nanoparticles.

	Base fluid	Nanoparticles	
	Water	SWCNT	MWCNT
$\rho\ (\text{kg/m}^3)$	4179	2600	1600
$C_p\ (\text{J/kg K})$	997.1	425	796
$\kappa\ (\text{W/mK})$	0.613	6600	3000

respectively. Let x − axis and y − axis be measured alongside the plate, whereas z − axis is measured alongside normal direction of plate as portrayed in Fig. 10.1. Further, u, v, w signify velocities component alongside x, y, z direction, respectively. It also presumed that velocity components of exterior potential flow or nonaxisymmetric free stream flow are $u_e(x) = (a+b)x$, $v_e(y) = (a-b)y$, $w_e(z) = -2az$, along (x, y, z)-direction, where a indicates strain rate of flow and b is shear rate. Also agree to take that heat conduction of the flow is encouraged through melting condition and nonlinear radiative heat flux. In mathematical sketch during our analysis, we remain with the succeeding hypothesis that absenteeism of chemical reaction, nonappearance of joule heating, negligible viscous dissipation, no-slip condition, and all body forces are supposed to be ignored, and nanofluid is in thermal equilibrium.

Based on aforementioned deliberation, prevailing equations are [13,15]:

$$\frac{\partial u}{\partial x} + \frac{\partial v}{\partial y} + \frac{\partial w}{\partial x} = 0 \qquad (10.1)$$

$$\rho_{nf}\left(u\frac{\partial u}{\partial x} + v\frac{\partial u}{\partial y} + w\frac{\partial u}{\partial z}\right) = -\frac{\partial p}{\partial x} + \mu_{nf}\left(u\frac{\partial^2 u}{\partial x^2} + \frac{\partial^2 u}{\partial y^2} + \frac{\partial^2 u}{\partial z^2}\right) \qquad (10.2)$$

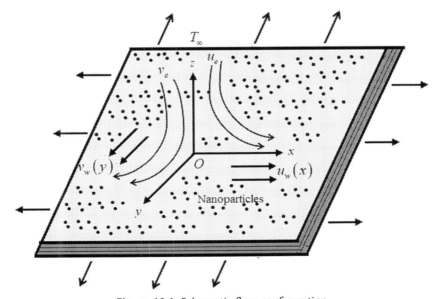

Figure 10.1 Schematic flow configuration.

$$\rho_{nf}\left(u\frac{\partial v}{\partial x}+v\frac{\partial v}{\partial y}+w\frac{\partial w}{\partial z}\right)=-\frac{\partial p}{\partial y}+\mu_{nf}\left(\frac{\partial^2 v}{\partial x^2}+\frac{\partial^2 v}{\partial y^2}+\frac{\partial^2 v}{\partial z^2}\right) \tag{10.3}$$

$$\rho_{nf}\left(u\frac{\partial w}{\partial x}+v\frac{\partial w}{\partial y}+w\frac{\partial w}{\partial z}\right)=-\frac{\partial p}{\partial z}+\mu_{nf}\left(\frac{\partial^2 w}{\partial x^2}+\frac{\partial^2 w}{\partial y^2}+\frac{\partial^2 w}{\partial z^2}\right) \tag{10.4}$$

$$\left(\rho c_p\right)_{nf}\left(u\frac{\partial T}{\partial x}+v\frac{\partial T}{\partial y}+w\frac{\partial T}{\partial z}\right)=\kappa_{nf}\left(\frac{\partial^2 T}{\partial x^2}+\frac{\partial^2 T}{\partial y^2}+\frac{\partial^2 T}{\partial z^2}\right)$$
$$+\frac{16\sigma^*}{k^*}\frac{\partial}{\partial y}\left(T^3\frac{\partial T}{\partial y}\right) \tag{10.5}$$

With boundary conditions [13,15]:

$$\left.\begin{array}{c} u=u_w(x), v=v_w(y), w=w_0 \quad \text{at } z=0 \\ u\to u_e, v\to v_e, T\to T_\infty \quad \text{as } z\to\infty \end{array}\right\} \tag{10.6}$$

Melting heat transmission occurrence is signified by:

$$\kappa_{nf}\left(\frac{\partial T}{\partial z}\right)=p_{nf}(\lambda+c_s(T_m-T_0))w \quad \text{at } z=0 \tag{10.7}$$

p signifies pressure and T signifies temperature. Here λ is used for nonuniform latent heat of the fluid. Eq. (10.7) exhibits that heat conducted on melting surface is equivalent to melting heat in addition to sensible heat essential to elevation of solid temperature from T_0 to the melting temperature T_m.

And $\mu_{nf}, \rho_{nf}, \left(\rho c_p\right)_{nf}$ are, respectively, viscosity, density, and specific heat of nanofluid and are defined by:

$$\mu_{nf}=\mu_f(1-\phi)^{-2.5}, \rho_{nf}=(1-\phi)\rho_f+\phi\rho_s, \left(\rho c_p\right)_{nf}=(1-\phi)\left(\rho c_p\right)_f+\phi\left(\rho c_p\right)_s\Big\} \tag{10.8}$$

Thermal conductivity κ_{nf} is given by [24]:

$$\kappa_{nf}=\frac{\phi\kappa_{lr}(\kappa_s-\kappa_{lr})\left(\chi_1^2-\chi_2^2+1\right)+\chi_1^2(\kappa_s-\kappa_{lr})\left\{\phi\chi_2^2\left(\kappa_{lr}-\kappa_f\right)+\kappa_f\right\}}{\chi_1^2(\kappa_s-\kappa_{lr})-\phi(\kappa_s-\kappa_{lr})\left(\chi_1^2+\chi_2^2-1\right)} \tag{10.9}$$

Here ϕ is nanofluid concentration, $\chi_1=1+\frac{h}{2\gamma_p}, \chi_2=1+\frac{h}{\gamma_p}$ in which γ_p indicates nanoparticle diameter and h is interfacial-layer thickness given by $h = 2\pi\sigma$ where σ is diffuseness of interfacial-layer and lies in 0.2—0.8 nm. There is no precise scheme to measure h, for cylindrical designed nanoparticles $h = 2$ [27]. And κ_{lr} symbolizes thermal conductivity of solid—liquid layer, $\kappa_{lr} = \delta\kappa_f$ and $0 \le \delta \le \frac{\kappa_s}{\kappa_f}$.

Hence thermal conductivity is reframed as:

$$\kappa_{nf} = \frac{\phi\kappa_{lr}\left(\kappa_s - \delta\kappa_f\right)\left(\chi_1^2 - \chi_2^2 + 1\right) + \chi_1^2\left(\kappa_s + \delta\kappa_f\right)\left\{\phi\chi_2^2\left(\delta\kappa_f - \kappa_f\right) + \kappa_f\right\}}{\chi_1^2\left(\kappa_s + \delta\kappa_f\right) - \phi\left(\kappa_s - \delta\kappa_f\right)\left(\chi_1^2 + \chi_2^2 - 1\right)}$$

$$(10.10)$$

Let us introduce the similarity variables as:

$$\left. \begin{array}{c} u = cxf'(\eta), v = cyg'(\eta), w = -\sqrt{cv_f}(f(\eta) + g(\eta)), \\[2mm] \theta(\eta) = \dfrac{T - T_\infty}{T_m - T_\infty}, \eta = \sqrt{\dfrac{c}{v_f}}z \end{array} \right\}$$

$$(10.11)$$

Using (8) in Eqs. (10.1) and (10.5) and eliminating the pressure gradient p, we get:

$$f'' + (1 - \phi)^{2.5}\left((1 - \phi) + \phi\frac{\rho_s}{\rho_f}\right)\left((f + g)f'' - f'^2 + (\delta_1 + \delta_2)^2\right) = 0$$

$$(10.12)$$

$$g'' + (1 - \phi)^{2.5}\left((1 - \phi) + \phi\frac{\rho_s}{\rho_f}\right)\left((f + g)g'' - g'^2 + (\delta_1 - \delta_2)^2\right) = 0$$

$$(10.13)$$

$$\frac{\kappa_{nf}}{\kappa_f}\theta'' + \mathrm{Pr}.\left((1 - \phi) + \phi\frac{\left(\rho c_p\right)_s}{\left(\rho c_p\right)_f}\right)(f + g)\theta' + \frac{4}{3}\mathrm{N}\left\{(C_T + \theta)^3\theta'\right\}' = 0$$

$$(10.14)$$

Similarly, boundary conditions (10.6) and (10.7) change to:

$$\left. \begin{array}{c} f'(0) = 1, g'(0) = 1, f(0) + g(0) = S, \\[2mm] \dfrac{\kappa_{nf}}{\kappa_f}Me\theta'(0) + \mathrm{Pr}\left((1 - \phi) + \phi\dfrac{\rho_s}{\rho_f}\right)(f(0) + g(0)) = 0 \\[2mm] f'(\eta) \to \delta_1 + \delta_2, g'(\eta) \to \delta_1 - \delta_2, \theta(\eta) \to 0 \text{ as } \eta \to \infty \end{array} \right\}$$

$$(10.15)$$

Lok et al. [12] reported that without loss of any generality, the condition $f(0) + g(0) = S$ can be restored by $f(0) = S$, $g(0) = 0$.

Skin friction coefficients C_{fx} and C_{fy} along x axis and y axis and Nusselt number are well defined as follows:

$$\left.\begin{array}{c} C_{fx} = \dfrac{2\mu_{nf}}{\rho_f u_w^2}\left(\dfrac{\partial u}{\partial z}\right)_{z=0}, \quad C_{fy} = \dfrac{2\mu_{nf}}{\rho_f u_w^2}\left(\dfrac{\partial v}{\partial z}\right)_{z=0}, \\ Nu_x = -\dfrac{x}{\kappa_f(T_m - T_w)}\left(\kappa_{nf}\left(\dfrac{\partial T}{\partial z}\right)_{z=0} + \dfrac{4\sigma^*}{3k^*}\left(\dfrac{\partial T^4}{\partial z}\right)_{z=0}\right) \end{array}\right\} \quad (10.16)$$

By expending (10.11) in (10.16), we acquire:

$$\left.\begin{array}{c} C_{frx} = C_{fx}\text{Re}_x^{\frac{1}{2}} = 2(1-\phi)^{-2.5}f''(0), \quad C_{fry} = C_{fy}\text{Re}_y^{\frac{1}{2}} = 2(1-\phi)^{-2.5}g''(0), \\ Nu_r = Nu_x\text{Re}_x^{-\frac{1}{2}} = -\left\{\dfrac{\kappa_{nf}}{\kappa_f} + \dfrac{4N}{3}\{C_T + \theta(0)\}^3\right\}\theta'(0) \end{array}\right\}$$

(10.17)

where $\text{Re}_x = \dfrac{xu_w(x)}{\nu_f}$, $\text{Re}_y = \dfrac{yv_w(y)}{\nu_f}$ are local Reynolds numbers.

3. Numerical experiment

3.1 Numerical methodology

The leading Eqs. (10.12) and (10.14) are nonlinear in nature. So the Eqs. (10.12)−(10.14) accompanying boundary criterion (10.15) have been solved by shooting approach based on Runge−Kutta−Fehlberg fourth-fifth order scheme. To accomplish complete numerical solutions, robust Maple-18 computational software has been used. Boundary-criterion (10.15) at $\eta \to \infty$ has been transmuted to alike at $\eta = \eta_\infty$. Interior iteration is completed till we attain the outcome to preferred precision 10^{-6} in all cases.

3.2 Testing of code

To ensure the precision of mathematical computations and numerical coding accomplished in contemporary studies, we have compared values of $f''(0)$ for diverse values of δ_1 with Mahapatra and Sidui [13], Khashi'ie et al. [15] in the absence of melting heat and $\delta_2 = \phi = N = S = 0$, Pr = 6.2. These values of $f''(0)$ are enlisted in Table 10.2 and acknowledged that values of $f''(0)$ are agreed with preceding literature.

Table 10.2 Variation of $f''(0)$ for different values of δ when $\delta = \phi = S = N = 0$, $Pr = 6.2$.

δ_1	Mahapatra and Sidui [13]	Khashi'ie et al. [15]	Present result
0.1	−1.1246	−1.124605	−1.124605398
0.2	−1.0544	−1.055622	−1.0556220310
0.5	−0.7534	−0.753446	−0.7534458096
1.0	0.0000	0.000000	0.0000000000
2.0	2.1902	2.207088	2.2070877166

4. Results and discussion

This article is passionate to discover the parametric execution of factors on velocity outlines and thermal outlines. We have revealed such effects via several charts and diagrams. We have assigned the following values $Pr = 6.2$, $\phi = 0.05$, $Me = 2.0$, $\gamma_P = 1.0$, $\delta = 10.0$, $\delta_1 = 1.0$, $\delta_2 = 1.0$, $S = 0.2$, $N = 0.5$, and $C_T = 1.0$ for numerical computations.

4.1 Effect of nanoparticle diameter

Consequence of γ_p on thermal outlines is exemplified in Fig. 10.2. Graphical look reveals that temperature outlines of flow system reduce when γ_p

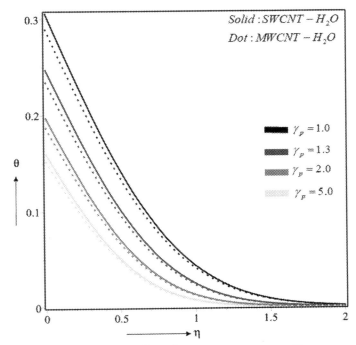

Figure 10.2 The effect of γ_p on temperature profiles.

enhances. The consequence is well distinct inside boundary layer. Since nano-sized particles with smaller diameters can transport supplemented thermal conductivity within flow system, hence progressive thermal outcomes are observed for lesser diameters. An impartial comparative investigation for two dissimilar carbon nanotubes viz. SWCNTs and MWCNTs is portrayed in the graph. We witnessed that temperature outlines are advanced in SWCNT-H_2O than the same in MWCNT-H_2O. Since thermal conductivity of SWCNTs is advanced concerning the identical of MWCNTs, progressive outlines of θ (η) are revealed for flow accompanied by SWCNTs particles than MWCNTs. When γ_p enhances, Table 10.3 acknowledged that values of Nu_r are reduced in each case, that is, values *of Nu_r* are identical in nature for both nanofluids.

Table 10.3 Values of Nu_r for different parameters.

γ_p	δ	Me	N	δ_1	δ_2	Nu_r SWCNT-H_2O	MWCNT-H_2O
1.0						1.198330	1.124477
1.3						1.118823	1.054753
2.0						1.007528	0.953961
5.0						0.970504	0.920286
	5.0					1.110305	1.046238
	10.0					1.198330	1.124477
	15.0					1.283887	1.200235
	20.0					1.370256	1.276446
		1.5				1.814010	1.686330
		2.0				1.198330	1.124477
		2.5				0.896925	0.845367
		3.0				0.717421	0.677909
			0.2			0.868240	0.821989
			0.5			1.198330	1.124477
			0.8			1.574100	1.465176
			1.0			1.854691	1.717010
				0.3		1.256846	1.173882
				0.5		1.240184	1.160005
				0.7		1.222561	1.145140
				1.0		1.198330	1.124477
					0.3	1.206223	1.131366
					0.5	1.204795	1.130120
					0.7	1.202685	1.1282789
					1.0	1.198330	1.124477

4.2 Effect of interfacial layer

An upshot of δ on $\theta(\eta)$ is exposed in Fig. 10.3. It is established that outlines of $\theta(\eta)$ enhance together with δ. Mathematically $\delta = \frac{K_{lr}}{K_f}$; therefore, escalating in δ proposes augmentation of thermal intake ability of nanolayer, and consequently outlines of $\theta(\eta)$ increase. We detected that outlines of $\theta(\eta)$ are more advanced in SWCNT-H_2O than the same in MWCNT-H_2O and physical analysis behind this impact discussed in former section. Table 10.3 reproduced that values of Nu_r are enhanced with δ.

4.3 Effect of melting parameter

Fig. 10.4 illustrates the encouragement of Me on $\theta(\eta)$. It is established that supplementation in Me implies descending in $\theta(\eta)$. The incidence of melting heat at surface always performs as blowing boundary conditions there. Therefore intensifying in Me implies intensification in temperature alteration among ambient surface and melting surface which in turn declines thermal gradient. When Me enhances, a noteworthy reduction in values of

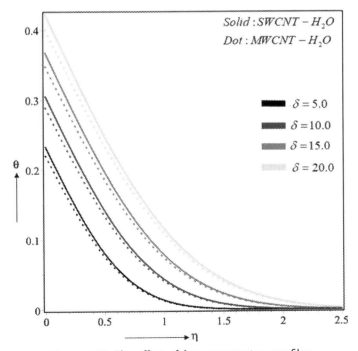

Figure 10.3 The effect of δ on temperature profiles.

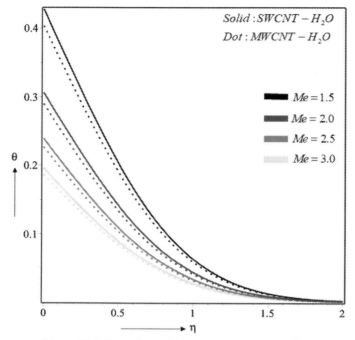

Figure 10.4 The effect of Me on temperature profiles.

Nu_r is conveyed in Table 10.3. And reduction rate in Nu_r is maximum for both cases when Me augments from 1.5 to 2.0 in extant assessment. Rate of detraction Nu_r are 33.94% and 33.31% for of SWCNT-H$_2$O and MWCNT-H$_2$O, respectively.

4.4 Effect of radiation parameter

Stimulus of N on thermal outlines is uncovered in Fig. 10.5. We revealed that thermal outlines are monotonically intensified with accumulative values of N. Incidence of nonlinear solar radiation in flow system implies continuous augments of the Brownian motion of nano-sized particles plus fluid molecules within flow structure. Subsequently, their kinetic energy is amplified within flow structure which exchanges into expansions in thermal outlines. It is perceived from Table 10.3 that if N upsurges from 0.2 to 0.5, Nu_r enhances 38.02% and 36.80% for SWCNT-H$_2$O and MWCNT-H$_2$O, respectively. Therefore when N improves, heat conveyance rate is augmented in flow. Rate of change in Nu_r is lower in MWCNT-H$_2$O comparison to SWCNT-H$_2$O.

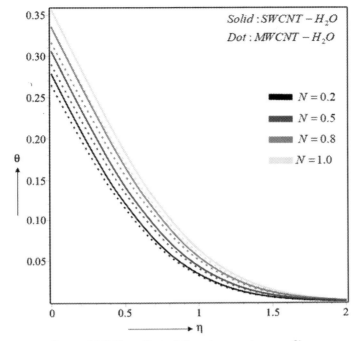

Figure 10.5 The effect of N on temperature profiles.

4.5 Effect of the parameter δ_1

Figs. 10.6 and 10.7 reveal that as the value of δ_1 is increasing, velocity outlines $f'(\eta)$ and its transverse component $g'(\eta)$ are enhanced gradually. Primarily very close to surface, outlines of $g'(\eta)$ have negligible impact. But marginally away from it, separable impression takes our attention. Since $\delta 1 = \frac{a}{c}$ and as for graphical description, we have deliberated all values of $\frac{a}{c} \leq 1$; therefore, decrease of a correlated to c occurs. Consequently, flow experienced a lesser amount of straining motion neighboring the stagnation region which can lessen the acceleration of the external stream. We witnessed that outlines of $f'(\eta)$ are advanced in SWCNT-H$_2$O than the same in MWCNT-H$_2$O, but a converse effect is perceived for the outlines of $g'(\eta)$. Fig. 10.8 displays that thermal outlines are augmented with parameter δ_1, and in this, tactic escalation of thermal boundary layer of the flow is explored. Tables 10.3 and Table 10.4 authorize that values of C_{fx}, C_{fy}, and Nu_r are identical in nature whenever δ_1 augments. Rate of change in C_{fx} is higher in comparison to both C_{fy} and Nu_r with the increased value of δ_1. Also realized that when $\delta 1$ changes from 0.3 to 0.5, values of C_{fx} enhanced

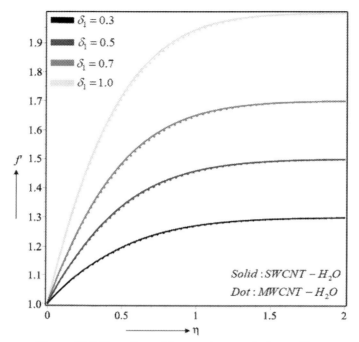

Figure 10.6 The effect of δ_1 on primary velocity profiles.

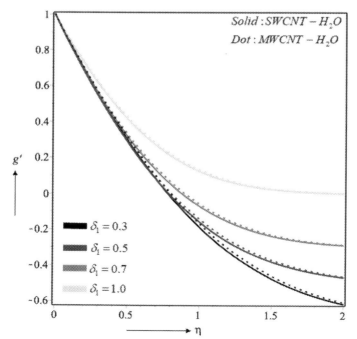

Figure 10.7 The effect of δ_1 on secondary velocity profiles.

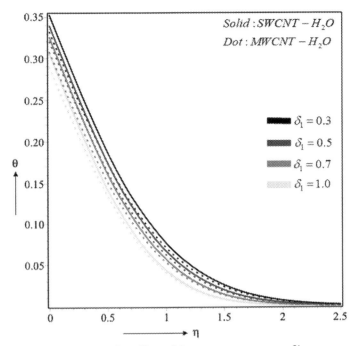

Figure 10.8 The effect of δ_1 on temperature profiles.

Table 10.4 Values of C_{frx} and C_{fry} for different parameters.

		C_{frx}		C_{fry}	
δ_1	δ_2	SWCNT-H$_2$O	MWCNT-H$_2$O	SWCNT-H$_2$O	MWCNT-H$_2$O
0.3		1.294853	1.262765	−3.617696	−3.522735
0.5		2.249753	2.194091	−3.810618	−3.712159
0.7		3.277165	3.196212	−3.754639	−3.658641
1.0		4.947632	4.825689	−3.336074	−3.251690
	0.3	1.325464	1.2925834	−1.180714	−1.151206
	0.5	2.286476	2.229872	−1.884233	−1.837001
	0.7	3.307391	3.225667	−2.518554	−2.455211
	1.0	4.947632	4.825689	−3.336074	−3.251690

73.74% and 73.75% for SWCNT-H$_2$O and MWCNT-H$_2$O, respectively, which is utmost rate of change in C_{frx} present numerical investigations.

4.6 Effect of the parameter δ_2

Figs. 10.9 and 10.10 enlighten the effects of δ_2 on $f'(\eta)$ and $g'(\eta)$, respectively, and we detected that both outlines are opposing in nature. Outlines

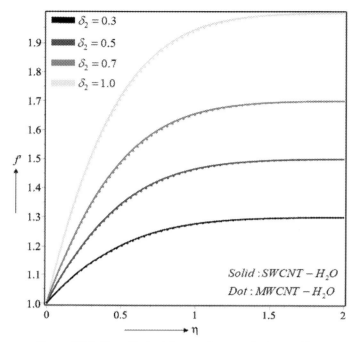

Figure 10.9 The effect of δ_2 on primary velocity profiles.

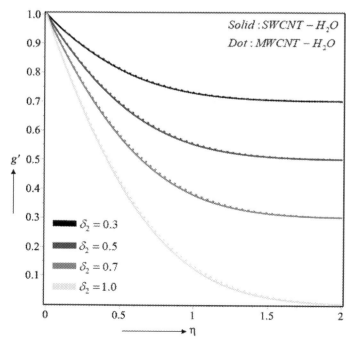

Figure 10.10 The effect of δ_2 on secondary velocity profiles.

of $f'(\eta)$ are enhanced, whereas outlines of $g'(\eta)$ are declined in nature for enhanced values of $\delta 2$. We witnessed that sketches of $f'(\eta)$ are advanced in SWCNT-H$_2$O than the same in MWCNT-H$_2$O, but for $g'(\eta)$, reverse effect is observed. Consequence of δ_2 on $\theta(\eta)$ reflects in Fig. 10.11 and witnessed that outlines of $\theta(\eta)$ decline when δ_2 improves. Because δ_2 is the ratio of shear and strain rate consequently with escalation of δ_2, shear stress becomes larger, which leads to decline in thermal outlines. From Table 10.3 and Table 10.4, it is accepted that as δ_2 intensifies, values of C_{fx} and Nu_r augment, whereas values of C_{fry} decline. Also perceived that when δ_2 upsurges, rates of change in C_{fx} and C_{fry} are very much noteworthy, whereas the nominal effect is perceived in rate of change in Nu_r. Maximum rate of change in C_{fry} is attained when δ_2 enhances from 0.3 to 0.5, and the rate of change is 58.58% and 58.57% for SWCNT-H$_2$O and MWCNT-H$_2$O, respectively

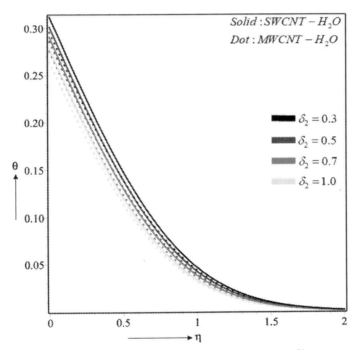

Figure 10.11 The effect of δ_2 on temperature profiles.

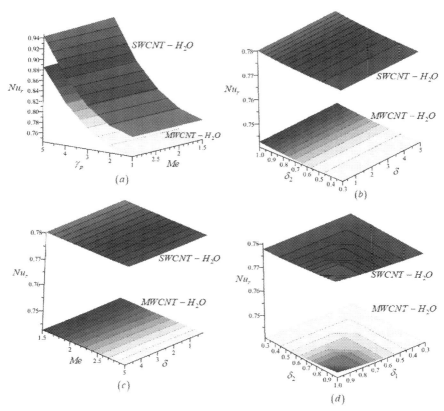

Figure 10.12 Surface plot (A) Nu_r for Me versus γ_p, (B) Nu_r for δ versus $\delta 2$, (C) Nu_r for δ versus Me, and (D) Nur for $\delta 1$ versus $\delta 2$.

4.7 Effect of parameters on engineering coefficient

Fig. 10.12 reveals 3D view of Nu_r for some nondimensional parameters. Surface plot of Nu_r conveys for parameters γ_p and Me conveys in Fig. 10.12A. It is detected that the smallest value of γ_p and uppermost value of Me together agree to the supreme value of Nu_r. Fig. 10.12B portrayed variation of δ and δ_1 on Nu_r. Fig. 10.12C expresses a surface plot of Nu_r against the parameters δ and Me, and established that the maximum values attained for Nu_r are the maximum values of δ and Me. Fig. 10.12D illustrates the surface plot of Nur for parameters δ_1 and δ_2. Fig. 10.13A delivers a surface plot of C_{frx} alongside the parameters δ_1 and δ_2. It is detected that C_{frx} achieves its uppermost value for utmost values of δ_1 and δ_2. Fig. 10.13B portrays the surface plot of C_{fry} for parameters δ_1 and δ_2.

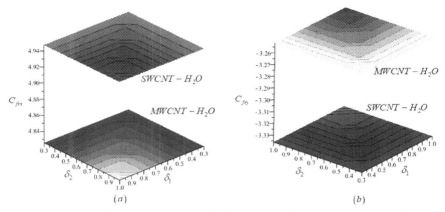

Figure 10.13 Surface plot (A) C_{frx} for δ1 versus δ2, (B) C_{fry} for δ1 versus δ2.

5. Conclusions

In this article, we have deliberated a 3D nonaxisymmetric stagnation-point viscous flow over a stretched flat plate. Water-based nanofluids comprising SWCNTs and MWCNTs as nanoparticles are considered here accompanied by nonlinear radiative heat flux. Thermal conductivity of nanofluids incorporating the consequences of diameter of nanoparticles and nanolayer is accounted here. Accept that heat transference of flow system is encouraged through melting condition. By implementing similarity analysis, we renovate the foremost partial differential equations (PDEs) of accepted nanofluid model into ODEs. The resulting ODEs are unravelled numerically by retaining Runge–Kutta–Fehlberg fourth-fifth order scheme through shooting system to execute flow analysis. The effect of exciting flow parameters on the flow specification is accomplished perfectly through figures and charts. Based on overhead analysis, noteworthy features are:

❖ Temperature circulation and Nu_r decline as γ_p enhanced.
❖ Outlines of $\theta(\eta)$ and values of Nu_r are in identical nature as δ and N are upswings.
❖ Thermal outlines $\theta(\eta)$ and Nu_r reduce as Me enhanced, and the rate of reduction in Nu_r is maximum when Me enhances.
❖ Outlines of $\theta(\eta)$ are identical in nature when δ_1 and δ_2 amplify.
❖ Outlines of $f'(\eta)$ and $g'(\eta)$ are in identical nature when δ_1 augmented, but $f'(\eta)$ and $g'(\eta)$ are conflicting in nature when δ_2 enhances.

Nomenclature

(u, v, w) Velocity components

(u_e, v_e, w_e) Velocity components of external potential flow

a, b Stretching strain rate and shear rate

C_{frx}, C_{fry} Reduced skin friction coefficient

C_{fx}, C_{fy} Skin friction number

$CT = \frac{Th}{Tm - Th}$ Temperature Ratio

h Interfacial-layer thickness

K Radius of curvature

$Me = \frac{(c_p)_f (T_m - T_\infty)}{\lambda + c_s (T_m - T_0)}$ Melting parameter

$N = \frac{4\sigma_1 (T_m - T_h)^3}{k^* \kappa_f}$ Solar radiation parameter

Nu_r Reduced Nusselt number

p Pressure

$Pr = \frac{\mu_r (c_p)_f}{\kappa_f}$ Prandtl number

Re_x, Re_y Local Reynolds numbers

$S = \frac{-w0}{\sqrt{cvf}}$ is suction/injection parameter

T Temperature

T_m Melting temperature

T_∞ Temperature away from surface

u_w, v_w Stretching velocity

λ Nonuniform latent heat

ρ Density

μ Dynamic viscosity

ν Kinematics viscosity

κ Thermal conductivity

ϕ Nanofluid concentration

σ Diffuseness of interfacial layer

γ_p Nanoparticle diameter

$\delta_1 = \frac{a}{c}$ Ratio of strain rate of ambient fluid and plate

$\delta_2 = \frac{b}{c}$ Ratio of shear rate of ambient fluid and plate

ρC_p Heat capacitance

κlr Thermal conductivity of the solid—liquid layer

References

[1] K. Hiemenz, Die Grenzschicht an einem in den gleichförmigenFlüssigkeitsstromeingetauchtengeradenKreiszylinder, Dingler'sPolytech J. 326 (1911) 321—410.

[2] F. Homann, Der einfluss grosser zähigkeitbei der strömung um den zylinder und um die kugel, ZAMM - ZeitschriftFürAngewandteMathematik Und Mechanik 16 (3) (1936) 153—164.

[3] L. Howarth, CXLIV; the boundary layer in three dimensional flow — part II. The flow near a stagnation point, London, Edinburgh Dublin Phil. Mag. J. Sci. 42 (335) (1951) 1433—1440.

[4] A. Davey, Boundary-layer flow at a saddle point of attachment, J. Fluid Mech. 10 (1961) 593—610.

[5] A. Davey, D. Schofield, Three-dimensional flow near a two dimensional stagnation point, J. Fluid Mech. 28 (1967) 149—151.

[6] N. Bachok, A. Ishak, I. Pop, Stagnation-point flow over a stretching/shrinking sheet in a nanofluid, Nanoscale Res. Lett. 6 (2011) 623.

[7] N. Bachok, A. Ishak, I. Pop, Boundary layer stagnation-point flow and heat transfer over an exponentially stretching/shrinking sheet in a nanofluid, Int. J. Heat Mass Tran. 55 (25—26) (2012) 8122—8128.

[8] X.H. Zhang, A. Abidi, A.E.S. Ahmed, M.R. Khan, M.A.E. Shorbagy, M. Shutaywi, A. Issakhov, A.M. Galal, MHD stagnation point flow of nanofluid over a curved stretching/shrinking surface subject to the influence of Joule heating and convective condition, Case Stud. Therm. Eng. 26 (2021) 101184.

[9] S. Rajput, K. Bhattacharyya, A.K. Verma, M.S. Mandal, A.J. Chamkha, D. Yadav, Unsteady stagnation-point flow of CNTs suspended nanofluid on a shrinking/expanding sheet with partial slip: multiple solutions and stability analysis, Waves Random Complex Media (2022), https://doi.org/10.1080/17455030.2022.2063986.

[10] A. Dawar, Z. Shah, S. Islam, W. Deebani, M. Shutaywi, MHD stagnation point flow of a water-based copper nanofluid past a flat plate with solar radiation effect, J. Petrol. Sci. Eng. 220 (A) (2023) 111148.

[11] P.D. Weidman, Non-axisymmetric homann stagnation-point flows, J. Fluid Mech. 702 (2012) 460—469.

[12] Y.Y. Lok, J.H. Merkin, I. Pop, Mixed convection non-axisymmetric Homann stagnation-point flow, J. Fluid Mech. 812 (2017) 418—434.

[13] T.R. Mahapatra, S. Sidui, Heat transfer in non-axisymmetric Homann stagnation-point flows towards a stretching sheet, Eur. J. Mech. B Fluid 65 (2017) 522—529.

[14] R.B. Kudenattia, S.R. Kirsur, Numerical and asymptotic study of non-axisymmetric magnetohydrodynamic boundary layer stagnation-point flows, Math. Methods Appl. Sci. 40 (16) (2017) 5841—5850.

[15] N.S. Khashi'ie, N.Md Arifin, I. Pop, R. Nazar, E.H. Hafidzuddin, N. Wahi, Non-axisymmetric Homann stagnation point flow and heat transfer past a stretching/shrinking sheet using hybrid nanofluid, Int. J. Numer. Methods Heat Fluid Flow (2020), https://doi.org/10.1108/HFF-11-2019-0824.

[16] T.R. Mahapatra, S. Sidui, Non-axisymmetric Homann stagnation-point flow of a viscoelastic fluid towards a fixed plate, Eur. J. Mech. B Fluid 79 (2020) 38—43.

[17] A.U. Khan, S. Saleem, S. Nadeem, A.A. Alderremy, Analysis of unsteady non-axisymmetric Homann stagnation point flow of nanofluid and possible existence of multiplesolutions, Physica A 554 (2020) 123920.

[18] M. Khan, M. Sarfraz, J. Ahmed, L. Ahmed, C. Fetecau, Non-axisymmetric Homann stagnation-point flow of Walter's B nanofluid over a cylindrical disk, Appl. Math. Mech. 41 (2020) 725—740.

[19] M.R. Jagwal, I. Ahmad, M. Sajid, Non-axisymmetric Homann stagnation point flow of Maxwell nanofluid towards fixed surface, Int. J. Mod. Phys. C 32 (06) (2021) 2150076, https://doi.org/10.1142/S0129183121500765.

[20] Q.-Z. Xue, Model for effective thermal conductivity of nanofluids, Phys. Lett. 307 (2003) 313—317.

[21] W. Yu, S.U.S. Choi, The role of interfacial layers in the enhanced thermal conductivity of nanofluids: a renovated Maxwell model, J. Nanoparticle Res. 5 (2003) 167—171.

[22] H. Xie, M. Fujii, X. Zhang, Effect of interfacial nanolayer on the effective thermalconductivity of nanoparticle-fluid mixture, Int. J. Heat Mass Tran. 48 (2005) 2926—2932.

[23] K.C. Leong, C. Yang, S.M.S. Murshed, A model for the thermal conductivity of nanofluids — the effect of interfacial layer, J. Nanoparticle Res. 8 (2006) 245—254.

[24] S.M.S. Murshed, K.C. Leong, C. Yang, Investigations of thermal conductivity and viscosity of nanofluids, Int. J. Therm. Sci. 47 (2008) 560—568.

[25] P. Rana, O.A. Beg, Mixed convection flow along an inclined permeable plate: effect of magnetic field, nanolayer conductivity and nanoparticle diameter, Appl. Nanosci. 5 (2015) 569–581.

[26] P. Rana, R. Dhanai, L. Kumar, MHD slip flow and heat transfer of Al_2O_3 -water nanofluid over a horizontal shrinking cylinder using Buongiorno's model: effect of nanolayer and nanoparticle diameter, Adv. Powder Technol. 28 (7) (2017) 1727–1738.

[27] M.T. Sk, K. Das, P.K. Kundu, Consequences of nanoparticle diameter and solid–liquid interfacial layer on the SWCNT/EO nanofluid flow over various shaped thin slendering needles, Chin. J. Phys. 56 (2018) 2439–2447.

[28] S.S. Giri, K. Das, P.K. Kundu, Influence of nanoparticle diameter and interfacial layer on magnetohydrodynamic nanofluid flow with melting heat transfer inside rotating channel, Math. Methods Appl. Sci. (2020), https://doi.org/10.1002/mma.6818.

[29] N. Acharya, Spectral simulation to investigate the effects of nanoparticle diameter and nanolayer on the ferrofluid flow over a slippery rotating disk in the presence of low oscillating magnetic field, Heat Trans. (2021), https://doi.org/10.1002/htj.22157.

[30] N. Acharya, F. Mabood, S.A. Shahzad, I.A. Badruddin, Hydrothermal variations of radiative nanofluid flow by the influence of nanoparticles diameter and nanolayer, Int. Commun. Heat Mass Tran. 130 (2022) 105781.

[31] R.S.R. Gorlaa, A. Chamkhab, A. Aloraierb, Melting heat transfer in a nanofluid flow past a permeable continuous moving surface, J. Nav. Architect. Mar. Eng. 8 (2) (2011) 83–92.

[32] K. Das, Radiation and melting effects on MHD boundary layer flow over a moving surface, Ain Shams Eng. J. 5 (2014) 1207–1214.

[33] T. Hayat, M. Farooq, A. Alsaedi, Melting heat transfer in the stagnation-point flow of Maxwell fluid with double-diffusive convection, Int. J. Numer. Methods Heat Fluid Flow 24 (3) (2014) 760–774.

[34] B.J. Gireesha, B. Mahanthesh, I.S. Shivakumara, K.M. Eshwarappa, Melting heat transfer in boundary layer stagnation-point flow of nanofluid toward a stretching sheet with induced magnetic field, Eng. Sci. Technol. Int. J. 19 (2016) 313–321.

[35] M. Sheikholeslami, H.B. Rokni, Effect of melting heat transfer on nanofluid flow in existence of magnetic field considering Buongiorno Model, Chin. J. Phys. 55 (4) (2017) 1115–1126.

[36] M. Imtiaz, F. Shahid, T. Hayat, A. Alsaedi, Melting heat transfer in Cu-water and Ag-water nanofluids flow with homogeneous-heterogeneous reactions, Appl. Math. Mech. 40 (2019) 465–480.

[37] C. Liu, H. Akhter, M. Ramzan, M.K. Alaoui, Impact of melting heat transfer in the bioconvective casson nanofluid flow past a stretching cylinder with entropy generation minimization analysis, Int. J. Mod. Phys. B 35 (31) (2021) 2150315.

[38] K. Singh, A.K. Pandey, M. Kumar, Melting heat transfer assessment on magnetic nanofluid flow past a porous stretching cylinder, J Egypt Math Soc 29 (1) (2021).

[39] K. Das, Towards the understanding of the melting heat transfer in a Cu–water nanofluid flow, J. Eng. Phys. Thermophys. 95 (2022) 1207–1213.

CHAPTER ELEVEN

On the hydrothermal performance of radiative Ag–MgO–water hybrid nanofluid over a slippery revolving disk in the presence of highly oscillating magnetic field

Nilankush Acharya[1] and Kalidas Das[2]

[1]NCP Umasashi High School, Kolkata, West Bengal, India
[2]Department of Mathematics, Krishnagar Government College, Krishnanagar, West Bengal, India

Highlights

- The Ag–MgO–water hybrid nanofluidic transportation over a revolving slippery disk is studied.
- Highly oscillating magnetic field and Shliomis's theory are clutched to model the flow.
- Thermal radiation and heat source/sink are introduced within the flow regime.
- Fluidic friction seems to escalate for high field frequency.
- Thermal radiation, heat source, and temperature ratio accelerate the heat transport.

1. Introduction

A magnetic liquid that contains a stable mixture of tiny magnetic ingredients can significantly be controlled by a magnetic field. Such ingredients were manufactured during the 1960s, and the advancement of this metal-liquid dispersion discovers the preeminent potential for the research topic named ferrohydrodynamics (FHD) [1]. Several varieties of magnetic liquids exist; the leading combination is named colloidal ferrofluid [2,3] that can be categorized as finely distributed 1–100 nm-sized tiny particles within the base medium. Thermal agitation [4] assistances the small particles to be finely distributed within the liquid because coating with surfactants and Brownian migration prevents them from agglomeration. Natural fluids generally never respond

Advanced Materials-Based Fluids for Thermal Systems
ISBN: 978-0-443-21576-6
https://doi.org/10.1016/B978-0-443-21576-6.00012-1

Copyright © 2024 Elsevier Inc.
All rights are reserved, including those
for text and data mining, AI training,
and similar technologies.

to magnetic fields, thus separate chemistry and colloidal study are required to understand the hydrothermal variations of magnetized liquids. The study of such magnetized liquids attracted research scientists not only for their novel rheological background but also for their several applications in spinning shafts sealing [5], drug targeting [6], retinal detachment repair during eye surgery, controlling heat in a loudspeaker, etc. [7].

Massive efforts were contributed by the investigators to prepare liquid metals that don't go through strong changes in viscosity under applied magnetic fields. The controlled viscous magnetic fluids can be applied for magnetically controlled damping systems [8]. Viscosity changes affect the applications of magnetic fluids significantly. Strong physics supports the viscosity variation owing to magnetic fields [9]. Influenced by the shear flow, the dissolved magnetized particles start to spin about their own spinning axis parallel to the flow vorticity. When the magnetic scenario is introduced within the dispersed particles under shear, then two extreme cases may evolve. One corresponds to the circumstance when the introduced magnetic field collinearly acts with flow vorticity, then the magnetic moment tends to align with the magnetic fields' direction, which becomes identical with the particles' spinning axis. During such circumstances, no change in viscosity is perceived. Again, for the second situation, the flow vorticity and magnetic field act perpendicularly, then the magnetic field attempts to align the magnetic moment with the field's direction, while the viscous torques produced by the flow force the particles to rotate which creates misalignment among magnetic moment and magnetic fields. Consequently, viscous torque equilibriums the magnetic torque, and hence rotational viscosity depends on both the magnetic field's strength and its direction [10]. The mean angular speed $\boldsymbol{\omega_p}$ of suspended particles is finalized by the intermediate balance between magnetic torques and the viscosity [11]: $\mathbf{M} \times \mathbf{H} = -6\mu_f \phi (\Omega - \omega_p)$, where $\mathbf{M}$ *and* $\mathbf{H}$ epitomizes the fluid's magnetization and magnetization strength, μ_f signifies the viscosity, fluid's angular velocity is marked by Ω, and ϕ represents the nanoparticles' volume fraction. Therefore, $\mathbf{H}$, the magnetic field switches an extra viscosity, termed as rotational viscosity, and is symbolized by $\Delta\mu_f$. One can formulate the mathematical correlation as [12]: $\Delta\mu_f = \frac{3}{2}\mu_f \phi \frac{\Omega - \omega_p}{\Omega}$. Free particle spinning is hindered by the slow oscillating magnetic field, that is, $\omega_p < \Omega$ and hence $\Delta\mu_f > 0$. On the contrary, particles revolve faster than functioning liquid for a highly oscillated magnetized field, that is, $\omega_p > \Omega$. Thus, the second scenario initiates the "negative viscosity" and declines the resulting viscosity [13]. Andhariya et al. [14] deliberated the conduct of rotational viscosity of kerosene or water-based ferrofluids owing to the capillary size effect and magnetized field. They marked a higher rotational viscous effect for water-ferrofluid compared to the kerosene-based ferrofluid. Sanchez and Rinaldi [15] examined the magnetized viscous features of magnetized fluids considering both oscillating plus rotating magnetic fields using Brownian simulation. They also observed the negative magneto-viscosity for oscillating and corotating magnetic fields. Papadopoulos et al. [16] investigated the magnetically driven ferrofluidic transport through a pipe. The report assures insignificant magneto-viscous changes for nonuniform magnetic fields. Yarahmdi et al. [17] studied experimentally ferrofluid heat transmission introducing both constant and oscillating magnetic fields. Various oscillation modes and magnetized arrangements were considered. The outcomes reveal that for oscillating fields, approximately 19.8% heat transmission can possibly be gained. Yu et al. [18] discussed nanoparticles' fluidization for oscillated magnetic effect. They noted silica nanoparticles' agglomeration can easily be fluidized with magnetic ingredients under oscillating magnetic environment. Lajvardi et al. [19] executed experimental laminar ferrofluid heat transference within warmed copper tube. Results ensure no enhanced heat transmission for magnetic absence, but significant heat transmission enhancement for magnetic effect. Schumacher et al. [20] and Ghofrani et al. [21] both discussed ferrofluid's behavior under alternating magnetic medium. Hassan [22] illustrated the high oscillating magnetic impact on ferrofluids and observed viscous reduction for rotational effect. Kimura et al. [23] investigated Farady diamagnetism for slow oscillating magnetized zone.

Fluidic transport over a spinning disk conveys various significant applications in turbomachinery, computer storage, lubrication, spin coating, etc. Karman [24] was the first to examine such flow. Krehkov et al. [25] investigated theoretically ferrofluidic transit through a pipe assuming the existence of oscillating magnetic fields. They noted non-Newtonian features of ferrofluid. Fang and Tao [26] investigated the slippery effects of unsteady motion running over a spinning disk. Rashidi et al. [27] introduced an artificial neural networking method to study the entropy variations of unsteady transport through the stretched swirling disk. Ram and Bhandari [28] inspected the phase variation impacts on highly oscillated magnetized unsteady ferrofluid motion owing to spinning disk and noted negative viscosity for higher angular frequency. Effects of cross-diffusion and heat source on Marangoni convected Casson liquid transport due to whirling disk was illustrated by Mahanthesh et al. [29]. Results assured mass transfer enhancement for Marangoni convection and thermodiffusion.

Hybrid nanofluids are the advanced version of classical mono nanofluids. Two or more metallic, oxide nanoparticles are dispersed within the base liquids to prepare hybrid nanofluids [30]. The presence of double nanoparticles improves the hybrid nanofluid's resulting thermal conductivity [31] and boosts the heat transmission capability compared to classical mono nanofluids [32]. Hybrid nanofluids are often utilized as an improved heat transport candidate in solar [33], automobiles [34], heat exchangers [35], defense equipment, electrocooling [36], etc. Ellahi et al. [37] discoursed low oscillated magnetic field impacts on nano-ferrofluidic transmission through stretched revolving disks and conclude that both heat transmission and thermal trajectory enriched for oscillated magnetization. Hassan et al. [38] exhibited tiny particles' shape influence on the nanofluidic transit under a low oscillated magnetic medium. They report velocity reduction for enhanced solid particle concentration. Hassan et al. [39] again considered the same scenario but considered a highly oscillated magnetic medium instead of a low oscillation as in Ref. [38]. Results convey that the thermal scenario gets improved for prolate-shaped nanoparticles. The impression of a low oscillated magnetized field on unsteady transport owing to a turning disk was communicated by Ram et al. [40]. Consequences address maximum radial velocity for magneto-viscosity. Shoaib et al. [41] investigated the magnetically driven $Cu-Al_2O_3-$water hybrid nanofluids transit caused by spinning disk considering thermal slip and heat absorption. Results marked thermal profile reduction for thermal slip factor. Waqas et al. [42] illustrated the hybrid radiative flow through a rotary disk and noticed enhanced heat transport for the temperature ratio parameter. Saeed et al. [43] examined the water-based $Al_2O_3-TiO_2$ hybrid nanofluid flow through a slippery whirling disk considering Darcy$-$Forchheimer's porous medium. They observed velocity reduction for nanoparticle concentrations. Imtiaz et al. [44] demonstrated fluidic magnetized transportation over variably thicker spinning disk, and spotted velocity reduction is negatively correlated with thickness. Acharya [45] introduced nanolayer and particles' diameter effects for ferrofluidic motion considering both highly oscillated magnetic zone absence-presence and received relative magnetization declination for enhanced field frequency. Gul et al. [46] simulated Fe_3O_4–Cu–water flow through the conical space of the turning disk and cone and perceived impressive heat transference for nanoparticles' concentrations. Acharya [47] introduced the spectral technique to reveal low oscillated magnetized impact for ferrofluidic motion through a slippery whirling disk covering particles' diameter and related nanolayer. Results indicate amplified heat transmission for oscillated cases compared to oscillated magnetic absence. Acharya et al. [48] portrayed the irreversibility scrutinization of radiative unsteady Fe_3O_4–GO–water stream running through a slippery circling disk and obtained lower entropy for hybrid medium compared to classic mono nanofluids. Lei et al. [49] analyzed magnetically driven GO–MoS_2–water flow through a whirling disk and perceive velocity drop-off for magnetic strength. Acharya [50] disclosed Ag–MgO–water flow within a cylinder-fitted cube and investigated the simultaneous impacts of the cylinder's thermal mode on the hybrid flow scenario.

Being motivated by the above-stated literature, this investigation unfolds the hydrothermal changes of radiative water-based Ag–MgO hybrid nanofluidic transport caused by a slippery swirling disk assuming the existence of a highly oscillated magnetized field. The presence of a heat sink/source is also presumed to model the simulation. Shliomis's theory is clutched to model the flow. The pioneering utilization of Ag nanoparticles lies in as drug carriers [51], anticancer agents, heat exchangers [52], biosensing [53], etc., while the tiny MgO nanoparticles are introduced in electronics [54], ceramics, catalysis, petrochemical products [55], oil recovery [56], coatings, and numerous additional fields. The shooting-based RK-4 procedure is operated to execute the simulation. Prior to this numeric investigation, no other attempts were executed to disclose the hydrothermal changes of the water-based Ag–MgO hybrid nanofluids considering the above circumstances. So, this numeric approach is unique and novel. The core questions that are of keen interest and explored throughout the investigation are:

- How do high field frequency, velocity slip, unsteadiness, and nanoparticle concentrations affect the velocity profiles?
- How do hybrid nanofluid streamlines behave under a highly oscillated magnetic field?
- How do unsteadiness, nonlinear radiation, heat source/sink, temperature ratio, and solid particle concentration affect the thermal profiles?
- What are the impacts of interrelated parameters on skin friction and Nusselt numbers?

2. Mathematical formulation
2.1 Flow modeling

The axially symmetric, incompressible, unsteady, nonconducting hybrid nanofluidic motion is presumed to flow over a revolving disk in presence of the highly oscillating magnetic field. The spinning disk is set to $z = 0$ and supposed to have angular velocity $\frac{\Omega_v}{1-\lambda t}$ w.r.t, where λ is of dimension Sec^{-1} and $1 - \lambda t > 0$. The graphic is portrayed in Fig. 11.1. The dissolved tiny nanoparticles are nonchemically reactive with the base liquid. A cylindrical frame of reference (r, θ, z) is considered to model the hybrid nanofluidic transport. Now, the essential bunch of equations related to continuity, momentum, magnetization, rotational motion, and energy is presented as follows [1–5,9]:

$$\nabla \cdot \mathbf{V} = 0 \qquad (11.1)$$

$$\rho_{hnf} \frac{d\mathbf{V}}{dt} = -\nabla p + \mu_{hnf}\nabla^2\mathbf{V} + \mu_0(\mathbf{M}.\nabla)\mathbf{H} + \frac{\mathbf{I}}{2\tau_s}\nabla \times (\omega_p - \Omega) \qquad (11.2)$$

$$\frac{d\mathbf{M}}{dt} = \omega_\mathbf{p} \times \mathbf{M} - \frac{1}{\tau_B}(\mathbf{M} - \mathbf{M}_0) \qquad (11.3)$$

$$\mathbf{I}\frac{d\omega_\mathbf{p}}{dt} = (\mathbf{M} \times \mathbf{H}) - \frac{\mathbf{I}}{\tau_S}(\omega_\mathbf{p} - \Omega) \qquad (11.4)$$

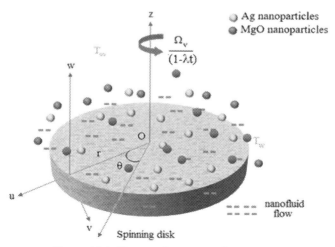

Figure 11.1 Flow configuration of the model.

On the hydrothermal performance of radiative 261

$$\left(\rho C_p\right)_{hnf}\frac{dT}{dt} = \kappa_{hnf}\nabla^2 T - \frac{\partial q_r}{\partial z} + Q(T - T_\infty) \tag{11.5}$$

where $\frac{d}{dt} = \frac{\partial}{\partial t} + \mathbf{V}.\nabla$, $\mathbf{V} = (u, v, w)$ in Eqs. (11.1)–(11.5) symbolizes the velocity, Q assures the heat source/sink parameter according as $Q > 0$ and $Q < 0$. Again, for q_r, the Rosseland approximation in consideration with an optically thick medium can be formulated as [42,43]:

$$q_r = -\frac{4}{3k^*}grad(e_b) \tag{11.6}$$

in which, the Stefan–Boltzmann law related to radiation expresses e_b as $e_b = \sigma^* T^4$. Thus, the nonlinear radiative zone q_r in Eq. (11.6) is redefined as [42,43]:

$$q_r = -\frac{4\sigma^*}{3k^*}\frac{\partial T^4}{\partial z} = -\frac{16\sigma^*}{3k^*}T^3\frac{\partial T}{\partial z} \tag{11.7}$$

If one includes T nondimensionally as $\Theta(\eta) = \frac{T-T_\infty}{T_w-T_\infty}$, then T changed as $T = T_\infty\{1 + (\Theta_w - 1)\Theta\}$, where $\Theta_w = \frac{T_w}{T_\infty}$.

Whole equations must contain Maxwell's equations:

$$\nabla \times \mathbf{H} = 0 \tag{11.8}$$

$$\nabla.(\mathbf{H} + 4\pi\mathbf{M}) = 0 \tag{11.9}$$

Now at $\tau_B = 0$, $\mathbf{M}_0$ in the form of Langevin functional can be outlined as:

$$\mathbf{M}_0 = nmL(\xi)\frac{\mathbf{H}}{H}, \xi = \frac{mH(t)}{k_B T_a}, L(\xi) = \coth \xi - \frac{1}{\xi}\} \tag{11.10}$$

where $L(\xi)$ demonstrates the Langevin function and ξ represents the proportion of magnetization energy to thermal energy.

Again, compared to the relaxation term, the inertial term seems minor, that is, $\mathbf{I}\frac{d\omega_p}{dt} \ll \mathbf{I}\frac{\omega_p}{\tau_S}$; consequently, one can reconstruct Eq. (11.4) as:

$$\omega_\mathbf{p} = \Omega + \frac{\tau_S}{\mathbf{I}}(\mathbf{M} \times \mathbf{H}) \tag{11.11}$$

Eliminating the term $(\omega_\mathbf{p} - \Omega)$ and using (11), Eqs. (11.2) and (11.3) are reshaped as:

$$\rho_{hnf}\frac{d\mathbf{V}}{dt} = -\nabla p + \mu_{hnf}\nabla^2\mathbf{V} + \mu_0(\mathbf{M}.\nabla)\mathbf{H} + \frac{1}{2}\nabla \times (\mathbf{M} \times \mathbf{H}) \tag{11.12}$$

$$\frac{d\mathbf{M}}{dt} = \Omega \times \mathbf{M} - \frac{1}{\tau_B}(\mathbf{M} - \mathbf{M}_0) - \frac{\tau_S}{\mathbf{I}}\mathbf{M} \times (\mathbf{M} \times \mathbf{H}) \tag{11.13}$$

Radially, the magnetic field can be components-wise decomposed as:

$$H_r = H_0 \cos(\omega_0 t), H_\theta = 0, H_z = 0 \tag{11.14}$$

The assumption of having negligible amplitude, that is, $mH_0 \leq k_B T_a$, one can easily overlook the nonlinearity of the Langevin function. So, Eq. (11.10) can be redefined as:

$$\left. \mathbf{M}_0 = \chi \mathbf{H}, \chi = \frac{nm^2}{3k_B T_a} \right\} \tag{11.15}$$

The acting magnetic field is assumed to be the two revolving fields' superposition, one is noted as right-hand polarized with subscript "-" and the other as left-hand polarized with subscript "+". Hence,

$$\mathbf{H}_\pm = (H_0 \cos \omega_0 t, \pm H_0 \sin \omega_0 t, 0), \quad \mathbf{H} = \frac{1}{2}(\mathbf{H}_+ + \mathbf{H}_-) \tag{11.16}$$

The field Eqs. (11.2)−(11.4) reveal the solution for quiescent liquid ($\Omega = 0$), and the subsequent magnetization spins having an angular frequency lagging behind the angle α. Thus,

$$\mathbf{M}_\pm = (M \cos(\omega_0 t - \alpha), \pm M \sin(\omega_0 t - \alpha), 0), \quad \omega_{\mathbf{p}} = (0, 0, \pm \omega_p) \tag{11.17}$$

Replacing Eqs. (11.14)−(11.17) with Eqs. (11.3) and (11.4), one obtains:

$$M = \chi H_0 \cos \alpha, \quad \omega_p = \frac{\tau_S}{\mathbf{I}} M H_0 \sin \alpha, \quad \tan \alpha = (\omega_0 - \omega_p)\tau_B \tag{11.18}$$

Now, $\xi = \frac{mH_0}{k_B T_a \sqrt{2}}$ is small, and $\frac{1}{\tau_S} = 6\mu_f \phi_{hnf}$ removing the angle α and considering terms up to ξ^2, one receives from Eq. (11.18):

$$\left. M = \frac{\xi H_0}{\sqrt{1 + \omega_0^2 \tau_B^2}}, \quad \omega_p = \frac{\frac{\xi^2}{3}\omega_0}{1 + \omega_0^2 \tau_B^2}, \quad \tan \alpha = \omega_0 \tau_B \left(1 - \frac{\frac{\xi^2}{3}\omega_0}{1 + \omega_0^2 \tau_B^2}\right) \right\} \tag{11.19}$$

where $\frac{H_0}{\sqrt{2}}$ links the root mean square quantity of $H_0 \cos \omega_0 t$. Assuming the existence of two revolving fields, one can reconstruct Eq. (11.18) in the presence of a hydrodynamical vortex $\Omega = (0, 0, \Omega)$ as:

$$M_+ = \chi H_0 \cos \alpha_+, \quad \omega_{p_+} = \frac{\tau_S}{I} \mathbf{M}_+ H_0 \sin \alpha_+ + \Omega, \quad \tan \alpha_+ = \left(\omega_0 - \omega_{p_+}\right)\tau_B$$

$$(11.20)$$

$$M_- = \chi H_0 \cos \alpha_-, \quad \omega_{p_-} = \frac{\tau_S}{I} \mathbf{M}_- H_0 \sin \alpha_- + \Omega, \quad \tan \alpha_- = \left(\omega_0 - \omega_{p_-}\right)\tau_B$$

$$(11.21)$$

It is echoed from the above correlations that faster rotation of magnetized particles is assured when together ω_0 and Ω possess the same sign. Henceforth, the last terms present in Eqs. (11.20) and (11.21) are:

$$\tan \alpha_+ = (\omega_0 - \Omega)\tau_B, \quad \tan \alpha_- = (\omega_0 + \Omega)\tau_B\}$$

$$(11.22)$$

Owing to rotary magnetic fields, the $\theta - components$ of magnetization are extracted as:

$$\begin{aligned}\mathbf{M}_\theta^+ &= \chi H_0 \cos \alpha_+ \sin(\omega_0 t - \alpha_+) \\ &= \frac{\chi H_0}{1 + (\omega_0 - \Omega)^2 \tau_B^2} (\sin \omega_0 t - (\omega_0 - \Omega)\tau_B \cos \omega_0 t)\end{aligned}$$

$$(11.23)$$

$$\begin{aligned}\mathbf{M}_\theta^- &= -\chi H_0 \cos \alpha_- \sin(\omega_0 t - \alpha_-) \\ &= \frac{\chi H_0}{1 + (\omega_0 - \Omega)^2 \tau_B^2} (\sin \omega_0 t - (\omega_0 + \Omega)\tau_B \cos \omega_0 t)\end{aligned}$$

$$(11.24)$$

The magnetization tangential component along radial direction for a linearly polarized field is exhibited as following Eqs. (11.22)–(11.24):

$$\mathbf{M}_\theta = \frac{1}{2}\left(\mathbf{M}_\theta^+ + \mathbf{M}_\theta^-\right) = \Omega\tau_B\chi H_0 \cos^2 \alpha \cos(\omega_0 t - 2\alpha)$$

$$(11.25)$$

After deserting ξ^2 from Eq. (11.19), it only exists $\tan \alpha = \omega_0\tau_B$. The magnetization part's presence in Eq. (11.25) guarantees magnetic torque presence along $z - axis$ and thus renders the density as:

$$\begin{aligned}\mathbf{M} \times \mathbf{H} &= -\Omega\tau_B\chi H_0^2 \cos^2 \alpha \cos \omega_0 t \cos(\omega_0 t - 2\alpha) \\ &= -2\Omega\mu_f\phi_{hnf}\xi^2 \cos^2 \alpha\left(\cos^2\omega_0 t \cos 2\alpha + \sin \omega_0 t \cos \omega_0 t \sin 2\alpha\right)\}\end{aligned}$$

$$(11.26)$$

On averaging Eq. (11.26) having a period of variation $\frac{2\pi}{\omega_0}$, the second term dissolves, and the remaining first term renders as $\frac{1}{2}\cos 2\alpha$. Thus, the subsequent form can be obtained:

$$\overline{\mathbf{M} \times \mathbf{H}} = -\Omega\mu_f\phi_{hnf}\xi^2 \cos^2 \alpha \cos 2\alpha$$

$$(11.27)$$

Applying both Eqs. (11.27) and (11.12) and following Eq. (11.13), the equation of motions is addressed as:

$$\rho_{hnf}\frac{d\mathbf{V}}{dt} = -\nabla p + \left(\mu_{hnf} + \frac{1}{4}\mu_f\phi_{hnf}\xi^2\cos^2\alpha\cos2\alpha\right)\nabla^2\mathbf{V} + \mu_0(\mathbf{M}.\nabla)\mathbf{H}$$

$$(11.28)$$

Let us symbolize the reduced pressure as $\nabla\widetilde{p}$ caused by magnetic force, hence:

$$-\nabla\widetilde{p} = -\nabla p + \mu_0(\mathbf{M}.\nabla)\mathbf{H}$$

$$(11.29)$$

Introducing Eq. (11.29), Eq. (11.28) is formulated as:

$$\rho_{hnf}\frac{d\mathbf{V}}{dt} = -\nabla\widetilde{p} + \left(\mu_{hnf} + \frac{1}{4}\mu_f\phi_{hnf}\xi^2\cos^2\alpha\cos2\alpha\right)\nabla^2\mathbf{V}$$

$$(11.30)$$

Thus, with the classic boundary layer approximations, the vital continuity, momentum, and energy, Eqs. (11.7)—(11.9), (11.30) can be reaffirmed as:

$$\frac{\partial u}{\partial r} + \frac{u}{r} + \frac{\partial w}{\partial z} = 0$$

$$(11.31)$$

$$\rho_{hnf}\left(\frac{\partial u}{\partial t} + u\frac{\partial u}{\partial r} - \frac{v^2}{r} + w\frac{\partial u}{\partial z}\right) + \frac{\partial\widetilde{p}}{\partial r} = \left(\mu_{hnf} + \frac{1}{4}\mu_f\phi_{hnf}\xi^2\cos^2\alpha\cos2\alpha\right)$$

$$\left(\frac{\partial^2 u}{\partial r^2} + \frac{1}{r}\frac{\partial u}{\partial r} - \frac{u}{r^2} + \frac{\partial^2 u}{\partial z^2}\right)$$

$$(11.32)$$

$$\rho_{hnf}\left(\frac{\partial v}{\partial t} + u\frac{\partial v}{\partial r} + \frac{uv}{r} + w\frac{\partial v}{\partial z}\right) = \left(\mu_{hnf} + \frac{1}{4}\mu_f\phi_{hnf}\xi^2\cos^2\alpha\cos2\alpha\right)$$

$$\left(\frac{\partial^2 v}{\partial r^2} + \frac{1}{r}\frac{\partial v}{\partial r} - \frac{v}{r^2} + \frac{\partial^2 v}{\partial z^2}\right)$$

$$(11.33)$$

$$\rho_{hnf}\left(\frac{\partial w}{\partial t} + u\frac{\partial w}{\partial r} + w\frac{\partial w}{\partial z}\right) + \frac{\partial\widetilde{p}}{\partial z} = \left(\mu_{hnf} + \frac{1}{4}\mu_f\phi_{hnf}\xi^2\cos^2\alpha\cos2\alpha\right)$$

$$\left(\frac{\partial^2 w}{\partial r^2} + \frac{1}{r}\frac{\partial w}{\partial r} + \frac{\partial^2 w}{\partial z^2}\right)$$

$$(11.34)$$

$$\left(\rho C_p\right)_{hnf}\left(\frac{\partial T}{\partial t}+u\frac{\partial T}{\partial r}+w\frac{\partial T}{\partial z}\right)=\kappa_{hnf}\left(\frac{\partial^2 T}{\partial r^2}+\frac{1}{r}\frac{\partial T}{\partial r}+\frac{\partial^2 T}{\partial z^2}\right)$$
$$-\frac{\partial q_r}{\partial z}+Q(T-T_\infty)$$

(11.35)

2.2 Thermophysical properties

For a hybrid nanofluid, a stable nanodispersion containing 25 nm Ag and 40 nm MgO nanoparticles is mixed with water. Related thermophysical features are addressed in Table 11.1. Also, the related thermophysical correlations are illustrated in Table 11.2 at reference temperature 20–30°C. To available reliable outcomes, in simulation experimentally assured correlations of dynamic viscosity and thermal conductivity provided by Esfe et al. [33] are merged.

2.3 Boundary conditions

The boundary conditions are:

$$\left. \begin{array}{l} u=L\dfrac{\partial u}{\partial z}, v=\dfrac{\Omega_v r}{1-\lambda t}+L\dfrac{\partial v}{\partial z}, w=0, T=T_w \ at \ z=0, \\[2mm] and \\[2mm] u\rightarrow 0, v\rightarrow 0, T\rightarrow T_\infty \ at \ z\rightarrow\infty. \end{array} \right\}$$

(11.36)

where $L=L'\sqrt{1-\lambda t}$.

2.4 Similarity transformation

The introduced nondimensional functions [26,27] for Eqs. (11.31)−(11.35) are:

$$\left. \begin{array}{l} \eta=\sqrt{\dfrac{\Omega_v}{\nu_f(1-\lambda t)}}z, u=\dfrac{r\Omega_v}{1-\lambda t}f(\eta), v=\dfrac{r\Omega_v}{1-\lambda t}g(\eta), \\[4mm] w=\sqrt{\dfrac{\nu_f\Omega_v}{(1-\lambda t)}}h(\eta), \Theta(\eta)=\dfrac{T-T_\infty}{T_w-T_\infty}, \bar{p}=-\dfrac{\mu_f\Omega_v}{1-\lambda t}P(\eta). \end{array} \right\}$$

(11.37)

Table 11.1 Thermophysical properties of the base liquid and Ag, MgO nanoparticles [33].

Physical properties	H₂O	Ag	MgO
C_p (J/Kg K)	4179	235	955
ρ (Kg/m^3)	997.1	10,500	3560
κ (W/mK)	0.613	429	45

Table 11.2 Thermophysical models of mono nanofluid and hybrid nanofluid [30,33].

Properties	Mono nanofluid (Ag–H$_2$O)
Density	$\rho_{nf} = (1-\phi)\rho_f + \phi\rho_s$
Heat capacity	$\left(\rho C_p\right)_{nf} = (1-\phi)\left(\rho C_p\right)_f + \phi\left(\rho C_p\right)_s$
Viscosity	$\mu_{nf} = \dfrac{\mu_f}{(1-\phi)^{2.5}}$
Thermal conductivity	$\dfrac{\kappa_{nf}}{\kappa_f} = \dfrac{\kappa_s + 2\kappa_f - 2\phi\left(\kappa_f - \kappa_s\right)}{\kappa_s + 2\kappa_f + \phi\left(\kappa_f - \kappa_s\right)}$

Properties	Hybrid nanofluid (Ag–MgO–H$_2$O)
Density	$\rho_{hnf} = \left(1-\phi_{hnf}\right)\rho_f + \phi_1\rho_1 + \phi_2\rho_2$
Heat capacity	$\left(\rho C_p\right)_{hnf} = \left(1-\phi_{hnf}\right)\left(\rho C_p\right)_f + \phi_1\left(\rho C_p\right)_1 + \phi_2\left(\rho C_p\right)_2$
Viscosity	$\mu_{hnf} = \mu_f\left(1 + 32.795\phi_{hnf} - 7214\phi_{hnf}^2 + 714600\phi_{hnf}^3 - 0.1941 \times 10^8\phi_{hnf}^4\right)$ $0.0 \leq \phi_{hnf} \leq 0.02$
Thermal conductivity	$\kappa_{hnf} = \kappa_f\left(\dfrac{0.1747\times10^5 + \phi_{hnf}}{0.1747\times10^5 - 0.1498\times10^6\times\phi_{hnf} + 0.1117\times10^7\times\phi_{hnf}^2 + 0.1997\times10^8\times\phi_{hnf}^3}\right)$ $0.0 \leq \phi_{hnf} \leq 0.03$

The current flow profile with Eq. (11.37) reframes as:

$$2f + h' = 0 \tag{11.38}$$

$$\Sigma_1 \left\{ f^2 - g^2 + hf' + S\left(f + \frac{\eta}{2} f' \right) \right\}$$

$$= \left(\Sigma_2 + \frac{1}{4} \phi_{hnf} \xi^2 \frac{\left(2 \left(1 + \omega_0^2 \tau_B^2 \left(1 - \frac{\frac{\xi^2}{3}}{1 + \omega_0^2 \tau_B^2} \right) \right)^2 \right)^{-1} - 1}{\left\{ 1 + \omega_0^2 \tau_B^2 \left(1 - \frac{\frac{\xi^2}{3}}{1 + \omega_0^2 \tau_B^2} \right) \right\}^2} \right) f'' \tag{11.39}$$

$$\Sigma_1 \left\{ 2fg + hg' + S\left(g + \frac{\eta}{2} g' \right) \right\}$$

$$= \left(\Sigma_2 + \frac{1}{4} \phi_{hnf} \xi^2 \frac{\left(2 \left(1 + \omega_0^2 \tau_B^2 \left(1 - \frac{\frac{\xi^2}{3}}{1 + \omega_0^2 \tau_B^2} \right) \right)^2 \right)^{-1} - 1}{\left\{ 1 + \omega_0^2 \tau_B^2 \left(1 - \frac{\frac{\xi^2}{3}}{1 + \omega_0^2 \tau_B^2} \right) \right\}^2} \right) g'' \tag{11.40}$$

$$\Sigma_1 \left\{ hh' + \frac{S}{2} (h + \eta h') \right\}$$

$$= -P' + \left(\Sigma_2 + \frac{1}{4} \phi_{hnf} \xi^2 \frac{\left(2 \left(1 + \omega_0^2 \tau_B^2 \left(1 - \frac{\frac{\xi^2}{3}}{1 + \omega_0^2 \tau_B^2} \right) \right)^2 \right)^{-1} - 1}{\left\{ 1 + \omega_0^2 \tau_B^2 \left(1 - \frac{\frac{\xi^2}{3}}{1 + \omega_0^2 \tau_B^2} \right) \right\}^2} \right) h'' \tag{11.41}$$

$$\frac{\Sigma_3}{\Sigma_4}\mathrm{Pr}\left(h\Theta' + S\frac{\eta}{2}\Theta'\right)$$
$$= \left(\Theta'' + \frac{4N}{3\Sigma_4}\frac{d}{d\eta}\left\{(1 + \Theta(\eta)(\Theta_w - 1))^3\frac{d\Theta(\eta)}{d\eta}\right\}\right) + \frac{\mathrm{Pr}}{\Sigma_4}\lambda_H\Theta$$

$$(11.42)$$

where
$$\Sigma_1 = \frac{\rho_{hnf}}{\rho_f}, \Sigma_2 = \frac{\mu_{hnf}}{\mu_f}, \Sigma_3 = \frac{(\rho C_p)_{hnf}}{(\rho C_p)_f},$$
$$\Sigma_4 = \frac{\kappa_{hnf}}{\kappa_f}, \phi_{hnf} = \phi_1 + \phi_2$$

The changed boundary conditions for Eq. (11.36) are:

$$\left.\begin{array}{l}f(0) = L_s f'(0), g(0) = 1 + L_s g'(0), h(0) = 0, \Theta(0) = 1 \\ f(\infty) = 0, g(\infty) = 0, \Theta(\infty) = 0\end{array}\right\}$$

$$(11.43)$$

where λ_H assures the heat source/sink parameter according as $\lambda_H > 0$ and $\lambda_H < 0$.

The skin friction and Nusselt number are:

$$\left.\begin{array}{l}C_f = \frac{\sqrt{\tau_{wr}^2 + \tau_{w\theta}^2}}{\rho_f\left(\frac{r\Omega_v}{1 - \lambda t}\right)^2}, Nu = \frac{r(q_w + q_r)}{\kappa_f(T_w - T_\infty)}\end{array}\right\}$$

$$(11.44)$$

where in Eq. (11.44):

$$\tau_{wr} = \left(\mu_{hnf} + \frac{1}{4}\phi_{hnf}\mu_f\xi^2\frac{2\left(1 + \omega_0^2\tau_B^2\left(1 - \frac{\frac{\xi^2}{3}}{1 + \omega_0^2\tau_B^2}\right)^2\right)^{-1} - 1}{\left\{1 + \omega_0^2\tau_B^2\left(1 - \frac{\frac{\xi^2}{3}}{1 + \omega_0^2\tau_B^2}\right)^2\right\}}\right)$$
$$\left(\frac{\partial u}{\partial z} + \frac{\partial w}{\partial r}\right)_{z=0}$$

$$(11.45)$$

On the hydrothermal performance of radiative

$$\tau_{w\theta} = \left\{ \left(\mu_{hnf} + \frac{1}{4}\phi_{hnf}\mu_f \xi^2 \frac{2\left(1+\omega_0^2\tau_B^2\left(1 - \frac{\frac{\xi^2}{3}}{1+\omega_0^2\tau_B^2}\right)\right) - 1}{\left\{1+\omega_0^2\tau_B^2\left(1 - \frac{\frac{\xi^2}{3}}{1+\omega_0^2\tau_B^2}\right)\right\}^2} \right) \right.$$

$$\left. \left(\frac{\partial v}{\partial z} + \frac{1}{r}\frac{\partial w}{\partial \theta}\right)_{z=0} \text{ and } q_w = -\kappa_{hnf}\left(\frac{\partial T}{\partial z}\right)_{z=0} \right\}$$

(11.46)

Inserting Eqs. (11.37), (11.45), (11.46), one obtains:

$$C_{fr} = \text{Re}_z^{\frac{1}{2}} C_f = \left\{ \left(\mu_{hnf} + \frac{1}{4}\phi_{hnf}\mu_f \xi^2 \frac{2\left(1+\omega_0^2\tau_B^2\left(1 - \frac{\frac{\xi^2}{3}}{1+\omega_0^2\tau_B^2}\right)\right) - 1}{\left\{1+\omega_0^2\tau_B^2\left(1 - \frac{\frac{\xi^2}{3}}{1+\omega_0^2\tau_B^2}\right)\right\}^2} \right) \sqrt{f'^2(0) + g'^2(0)} \right\}$$

(11.47)

$$Nu_r = \text{Re}_z^{-\frac{1}{2}} Nu = -\frac{K_{hnf}}{K_f}\left(1 + \frac{4N}{3\Sigma_4}\{(1+\Theta(0)(\Theta_w - 1))^3\}\right)\Theta'(0)$$

(11.48)

3. Numerical method and code validation
3.1 Numerical method

The classical shooting scheme with the Runge–-Kutta–Fehlberg technique is utilized to solve the foremost highly nonlinear ordinary differential equations (ODEs) (11.38)–(11.42) having the boundary restriction in (11.43). Primarily, the leading equations are turned into first-ordered ODEs, and after then shooting strategy is introduced to run the step-by-step

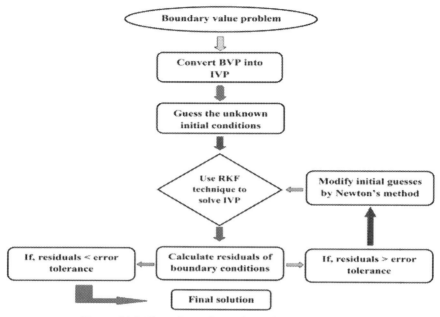

Figure 11.2 Flow chart of RK-4 base shooting method.

integration. MAPLE-18 software is used to run the whole simulation. The asymptotic restriction, that is, at $\eta \to \infty$ is replaced by $\eta = \eta_\infty$. The convergence limit of the intermediate iterative scheme is set to 10^{-6} in all cases. A flow chart is illustrated in Fig. 11.2.

3.2 Code validation

To compare the numeric outcomes with the preceding literature, the numeric values of $f'(0)$ and $g'(0)$ are extracted for several S in the absence of an oscillating magnetic zone. During numeric simulation, the parametric setup was arranged by choosing $\phi_1 = 0.0, \phi_2 = 0.0, \xi = 0.0, N = 0.0, \Theta_w = 0.0, \lambda_H = 0.0, L_s = 0.0$. It is apparent that the numeric outcomes are well satisfied in Table 11.3 with previous Fang and Tao [26] and Rashidi et al. [27].

4. Results and discussion

This segment discourses the parametric influence on the velocity and thermal distributions. Frictional effect and heat transference outcomes are

Table 11.3 Comparison of numeric outcomes with preceding literature.

	$f'(0)$			$g'(0)$		
	Fang and Tao [26]	Rashidi et al. [27]	Present	Fang and Tao [26]	Rashidi et al. [27]	Present
$S = -0.1$	0.5308	0.53077	0.530774	−0.5789	−0.57897	−0.578976
$S = -0.2$	0.5515	0.55152	0.551551	−0.5416	−0.54156	−0.541561
$S = -0.5$	0.6143	0.61433	0.614332	−0.4284	−0.42839	−0.428396
$S = -1.0$	0.7198	0.71981	0.719812	−0.2366	−0.23662	−0.236627

also merged within this section. The requisite number of plots, three-dimensional figures, and tables are clutched to enrich the section. Throughout the simulation, the default numeric set of values that were fitted against the parameter to generate the simulative graphs and tables is $S = 2, L_s = 0.5, \omega_0 \tau_B = 2, \xi = 0.75, \phi_1 = 0.01, \phi_2 = 0.01, \Pr = 6.2, N = 1.0, \Theta_w = 1.5, \lambda_H = 0.5$ unless specified.

4.1 Effect of high field frequency parameter $(\omega_0 \tau_B)$

Fig. 11.3 delineates the high-field frequency impact on radial velocity, tangential velocity, and axial velocity in Fig. 11.3A—C, respectively. The radial velocity declines with $\omega_0 \tau_B$. For highly oscillating magnetic fields, when the fluid particles' rotation is faster than that of the revolving magnetic

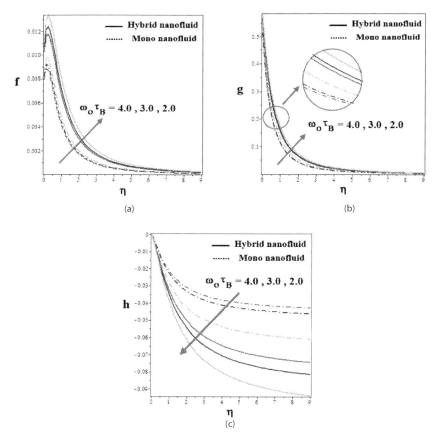

Figure 11.3 Effect of high field frequency on (A) radial velocity, (B) tangential velocity, and (C) axial velocity.

liquid is depicted for $\omega_0\tau_B = 2$, $\omega_0\tau_B = 3$, and $\omega_0\tau_B = 4$ in Fig. 11.3A. After acquiring the peak value, the decay of the radial velocity component toward the steady-state regime is quite faster. Also, one can mark the local impact of negative viscosity over the radial component because no significant change in peak value is traced for $\omega_0\tau_B \to \infty$. Hybrid nanofluids exhibit a high magnitude in radial velocity compared to mono nanofluids. Fig. 11.3B reveals variation in tangential velocity owing to high field frequency. A minor change in the trajectory was detected. Near the surface, that is, $0.0 \leq \eta \leq 0.5$(*not precisely marked*), the influence is negligible while slightly away from it, that is, $0.6 \leq \eta \leq 4.0$(*not precisely marked*), comparatively a slight variation is noted. Here, the tangential velocity moves to a steady-state regime faster for $\omega_0\tau_B = 2$ compared to $\omega_0\tau_B = 3$ as the orienting effect of the oscillating field declines with swelling frequency. Additionally, the well-known centrifugal force produced owing to the disk's spinning being carried by liquid particles and consequently away from the plate, its consequence drops off resulting in the tangential velocity reduction [3,9,28]. Hybrid nanofluids exhibit a high-velocity profile compared to classic mono nanofluids. Fig. 11.3C illustrates the axial velocity variation for the nondimensional parameter $\omega_0\tau_B$. Away from the disk's surface, the finite magnitudes acquired by axial velocity are 0.095, 0.08, and 0.07 for $\omega_0\tau_B = 2$, $\omega_0\tau_B = 3$, and $\omega_0\tau_B = 4$, respectively. For higher field frequency, the particle's angular speed is higher compared to magnetic fluid's angular speed, thus the rotational viscosity remains negative, and consequently, an impressive impact on axial speed is noted for the magnetic field [3,9,28]. Actually, the centrifugal force is caused by radial motion which leads to the axial motion. The negative magnitude of axial velocity discloses the hybrid nanofluid's flow toward the disk, and the high-field frequency effect on the axial motion is negligible near the disk; however, away from the plate, its impact can be detected as the disk's influence reduces and the liquid freely moves [3,9,28]. A three-dimensional view of the aforementioned velocity components is checked from Fig. 11.4A–C. Table 11.4 following Eq. (11.47) exhibits that frictional hindrance enhances for mono nanofluid most compared to hybrid one. The enhancement rate is 3.58% for hybrid nanofluids; however, it is 2.43% for mono nanofluids. Fig. 11.5A–B exhibit the variation three-dimensionally. Fig. 11.4D specifies that initially, the pressure remains high near the plate's surface. However, for $\omega_0\tau_B = 2$ and $\omega_0\tau_B = 4$, the negative viscosity originates from the magnetic field which diminishes the pressure acting on the hybridized magnetized nanofluid. During fluidic motion, all fluids are composed of energetic molecules. When such

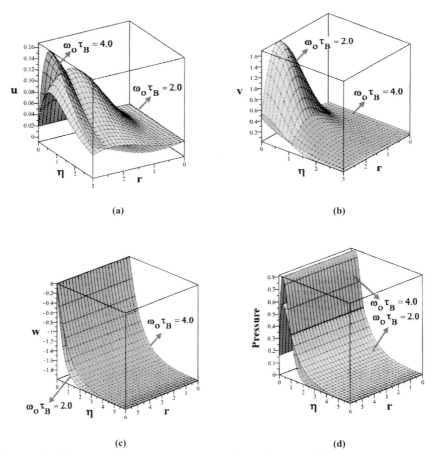

Figure 11.4 Three-dimensional view on the effect of high field frequency on (A) radial velocity, (B) tangential velocity, (C) axial velocity, and (D) pressure distribution.

energetic molecules strike the disk surface, a tangential and normal force is exerted on the plate's surface because of the momentum change of colliding molecules. When the magnetic field is absent, then fluid and particle angular velocities are equal and thus the total number of energetic fluid molecules colliding with a surface are maximum which results in having maximum pressure [3,9,28]. On the contrary, when the fluid's angular velocity is less compared to the colloidal particles owing to the alternating magnetic field, the pressure becomes less compared to the situation when the magnetic field does not exist. Since the negative viscosity appears during the oscillating magnetic field, thus the number of total molecules which hit the surface is decreased.

Table 11.4 Numeric outcomes of skin friction.

$\omega_0\tau_B$	L_s	S	ϕ_2	C_{fr} Hybrid nanofluid	C_{fr} Mono nanofluid
2.0	0.5	2.0	0.01	0.932244	1.026974
3.0				0.956681	1.044272
4.0				0.965636	1.051920
2.0	0.0			1.935846	2.135941
	0.5			0.990441	1.035344
	1.0			0.662084	0.681891
	0.5	2.0		0.536059	0.557967
		2.5		0.561413	0.583189
		3.0		0.582554	0.604101
		2.0	0.000	0.982164	1.026974
			0.014	1.013822	1.044272
			0.015	1.041061	1.051919

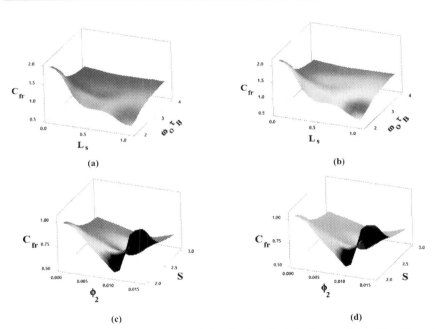

Figure 11.5 Three-dimensional view of skin friction for hybrid nanofluids (A)–(C) and mono nanofluids (B)–(D).

4.2 Effect of slip parameter (L_s)

The influence of the velocity slip parameter on velocity profiles is portrayed in Fig. 11. 6A–D. The radial speed in Fig. 11.6A seems to decline for the

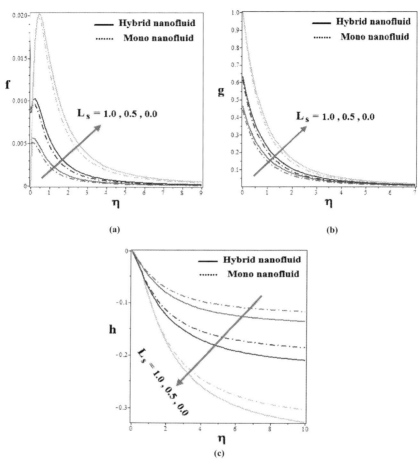

Figure 11.6 Effect of velocity slip on (A) radial velocity, (B) tangential velocity, and (C) axial velocity.

slippery parameter. Near the disk's surface radial speed is high for the no-slip regime, while the slip regime drops down the magnitude significantly [44,45]. Corresponding effects are prominent and significant. Hybrid nanofluids explore comparatively high magnitudes than mono nanofluids. A similar trajectory is noted for the tangential velocity profile in Fig. 11.6B. The massive outcome is marked within $0.0 \leq \eta \leq 4.0$ (*not precisely marked*). Both hybrid and mono nanofluids tend to reduce their corresponding magnitude away from the surface. An opposite scenario is tracked for axial velocity in Fig. 11.6C. Low-velocity magnitude is perceived near the surface; however, away from the surface, such magnitude

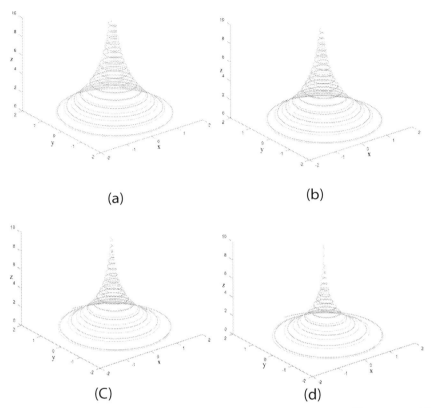

Figure 11.7 Three-dimensional streamlines pattern for (A) $L_s = 0.0$, (B) $L_s = 0.5$, (C) $L_s = 1.0$, and (D) $L_s = 1.5$.

is hiked. Fig. 11.7 describes the three-dimensional hybrid nanofluidic streamlines. The streamlines in Fig. 11.7A–D seemed to be spiral-shaped in the vicinity but tends to be parallel with the revolving axis. All the velocity components named radial, axial, and tangential do exist immediately at the plate's surface. The radial outward deflection is caused by the radial component, whereas the circumferential deflection is sourced from the tangential part. The existence of those two successive deflections causes the streamline to be spiral-shaped [26,27]. Enhancing the slip parametric input exhibits the significant propensity of the streamlines to be parallel with the spinning axis. Fig. 11.8A–D reveal two-dimensional streamlines variation for hybrid magnetized nanofluidic transport, and it is quite apparent that stream function values are reduced for the higher slippery

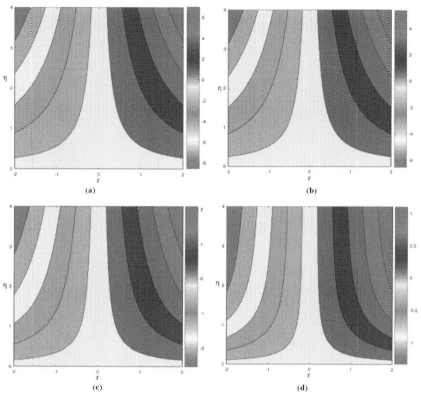

Figure 11.8 Streamlines pattern for (A) $L_s = 0.0$, (B) $L_s = 0.5$, (C) $L_s = 1.0$, and (D) $L_s = 1.5$.

regime. The frictional effect in Table 11.4 seems to reduce for a slippery surface. The reduction rate is 65.70% for hybrid nanofluids and 68.07% for mono nanofluids. Fig. 11.5A–B display the variation three-dimensionally.

4.3 Effect of unsteady parameter (S)

Fig. 11.9A depicts the radial velocity reduction for the unsteadiness factor. Near the surface, a hike in velocity is perceived, but after then, smooth decay is revealed [44]. Effects are prominent and distinct within $0.0 \leq \eta \leq 4.0$ (*not precisely marked*). Hybrid nanofluids explore higher velocity compared to mono nanofluids. Similar outcomes are traced out for

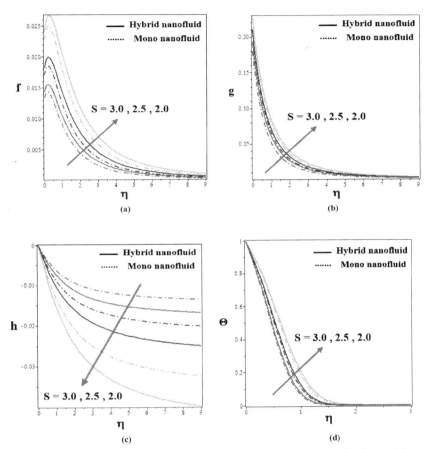

Figure 11.9 Effect of unsteadiness on (A) radial velocity, (B) tangential velocity, (C) axial velocity, and (D) temperature.

tangential velocity in Fig. 11.9B. Effects are minor compared to radial velocity profiles. Only near the plate, slight deviation is received. A totally reverse scenario is noted for axial velocity profiles in Fig. 11.9C. Here, mono nanofluids render higher velocity magnitude compared to hybrid nanofluids. Frictional output is displayed in Table 11.4 and seems to enlarge for both nanofluids. The enhancement rate is 8.26% for mono nanofluids, while it is 8.67% for hybrid nanofluids. The same is portrayed for hybrid nanofluids in Fig. 11.5C and mono nanofluids in Fig. 11.5D.

The thermal profile also drops off for unsteadiness in Fig. 11.9D. Slight away from the plate, the effects are prominent. Hybrid nanofluids disclose a

Table 11.5 Numeric outcomes of Nusselt number.

S	N	θ_w	$\lambda_H > 0.0$	$\lambda_H < 0.0$	ϕ_2	Nu_r Hybrid nanofluid	Mono nanofluid
2.0	1.0	1.5	0.5		0.01	0.856766	0.827119
2.5						0.718285	0.683427
3.0						0.541078	0.520395
2.0	0.5					0.303721	0.264950
	1.0					0.373923	0.368857
	1.5					0.505241	0.484950
	1.0	1.5				0.217402	0.186097
		2.0				0.313332	0.276195
		2.5				0.505242	0.484950
		1.5	0.3			0.041161	0.033241
			0.5			0.405909	0.388847
			0.7			0.680730	0.656087
				−0.3		1.475889	1.427084
				−0.5		1.298598	1.255364
				−0.7		1.095019	1.058102
			0.5		0.000	0.234693	... ...
					0.014	0.395712	... ...
					0.015	0.504498	... ...

high thermal profile compared to mono nanofluids because the existence of two nanoparticles boosts the thermal conductivity of the resulting fluids and thereby improves the temperature. Table 11.5 following Eq. (11.48) demonstrates the heat transference reduction for unsteadiness for both nanofluids, but hybrid nanofluids illustrate higher heat transmission magnitude. The same is revealed in Fig. 11.10A−B. The reduction rate is 36.74% for hybrid nanofluidic medium and 37.08% for mono nanofluidic medium.

4.4 Effect of radiation (N), temperature ratio (Θ_w), and heat source/sink (λ_H)

Fig. 11.11A describes thermal radiation's impression on the thermal profile. Temperature seems to escalate for radiation. Hybrid nanofluids explore a higher thermal profile compared to mono nanofluids. A heightening in thermal radiation speeds the frequent migration of nanoparticles, and thus random frictional collision among the metallic ingredients is increased. The continuous molecular friction translates the engendered kinetic energy into thermal energy. Consequently, temperature upsurges. On a similar note, the thermal trajectory swells for the temperature ratio factor in

On the hydrothermal performance of radiative 281

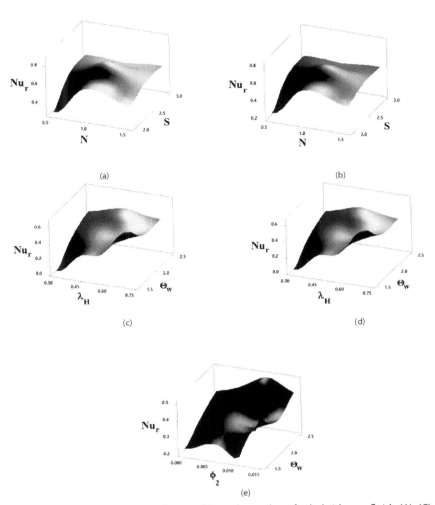

Figure 11.10 Three-dimensional view of Nusselt numbers for hybrid nanofluids (A), (C), and (E) and mono nanofluids (B) and (D).

Fig. 11.11B. Heat transference for radiation in Table 11.5 is marked to accelerate at the rate of 66.35% for hybrid nanofluids, while mono nanofluids exhibit a 53.37% increment. Fig. 11.10A–B support the above statement. The temperature ratio factor describes a hike in heat transport for both mono nanofluids and hybrid nanofluids in Table 11.5 and Fig. 11.10C–D. The heat source within the flow augments the thermal graph in Fig. 11.11C for both nanofluids, while the reduction is traced out in Fig. 11.11D for a

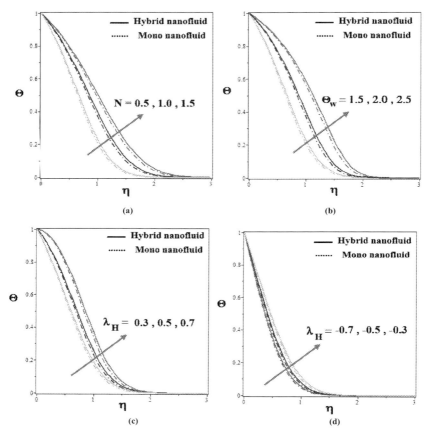

Figure 11.11 Effect of (A) thermal radiation, (B) temperature ratio, (C) heat source, and (D) heat sink on temperature.

heat sink. Heat transmission speeds up for the heat source in Table 11.5 and decreased for a heat sink. Fig. 11.10D explores the 3D view.

4.5 Effect of nanoparticles concentration (ϕ_2)

Fig. 11.12A depicts the radial velocity variations for nanoparticle concentration. Velocity seems to decline for higher concentrations. Higher concentrated nanofluids make themselves denser, and thus the resulting viscosity is enhanced which hinders their smooth transport over the specific geometrical texture. Consequently, mono nanofluids exhibit high velocity compared to hybrid nanofluids because existing double tiny ingredients make hybrid nanofluids comparatively much denser. Effects are clear and

On the hydrothermal performance of radiative 283

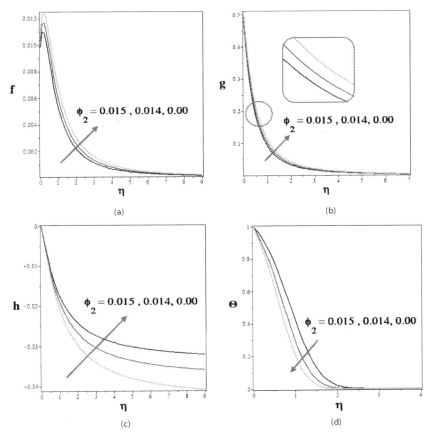

Figure 11.12 Effect of nanoparticles concentrations on (A) radial velocity, (B) tangential velocity, (C) axial velocity, and (D) temperature.

distinct. A similar conclusion can be penned down for tangential velocity from Fig. 11.12B. Here minor impacts are perceived. The opposite trajectory is revealed for axial velocity in Fig. 11.12C. Hybrid nanofluids acquire a high magnitude compared to mono nanofluids. Table 11.4 and Fig. 11.5C–D address increasing frictional hindrance for higher concentrations. Also, temperature shows an elevated scenario for nanoparticle concentrations. Double nanoparticles' existence maximizes the thermal effects and hence hybrid nanofluidic medium receives a high thermal magnitude compared to mono nanofluids in Fig. 11.12D. Heat transference also uplifts with higher concentrations as reflected in Table 11.5 and Fig. 11.10E. The heat transference enhancement rate is 27.49%.

5. Conclusion

The Ag—MgO—H_2O-based hybrid nanofluidic transport over a slippery swirling disk assuming the existence of highly oscillating magnetic fields is examined numerically throughout the study. Thermal radiation and heat sink/source are the external factors whose influence is considered during the modeling and simulation of the work. Shliomis's theory is clutched to model the flow. After reducing the foremost equations in dimensionless form, the shooting-based Runge—Kutta—Fehlberg technique is employed to run the simulation. Several streamlines, 3D plots, graphs, and tables are illustrated to enlighten the noteworthy fallouts of the investigation. The core conclusion that can be drawn from the study are as follows:

❖ The radial and tangential velocities are decreasing functions of high field frequency, velocity slip, nanoparticle concentrations, and unsteadiness. Hybrid nanofluids render high-velocity magnitude compared to classical nanofluids. However, axial velocity explores the reverse scenario for the aforementioned factors.

❖ Frictional effects seem to swell for high field frequency, unsteadiness, and concentrations, whereas slippery surface extracts the opposite outcome. Classical mono nanofluids reveal higher friction compared to hybrid ones.

❖ Thermal profiles decline for unsteadiness and heat sink, whereas radiation, temperature ratio, and heat source exhibit the opposite trend. Hybrid nanofluids depict high magnitude.

❖ Heat transference upsurges for radiation, temperature ratio, heat source, and nanoparticles concentrations, while the reverse trend is marked for unsteadiness and heat sink. Hybrid nanofluids explore the highest magnitude in heat transport and thus prove to be the better candidate in heat transmission.

Nomenclature

(u, v, w) Velocity components
r, θ, z Cylindrical coordinates
ω_p Particle's mean angular speed
Ω Flow vorticity
$\mathbf{H}$ Applied magnetic field strength
$\mathbf{M}$ Magnetization vector
τ_S Neel relaxation time
τ_B Brownian relaxation time
$\mathbf{I}$ Total moment of Inertia

$\mathbf{M_0}$ Instantaneous magnetization equilibrium
$\boldsymbol{\mu_0}$ Free space permeability
$\boldsymbol{n}$ Number of particles
$\boldsymbol{m}$ Magnetic moment of particles
$\boldsymbol{\xi}$ Langevin parameter
$\boldsymbol{k_B}$ Boltzmann constant
$\boldsymbol{T_a}$ Absolute temperature
$\boldsymbol{\omega_0}$ Magnetic field's angular frequency
$\boldsymbol{H_0}$ Amplitude of the magnetic field
$\boldsymbol{\chi}$ Magnetic susceptibility
$\boldsymbol{\alpha}$ Phase angle of magnetization and magnetic field
$\mathbf{Q}$ Heat source/sink
$\boldsymbol{T_w}$ Temperature of the surface
$\boldsymbol{T_\infty}$ Temperature away from surface
$\boldsymbol{T}$ Hybrid nanofluid temperature
$\boldsymbol{\rho}$ Density
$\boldsymbol{\mu}$ Dynamic viscosity
$\boldsymbol{\kappa}$ Thermal conductivity
$\boldsymbol{\rho C_p}$ Heat capacitance
$\boldsymbol{q_r}$ Radiative heat flux
$\boldsymbol{\sigma^*}$ Stefan Boltzmann constant
$\boldsymbol{k^*}$ Mean absorption coefficient
$\boldsymbol{\phi}$ Nanoparticle volume fraction
$\mathbf{Pr} = \frac{\mu_f (\rho C_p)_f}{\rho_f \kappa_f}$ Prandtl number
$\boldsymbol{S} = \frac{\lambda}{\Omega_v}$ Unsteadiness parameter
$\boldsymbol{L_s} = \boldsymbol{L'} \sqrt{\frac{\Omega}{\nu_f}}$ Slip parameter
$\boldsymbol{N} = \frac{4\sigma^* T_\infty^3}{3\kappa^* \kappa_f}$ Radiation parameter
$\boldsymbol{\Theta_w} = \frac{T_w}{T_\infty}$ Temperature ratio factor
$\boldsymbol{\lambda_H} = \frac{Q(1-\lambda t)}{\Omega_v (\rho C_p)_f}$ Heat source/sink
$\boldsymbol{Nu}$ Nusselt number
$\boldsymbol{C_f}$ Radial skin friction
$\boldsymbol{C_{fr}}$ Reduced skin friction
$\boldsymbol{Nu_r}$ Reduced Nusselt number
$\mathbf{Re}_z = \frac{r^2 \Omega}{\nu_f (1-\lambda t)}$ Local Reynold's number

Subscript
$\boldsymbol{f}$ Base fluid
$\boldsymbol{nf}$ Nanofluid
$\boldsymbol{hnf}$ Hybrid nanofluid
$\boldsymbol{s}$ Nanoparticle
$\mathbf{1}$ First nanoparticle
$\mathbf{2}$ Second nanoparticle

References

[1] R.E. Rosensweig, Ferrohydrodynamics, Cambridge University Press, Cambridge, 1985.

[2] S. Odenbach, Magnetic fluids, Adv. Colloid Interface Sci. 46 (1993) 263−282.

[3] P. Ram, A. Bhandari, K. Sharma, Effect of magnetic field-dependent viscosity on revolving ferrofluid, J. Magn. Magn Mater. 323 (2010) 3476−3480.

[4] A. Engel, H.W. Muller, P. Reimann, A. Jung, Ferrofluids as thermal ratchets, Phys. Rev. Lett. 91 (2003) 060602−060605.

[5] S. Odenbach, Magnetoviscous Effects in Ferrofluids, Springer, 2003.

[6] M. Sheikholeslami, R. Ellahi, Simulation of ferrofluid flow for magnetic drug targeting using lattice Boltzmann method, Z. Naturforsch. 70 (2015) 115−124.

[7] A. Zeeshan, A. Majeed, R. Ellahi, Unsteady ferromagnetic liquid flow and heat transfer analysis over a stretching sheet with the effect of dipole and prescribed heat flux, J. Mol. Liq. 223 (2016) 528−533.

[8] C. Rinaldi, A. Chaves, S. Elborai, X.T. He, M. Zahn, Magnetic fluid rheology and flows, Curr. Opin. Colloid Interface Sci. 10 (2005) 141−157.

[9] P. Ram, V. Kumar, Ferrofluid flow with magnetic field dependent viscosity due to rotating disk in porous medium, Int. J. Appl. Mech. 4 (2012) 1250041.

[10] M.I. Shliomis, K.I. Morozov, Negative viscosity of ferrofluid under alternating magnetic field, Phy. Fluids. 6 (1994) 2855−2861.

[11] J.C. Bacri, R. Perzynski, M.I. Shliomis, G.I. Burde, Negative viscosity effect in a magnetic fluid, Phys. Rev. Lett. 75 (11) (1995) 2128−2131.

[12] J.C. Bacri, A.O. Cebers, R. Perzynski, Behavior of a magnetic fluid microdrop in a rotating magnetic field, Phys. Rev. Lett. 72 (17) (1994) 2705−2708.

[13] R.E. Rosensweig, J. Popplewell, R.J. Johnston, Magnetic fluid motion in rotating field, J. Magn. Magn Mater. 85 (1990) 171−180.

[14] N. Andhariya, B. Chudasama, R. Patel, et al., Field induced rotational viscosity of ferrofluid: effect of capillary size and magnetic field direction, J. Colloid Interface Sci. 323 (2008) 153−157.

[15] J.H. Sanchez, C. Rinaldi, Magnetoviscosity of dilute magnetic fluids in oscillating and rotating magnetic fields, Phys. Fluids 22 (2010) 043304.

[16] P.K. Papadopoulos, P. Vafeas, P.M. Hatzikonstantinou, Ferrofluid pipe flow under the influence of the magnetic field of a cylindrical coil, Phys. Fluids 24 (2012) 122002.

[17] M. Yarahmadi, H.M. Goudarzi, M.B. Shafii, Experimental investigation into laminar forced convective heat transfer of ferrofluids under constant and oscillating magnetic field with different magnetic field arrangements and oscillation modes, Exp. Therm. Fluid Sci. 68 (2015) 601−611.

[18] Q. Yu, R.N. Dave, C. Zhu, J.A. Quevedo, R. Pfeffer, Enhanced fluidization of nanoparticles in an oscillating magnetic field, AIChE J. 51 (7) (2005) 1971−1979.

[19] M. Lajvardi, J. Moghimi-Rad, I. Hadi, A. Gavili, T. Dallali Isfahani, F. Zabihi, et al., Experimental investigation for enhanced ferrofluid heat transfer under magnetic field effect, J. Magn. Magn Mater. 322 (2010) 3508−3513.

[20] K.R. Schumacher, J.J. Riley, B.A. Finlayson, Effects of an oscillating magnetic field on homogeneous ferrofluid turbulence, Phys. Rev. E. 81 (2010) 016317.

[21] A. Ghofrani, M. Dibaei, A. Hakim Sima, M. Shafii, Experimental investigation on laminar forced convection heat transfer of ferrofluids under an alternating magnetic field, Exp. Therm. Fluid Sci. 49 (2013) 193−200.

[22] M. Hassan, Impact of iron oxide particles concentration under a highly oscillating magnetic field on ferrofluid flow, Eur. Phys. J. Plus 133 (2018) 230.

[23] T. Kimura, F. Kimura, Y. Kimura, Faraday diamagnetism under slowly oscillating magnetic fields, J. Magn. Magn Mater. 451 (2018) 65−69.

[24] T.V. Karman, Uber Laminare und turbulence Reibung, Z. Angew. Math. Mech. 1 (1921) 233−252.

[25] A.P. Krekhov, M.I. Shliomis, S. Kamiyama, Ferrofluid pipe flow in an oscillating magnetic field, Phys. Fluids 17 (2005) 033105.

[26] T. Fang, H. Tao, Unsteady viscous flow over a rotating stretchable disk with deceleration, Commun. Nonlinear Sci. Numer. Simul. 17 (12) (2012) 5064−5072.

[27] M.M. Rashidi, M. Ali, N. Freidoonimehr, F. Nazari, Parametric analysis and optimization of entropy generation in unsteady MHD flow over a stretching rotating disk using artificial neural network and particle swarm optimization algorithm, Energy 55 (2013) 497−510.

[28] P. Ram, A. Bhandari, Effect of phase difference between highly oscillating magnetic field and magnetization on the unsteady ferrofluid flow due to a rotating disk, Results Phys. 3 (2013) 55−60.

[29] B. Mahanthesh, B.J. Gireesha, N.S. Shashikumar, T. Hayat, A. Alsaedi, Marangoni convection in casson liquid flow due to an infinite disk with exponential space dependent heat source and cross-diffusion effects, Results Phys. 9 (2018) 78−85.

[30] B. Takabi, A.M. Gheitaghy, M. Tazraei, Hybrid water-based suspension of Al2O3 and Cu nanoparticles on laminar convection effectiveness, J. Thermophys. Heat Tran. 30 (3) (2016) 523−532.

[31] D. Madhesh, R. Parameshwaran, S. Kalaiselvam, Experimental investigation on convective heat transfer and rheological characteristics of Cu−TiO2 HNFs, Exp. Therm. Fluid Sci. 52 (2014) 104−115.

[32] Y. Xuan, H. Duan, Q. Li, Enhancement of solar energy absorption using a plasmonic NF based on TiO2/Ag composite nanoparticles, RSC Adv. 4 (2014) 16206−16213.

[33] M.H. Esfe, A.A.A. Arani, M. Rezaie, et al., Experimental determination of thermal conductivity and dynamic viscosity of Ag−MgO/water hybrid nanofluid, Int. Commun. Heat Mass Tran. 66 (2015) 189−195.

[34] H.K. Dawood, H.A. Mohammed, N.A.C. Sidik, et al., Forced, natural and mixed-convection heat transfer and fluid flow in annulus: a review, Int. Commun. Heat Mass Tran. 62 (2015) 45−57.

[35] F. Jamil, H.M. Ali, Chapter 6-Applications of hybrid nanofluids in different fields, in: Hybrid Nanofluids for Convection Heat Transfer, 2020, pp. 215−254, https://doi.org/10.1016/B978-0-12-819280-1.00006-9.

[36] R.A. Kumar, M. Kavitha, P.M. Kumar, et al., Numerical study of graphene-platinum hybrid nanofluid in microchannel for electronics cooling, Proc. IMechE. Part C: J. Mech. Eng. Sci. 25 (21) (2021) 5845−5857.

[37] R. Ellahi, M.H. Tariq, M. Hassan, K. Vafai, On boundary layer nano-ferrofluid flow under the influence of low oscillating stretchable rotating disk, J. Mol. Liq. 229 (2017) 339−345.

[38] M. Hassan, A. Zeeshan, A. Majeed, R. Ellahi, Particle shape effects on ferrofuids flow and heat transfer under influence of low oscillating magnetic field, J. Magn. Magn Mater. 443 (2017) 36−44.

[39] M. Hassan, C. Fetecau, A. Majeed, A. Zeeshan, Effects of iron nanoparticles' shape on convective flow of ferrofluid under highly oscillating magnetic field over stretchable rotating disk, J. Magn. Magn Mater. 465 (2018) 531−539.

[40] P. Ram, V.K. Joshi, V. Kumar, S. Sharma, Rheological effects due to oscillating field on time dependent boundary layer flow of magnetic nanofluid over a rotating disk, Proc. Natl. Acad. Sci., India Section A 89 (2018) 367−375, 2019.

[41] M. Shoaib, M.A.Z. Raja, M.T. Sabir, K.S. Nisar, W. Jamshed, B.F. Felemban, I.S. Yahia, MHD Hybrid nanofluid flow due to rotating disk with heat absorption and thermal slip effects: an application of intelligent computing, Coatings 11 (2021) 1554.

[42] H. Waqas, U. Farooq, R. Naseem, S. Hussain, M. Alghamdi, Impact of MHD radiative flow of hybrid nanofluid over rotating disk, Case Stud. Therm. Eng. 26 (2021) 101015.

[43] A. Saeed, M. Jawad, W. Alghamdi, et al., Hybrid nanofluid flow through a spinning Darcy—Forchheimer porous space with thermal radiation, Sci. Rep. 11 (2021) 16708.

[44] M. Imtiaz, T. Hayat, A. Alsaedi, et al., Slip flow by a variable thickness rotating disk subject to magnetohydrodynamics, Results Phys. 7 (2017) 503—509.

[45] N. Acharya, Framing the impacts of highly oscillating magnetic field on the ferrofluid flow over a spinning disk considering nanoparticle diameter and solid—liquid interfacial layer, J. Heat Tran. 142 (10) (2020) 102503.

[46] T. Gul, Kashifullah, M. Bilal, et al., Hybrid nanofluid flow within the conical gap between the cone and the surface of a rotating disk, Sci. Rep. 11 (2021) 1180.

[47] N. Acharya, Spectral simulation to investigate the effects of nanoparticle diameter and nanolayer on the ferrofluid flow over a slippery rotating disk in the presence of low oscillating magnetic field, Heat Transf. 50 (6) (2021) 5951—5981.

[48] N. Acharya, S. Maity, P.K. Kundu, Entropy generation optimization of unsteady radiative hybrid nanofluid flow over a slippery spinning disk, Proc. IMechE. Part C: J. Mech. Eng. Sci. 236 (11) (2022) 6007—6024.

[49] T. Lei, I. Siddique, M.K. Ashraf, S. Hussain, S. Abdal, B. Ali, Computational analysis of rotating flow of hybrid nanofluid over a stretching surface, Proc. IME E J. Process Mech. Eng. 236 (6) (2022) 2570—2579.

[50] N. Acharya, Magnetized hybrid nanofluid flow within a cube fitted with circular cylinder and its different thermal boundary conditions, J. Magn. Magn Mater. 564 (2022) 170167.

[51] J.C. Scaiano, C. Aliaga, S. Maguire, D. Wang, Magnetic field control of photoinduced silver nanoparticle formation, J. Phys. Chem. B 110 (26) (2006) 12856—12859.

[52] X.-F. Zhang, Z. Liu, W. Shen, et al., Silver nanoparticles: synthesis, characterization, properties, applications, and therapeutic approaches, Int. J. Mol. Sci. 17 (9) (2016) 1534.

[53] S. Zeroual, P. Estelle, D. Cabaleiro, et al., Ethylene glycol based silver nanoparticles synthesized by polyol process: characterization and thermophysical profile, J. Mol. Liq. 310 (2020) 113229.

[54] S. Abinaya, H.P. Kavitha, M. Prakash, et al., Green synthesis of magnesium oxide nanoparticles and its applications: a review, Sustain. Chem. Pharm. 19 (2021) 100368.

[55] I. Fongkaew, B. Yotburut, W. Sailuam, et al., Effect of hydrogen on magnetic properties in MgO studied by first-principles calculations and experiments, Sci. Rep. 12 (2022) 10063.

[56] F. Amrouche, M.J. Blunt, S. Iglauer, et al., Using magnesium oxide nanoparticles in a magnetic field to enhance oil production from oil-wet carbonate reservoirs, Mater. Today Chem. 27 (2023) 101342.

CHAPTER TWELVE

Application of nanofluids and future directions

Saeed Esfandeh
Department of Mechanical Engineering, Jundi-Shapur University of Technology, Dezful, Iran

Highlights
- Potential industrial application of nanofluids in electricity and energy sectors
- Potential industrial application of nanofluids in lubrication sector
- Potential industrial application of nanofluids in enhancement oil recovery sector
- Potential industrial application of nanofluids in water desalination sector
- Introduction of some industrialized nanofluid products in Iran

1. Application of nanofluids in energy and electricity sector

Nanofluids play important and growing role in various industries (Fig. 12.1). As the first this section take a look on application of nanofluid in energy and electricity sector. Applying nanofluids in electricity and energy sector can fix the problems like cooling problem of electrical and energy industry equipment, extending the life of some equipment due to high operating temperature, low efficiency of power generation and energy conversion equipment due to lack of proper heat transfer, high cost of power generation and energy conversion equipment, and reliability of power generation and energy conversion system. In fact, nanofluids can play two roles in optimizing the performance of heat conversion and electricity generation systems that are: (1). improving heat transfer in power and energy conversion equipment and (2). improving mass transfer in power and energy conversion equipment.

1.1 Nanofluids in power plant cooling towers

All mechanical equipment with moving parts, such as motors and compressors, generate heat during operation due to the friction between the moving

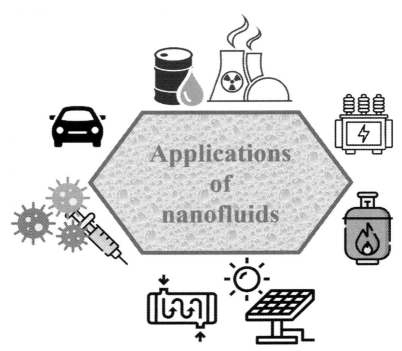

Figure 12.1 Application of nanofluids in various industries.

parts. The amount of produced heat is high, and if it is not controlled, it will cause damage and thermal stress to the equipment. Therefore, this heat must be removed from the system in some way. One equipment in power plant cycles is the cooling tower, which is responsible for removing the heat from other rotating equipment in the system and cycle. The cooling tower takes the heat from the applied operating fluids in the power plant cycle and transfers it to the surrounding environment. In a traditional cooling tower, the heat of the operating fluids of all equipment is removed by cool water, as a result of which water consumption in power plant cycles has been reported to be very high. For example, recondensing and changing the phase of the steam exiting the turbine to the liquid phase as an input of the boiler, cooling the coils in the electricity generator, and also cooling the oils that take the heat from the engine or turbine bearings or other parts are duties of water as a cooling fluid, which must take the heat from different operating fluids and return them to the power plant cycle.

In order to exchange heat between the operating fluid of the steam power plant cycle and water as a cooling fluid, it is necessary to create a structure such as heat exchangers. Although there are various methods for removing heat from power plant cycle equipment, as one of the promising solutions, nanofluids as a new generation of cooling fluids have a high potential to improve the performance and increase the efficiency of power plant cycles. Nanofluids have a much higher thermal conductivity coefficient than traditional cooling fluids such as water, oil, and ethylene glycol due to the presence of metal particles in nanodimensions. For example, the powered fluid containing copper nanoparticles has a thermal conductivity up to 700 times higher than that of water, and its thermal conductivity is up to 3000 times higher than that of oil.

In the past, cooling fluids containing particles in millimeters or maximum micrometers have been studied and used in many cases. Although the use of these suspensions improves the heat transfer characteristics of cooling fluids, due to the possibility of sedimentation of these particles, the possibility of corrosion on the surfaces due to the collision of particles with the surfaces, more pressure drop, and so need to spend more energy in transferring the cooling fluid, they are less welcomed.

Here, the importance of possessing nano-sized particles becomes more apparent, because the particles in very small sizes are less prone to sedimentation and clumping in the suspension, and due to their small size, they cause much less damage to equipment surfaces. It is also clear from the theoretical point of view that the lower sized particles will give the higher the total heat transfer surface, which increases the efficiency of heat exchange. Therefore, the design of cooling systems enriched with nanofluids in the cooling towers of steam power plants is one of the undeniable necessities and of course very beneficial. Of course, it should not be forgotten that the synthesis of a nanofluid with high stability and fixed natural characteristics under high temperature is one of the prerequisites for success in the optimization of the above-mentioned cooling system in steam power plants.

1.2 Nanofluids in transformer

Transformers are the most important part of a network. Improving the structure and performance of transformers has a great impact on the costs and reliability of the network. The application of nanotechnology in the field of electrical engineering has developed the way of manufacturing equipment such as transformers, electric motors, and other equipment. In fact, the use of nanotechnology is resulted in lower dimensions of transformers, volume, weight, and overall cost and also higher efficiency. Cooling in transformers, like many other equipment, plays an essential role in their performance, and on the other hand, the use of nanofluids can improve the thermal conductivity properties of the transformers working fluid and remove more heat from the transformer. Transformer oils are responsible for removing heat from this equipment, and the use of nanoparticles in working fluids can improve the process of removing heat from the transformer. Transformer oils lose their properties in a short period of time due to the high heat generated in the transformer, and their thermal performance is lower than expected. Therefore, it is necessary to improve the durability and thermal resilience of transformer oils, and one of the main solutions is to increase the thermal conductivity of these oils.

The heat transfer coefficient of oils used in transformers is around 0.130 W/m.K. Nanoparticles can be used to improve and increase this heat transfer coefficient. In the past years, the use of magnetic nanofluids containing suspended magnetic nanoparticles in electromagnetic equipment has been developed and positive results have also been obtained [1—3]. Due to that achievement, the use of nanoparticles and nanofluids in transformer oils has increased and developed significantly [4,5], especially since Segal and his colleagues' proposal in 1998 [3].

According to the conducted studies, the temperature of the transformer is one of the main factors in increasing the life and performance of the transformer. In the process of controlling and reducing the temperature of the transformer using nanofluids, there are some things to consider, including the type of nanoparticle and the volume percentage of nanoparticles. Laboratory methods or computer simulation methods are often used to obtain the optimal type and volume percentage of nanoparticles [6,7]. According to the results of Dhiaa et al. [7], the use of nanoparticles as dispersed particles in transformer's oil can decrease the temperature of the transformer by 6% and this temperature reduction can increase in the lifespan and overall operation time of the transformer by 100%. Although it should be noted that increasing the volume fraction of added nanoparticles does not always improve and reduce the temperature of the transformer. So finding the optimized volume fraction or volume concentration is essential.

1.3 Nanofluids in electrical generator

Generator is another equipment that is also used in power plants, and heat can be produced in its various parts. Like the transformer, the operating temperature of the generator is one of the most influential factors on its lifespan. Therefore, the higher amount of heat dissipation from the generators improves the quality of their performance and lifespan. The problem of

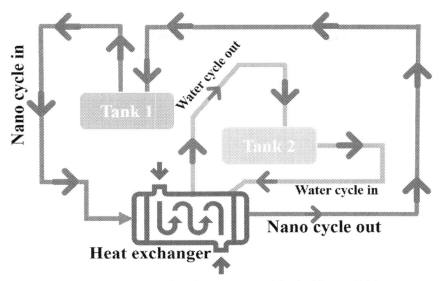

Figure 12.2 Improved cooling system enriched with nanofluids.

temperature increase in generators with high speed and high voltage and capacity is more evident than generators with low capacity, because with the increase of generator capacity, the ratio of heat produced to the surface unit increases. Low-capacity generators are often cooled with natural air or with cooling fans. Of course, the methods of generator cooling systems have been developed from direct to indirect cooling and from air cooling to hydrogen and water cooling. For high-capacity generators, cooling systems based on hydrogen and water are mostly used. Proper cooling systems will improve the performance, reduce the energy losses, and increase the lifespan of generators. One of the solutions to increase the efficiency of cooling systems is the use of nanofluids in these systems. So because of the high thermal conductivity coefficient of nanofluids, heat removal from generators can be increased. Fig. 12.2 shows a schematic of the cooling system enriched with nanofluids. Also, the flow rate of the nanoparticles affects the heat transfer coefficient, and as the concentration of nanoparticles increases, the heat transfer coefficient also increases.

1.4 Industrial products in Iran

According to reports, there are few industrial nanofluids that have been used in power plants with the aim of improving the energy and electricity efficiency of industrial equipment. For all industries, the heat exchangers play an undeniable role in production cost and quality. Nano Poshesh Felez Co. is a company in Iran that manufactured new age of cooling nanofluid for heat exchangers. As a new feature, the produced nanofluids create something like nanofins on the heat exchangers shell, in addition to enhancing the boiling point compared to conventional cooling fluids. This new feature increases the contact surface of fluid molecules with the metal wall of the heat exchanger and improves the heat transfer rate. Tarasht and Kahnuj are two power plants in Iran that applied the above-mentioned industrialized nanofluids in their cooling cycles and heat exchangers. The result in Tarasht and Kahnuj power plants was reduction of energy consumption by about 10% [8].

As one of the active companies in the field of producing industrial products based on nanotechnology, Nanoposhash Fellez Company has succeeded in producing cooling nanofluid. In 2008, this company received the confirmation of its industrial nanotechnology-based products from the Iran Nanotechnology Innovation Council (INIC). Nanoposhash Fellez Company has produced several other nanoproducts in addition to the nanocoolant product. By increasing the heat transfer coefficient of water, this technology increases the thermal efficiency of all devices that need cooling. Increasing the thermal conductivity coefficient of the base fluid, increasing the cooling ability of heat exchangers, increasing the power of generators, increasing the efficiency of all types of heat exchangers, and reducing the fuel consumption of cars are of the advantages of new produced nanofluid or nanocoolant [9].

2. Industrial application of nanoparticles in lubrication improvement

Lubrication plays an important and undeniable role in increasing the life of systems with moving components and moving surfaces in contact with each other. Due to the technology improvement, all mineral or synthetic base lubricants need additives to reach the highest efficiency in advanced systems with moving parts in contact with each other. Based on the studies, additives can add unique properties to conventional lubricants like the improvement of flash point and improvement of viscosity, pour point, and resistance to friction. Nanoparticles can be one of the types of additives of lubricants to improve their properties.

In general, it can be stated that the performance of a lubricant depends on the amount and type of additives used in the lubricant. The most important objective pursued in lubrication science is to facilitate relative motion of moving surfaces that are in contact with each other. In any instrument with the moving parts in contact with each other, the phenomenon of friction and abrasion occurs. In addition to causing wear and reducing moving parts life, friction will also increase energy consumption. Therefore, choosing a suitable lubricant has a great impact on the performance of systems and machines. Creating a suitable lubricant layer between surfaces with relative motion separates these surfaces and removes heat and wear particles.

Over the past years, nanoparticles have been mentioned as an efficient and innovative additive to improve the physical and chemical properties of conventional lubricants because of their especial surface properties and special thermal properties [10]. One of the most important features of nanoparticles that makes them superior to other additives is their very small dimensions, which result in less frictional reactions than other additives.

Despite the stated advantages, the use of nanoparticles as an additive in lubricants also has some obstacles. Placing nanoparticles in the lubricating-based fluid forms a colloidal solution, and colloidal solutions are inherently very weak in stability, and this can cause nanoparticles to sediment. This is an open challenge for researchers. So the effect of dispersing various types of nanoparticle in different volume fractions to lubricants has been studied in recent years [11].

Lubricants used in the industry can be divided into gas, liquid, semisolid, and solid categories. The main tasks of a lubricant are heat transfer, lubrication, and surfaces protection. All liquid lubricants consist of two parts: base oil and additives. Base oil additives can add various properties to lubricating oils. The following eight categories are the missions of additives after they are added to lubricants: (1). detergents and dispersants, (2). antiwear, (3). antioxidants, (4). viscosity index improver, (5). antirust and anticorrosion, (6). antifoam, (7). friction modifiers, and (8). pour point depressant [12].

2.1 Nanoadditives used in lubricants

WS2 nanoparticle is one of the widely used nanoparticles that is used as an additive to lubricants. This nanoparticle shows very good performance in unusual conditions such as high pressure and temperature as well as under high load and also has high resistance against oxidation [13,14]. Therefore, WS2 gives significant characteristics to lubricating oils like the ability to penetrate very well in very small pores, prevent the destruction of surfaces in contact with each other, and create self-lubricating surfaces. Fullerene nanoparticles and nanodiamonds [15,16] are also among the other widely used nanoparticles in the process of lubrication of moving surfaces in contact with each other. Fig. 12.3 shows a schematic of nanolubricant role between two moving parts in contact with each other.

Another category of nanoparticles used as additives in lubricants are metal nanoparticles, which, of course, have several major problems with the use of these nanoparticles, such as poor stability and solubility, as well as poor distribution ability of these nanoparticles in the base lubricant. In fact, metal nanoparticles are extremely unstable and quickly and easily aggregate. It is because of the adhesion of metal nanoparticles to each other, and as a result, the increase in the dimensions of the particles. This increase in the size of the particles causes the particles to go beyond the nanosize, and as a result, they lose their unique properties. It is also possible for the particles to sediment due to their size increase, and on the other hand, this increase in size will increase the possibility of particle deposition among moving surfaces in contact with each other. One of the ways to solve this problem is to apply a coating of organic molecules on the surface of metal nanoparticles, which has been discussed in some studies [16,17]. Nanoparticles of iron oxide [18], nickel oxide, copper oxide, and titanium are among the most widely used metal nanoparticles in the lubrication industry. In general, reducing friction and thus saving fuel, reducing operating temperature, increasing the life of the lubricant, high load tolerance, which means reducing the wear of parts, reducing SOx/NOx gases in internal combustion engines, increasing resistance to corrosion, the need to use nanoparticles in a very low amount in grams per liter, not destroying the filters due to the easy passage of particles through the filter are among the main advantages of nanolubricants.

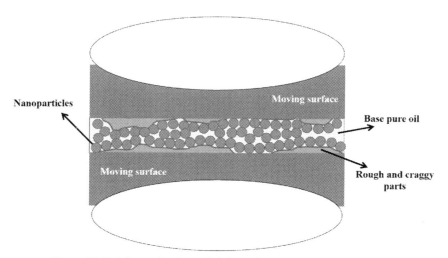

Figure 12.3 Schematic of nanolubricant between two moving parts.

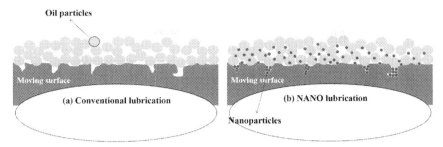

Figure 12.4 (A) Conventional lubricant and (B) nano-enriched lubricant.

Fig. 12.4 shows a schematic of conventional and nano-enriched lubricants performance facing with surface porosities and surface roughness.

2.2 Engine oils and nano-sized additives

One of the most widely used lubricants is engine oil, because many moving parts of an engine, all of engine parts require lubrication. Therefore, the use of nanolubricants in car engines has been of great interest to researchers in the past years. The use of nanoparticles in engine oils, in addition to reducing the friction between surfaces, which increases the life of the parts, will also reduce fuel consumption and reduce the sound of the car engine, because the lower the friction between the parts and the car can be expected to run faster. Another result of the use of nanolubricants is the reduction of greenhouse gases, which has a direct relationship with the reduction of car fuel consumption, and this helps car engines achieve a higher environmental standard. Nanodiamonds are one of the most widely applied nanoparticles in new engine oils. Among other applications of nanolubricants, we can refer to their application in the oil industry, specifically in turbines, compressors, and rotating oil equipment.

3. Applications of nanotechnology in solar water heaters

Technically, solar water-heating systems face with various challenges toward being commercial. One of the big challenges in this field is to increase the efficiency of the system, especially in cold seasons. Another challenge is improving the amount and rate of heat transfer to or from the working fluid, which can play an important role in achieving high efficiency. As the third obstacle, the dimensional compression of the solar water heater can be another positive and important factor in the development and acceptance of this product among society.

In general, the collector, the heat transfer fluid, and the storage tank are the main components of a solar water heater system, and the modification and optimization of each of the mentioned parts can have a great impact on improving the overall performance of the system. In the collectors sector, it is very important to modify and optimize the absorber plates to achieve the maximum absorption of solar radiation, especially on nonsunny days. Replacing the heat transfer fluid used in solar water heating systems with new fluids with improved thermal properties is another way to increase the efficiency of the system. To improve the storage tank, it is also possible to improve the thermal insulation applied in the tanks, and in this way, heat loss from the system can be minimized. Also, the use of old and new generation phase-change materials can increase the efficiency of the system.

There are various solutions to solve the problems and obstacles mentioned above, but in addition to all the solutions, there is a common solution to solve all the problems mentioned above, and that is the use of nanotechnology [19−24]. The use of nanofluids as one of the achievements of nanotechnology can have a tremendous effect on improving the thermal properties of the heat transfer fluid used in solar water heating systems. This fluid takes the heat from the collector and transfers it to the targeted water by using a heat exchanger. The heat transfer fluid should be selected based on various characteristics such as melting point, freezing point, viscosity, heat capacity, and thermal conductivity coefficient. Water, ammonia, methanol, and acetone are among the common fluids that have been used in solar water heaters, among which water has the highest efficiency. The main problem of conventional fluids is their poor heat transfer properties, which is an obstacle to ideal heat absorption and transfer from the solar radiation source. Also, this problem causes failure in dimensional compression process of the solar water heater system as one of the main future industries targets. One of the ways to improve the thermal properties of heat transfer fluids is to add solid particles to them, which of course will not be without problems, and the first problem in this way is the sedimentation of these suspended particles in the conventional base fluid. Therefore, the researchers started using particles in small dimensions, which include the synthesis of solid particles from metallic and nonmetallic materials in micro and nanodimensions. Solid particles can have thermal properties up to hundreds of times better than conventional fluids used in solar water heaters. Carbon nanotube, aluminum oxide, graphene, and copper oxide are among the nanoparticles that can be dispersed to heat transfer fluids in solar water heaters.

The increase in thermal conductivity of nanofluids depends on volume percentage, size, type of nanoparticles, and base fluid. Considering the constant

Reynolds number, the heat transfer of nanofluids increases with increasing volume percentage of nanoparticles and decreasing particle size [19,25].

In general, the main advantages of nanofluids are a high specific surface area and in result a higher heat transfer level between particles and fluids, high stability and unique dispersion of particle with the help of Brownian motion, reduction of energy consumption for pumping compared to pure fluids to achieve the equivalent heat transfer intensity, reducing particle clogging compared to conventional grouts, and adjustable properties including thermal conductivity and surface wettability by changing particle concentration for different applications.

Also metal and mineral nanoparticles can be used in phase-change materials to increase heat transfer and heat storage capacity as heat transfer fluid or solar heat storage materials. In some cases, the heat storage materials are in micron or millimeter size and usually have a core and shell structure. The core may be in the form of a nanocomposite of phase-change materials and a small fraction of nanoparticles with very high thermal conductivity and a higher melting point than phase-change materials. The shell is made of a thin layer of neutral nanoparticles with high thermal conductivity. The phase-change material in the core can absorb or release latent heat energy during the solid—liquid phase change. Carbon nanofibers, graphite with nanometer holes and nanometer oxides, and hydroxides are among the nanomaterials used in phase-change materials. Improving the insulating layer of storage tank in the solar water heater and the absorbent layer in the collector using nanoparticles is one of the other fields of application of nanotechnology in the solar water heater system. Using nanoparticles can create excellent thermal insulation without increasing the weight of the system.

4. Applications of nanotechnology in solar water desalination

Due to the fact that the available water sources for drinking purposes are scarce and most of them are saline and unusable for drinking, the need to produce fresh water in most regions of the world is quite clear. There are two general categories for water desalination methods: (1). thermal methods of producing fresh water (evaporation-distillation) and (2). membrane methods. With the help of renewable and free solar energy, it is possible to provide the needing thermal and electrical energy required for water desalination systems. The use of nanomaterials in both the abovementioned water desalination methods increases the production rate of fresh water and improves its quality, and ultimately increases efficiency and reduces energy loss in the system.

In recent years, the use of nanotechnology in the development process of solar water desalination has become more popular. The results of the studies have shown that the combination of two new technologies that are solar and nano can make great improvements in the way of applying solar water desalination. The main results of using nanotechnology in water desalination plants are: increasing the absorption of the sun's thermal energy, increasing the efficiency of converting solar energy into thermal energy (in thermal water desalination plants), increasing the water production rate, removing pollutants from the final produced water (in membrane-based water desalination systems), and cost reduction (due to reduction of surface area).

One of the favorite applications of researchers is the use of nanomaterials in the saline water entering the solar stills. In fact, added nanomaterials to saline water increase the percentage of sunlight absorption and increase the conversion efficiency of solar energy into thermal energy. Adding nanomaterials in the form of nanocomposites or nanoparticles to saline water increases the evaporation rate of water and as a result, increases the output rate of fresh water. To achieve the highest efficiency and increase the rate of water evaporation, nanoparticles or nanocomposites with the highest conductivity coefficient and the most powerful absorption of sunlight should be used in solar desalination systems. Gao et al. [26] were among the researchers who investigated the property of absorbing sunlight using gold nanoparticles in different concentrations of graphene, which obtained significant results. The use of graphene along with gold nanoparticles in their study is due to the fact that graphene has the highest absorption power and conductivity among existing materials.

Theron et al. [27] also investigated the effect of adding silver, gold, graphene, and the combination of graphene and gold nanoparticles to the sea salt water in the solar desalination basin on the rate of increase in water temperature and the increase in the evaporation rate of saline water. According to results of their research, after 60 min of sunlight, the temperature of saline water without nanoparticles in the desalination basin reached 50°C, while on the other hand, adding nanoparticles of silver, gold, graphene, and the combination of graphene and gold to the salt water increased the temperature of the water to 70, 70, 80, and 100°C, respectively. This shows that investing in the development of the application of nanofluids in different parts of a solar desalination plant can create a tremendous development in the desalination of saline water. In another research, Thembela Hillie & Mbhuti Hlophe [28] proposed suitable nanoparticles and nanocomposites to add to salt water to increase the heat transfer absorption rate of water in desalination system. Based on their studies, zeolite nanomaterials, carbon nanotubes, and metal oxide nanoparticles are among the best additions to the saline water in the desalination basin.

Of course, the application of nanotechnology and nanomaterials is not limited to their use in solar water desalination. One of the other applications is the use of membrane products with nanoscale holes. Combining the use of nanoscale membrane technology (nanofiltration) with one of the most widely used methods of water desalination (reverse osmosis) has made tremendous achievements in the development and optimization of water desalination systems. This makes the achievement of solar water desalination not only limited to water desalination but also brings virus and bacteria–free fresh water by applying nanofiltration [27]. It should be noted that some of nanoparticles such as silver nanoparticles also have antibacterial properties and adding them to saline water will give bacteria-free water next to increasing the speed and intensity of basin saline water evaporation of in water desalination systems to make more fresh water [29].

5. Application of nanotechnology in oil and gas wells

As stated in the previous sections, nanotechnology has created a new approach in all fields and sciences, and the oil and gas industry is not an exception. Although using nanotechnology in this industry is at the beginning of scientific developments in many countries, especially in developing countries like Iran, many hope for the successful application of nanotechnology, especially in the upstream sectors of oil and gas.

The application of this technology in the oil industry can result in many achievements, including facilitating and increasing the extraction of oil and gas. On the other hand, improving and increasing extraction of fossil fuels is equivalent to improving the economic condition of a country. Countries such as Iran and other Middle Eastern countries with many oil and gas resources can make one of the biggest profits by applying nanotechnology in oil industries, especially in upstream industries.

Upstream industries in oil are the set of activities of the oil industry from the exploration step to the exploitation and management of reservoirs. In general, the fields of application of nanotechnology in upstream oil industries can be divided into nine sectors: nanoparticles, nanosensors, nanocomposites, nanocoatings, nanocrystals, nanocalculators, nanofluids, nanofilters, and nanogels [30]. In the following, the upstream oil activities are presented in Table 12.1 [31]. The last column of this table is focused on the issue of

Table 12.1 Oil and gas Industry challenges classification and nanotechnology solution.

Industry challenges classification	Nanotechnology solution	Possible nano-enriched productions
Oil exploration	Nanosensors and imaging with the help of nanotechnology	
Oil drilling	Nanomaterials and nanocoating, nanofibers, and *nanofluids (nanoparticles)*	Drilling mud
Oil extraction	Nanomaterials and nanocoating, nanosensors, nanofibers, and *nanofluids (nanoparticles)*	Drilling cement
Oil reservoir management	Nanofibers, nanosensors, and *nanofluids (nanoparticles)*	EOR

applying nanoparticles in liquid or semisolid products like drilling mud and drilling cements and also adding nanoparticles or nanofluids in to the oil well for EOR goals.

One of the important subsections of reservoir management in which there is a possibility of developing the use of nanotechnology is the EOR process in reservoirs using nanoparticles and nanofluids. In oil drilling and extraction section, the role of nanoparticles and nanofluids can be very prominent. The production of optimized drilling mud as one of the subsectors of oil drilling and new generation drilling cement as one of the subsectors of oil extraction can be an arena to show the power of nanotechnology in the oil industry.

Drilling mud plays an important role in the speed of drilling operations. Drilling mud, which is also called drilling fluid, is responsible for several tasks like transferring of drilled fragments to the surface, controlling the temperature of the drill bit, controlling the pressure of the walls, and transferring the hydraulic power of the pump to the drill bit. In order to achieve the most optimized properties of the drilling fluid, additives must be used, among which nanoparticles can be mentioned. By using nanoparticles, important

Table 12.2 Enrich of drilling mud properties by nanomaterials [33].

Nanomaterials	Effect
Nanoclay	Improvement of mechanical properties of drilling mud
Nanopolymer	Fluid leakage control
Iron nanohydroxide	Fluid leakage control and oil well stability improvement
Calcium carbonate nanoparticles	Fluid leakage control and oil well stability improvement
Carbon nanostructures (Graphene and carbon nanotubes)	Improvement of heat transfer coefficient, improvement of rheological behavior and mud stability, fluid leakage control, and reduction of the corrosion of drilling tools

properties of the drilling fluid such as density and viscosity can be adjusted in the most optimal possible state. Generally, the purpose of adding nanoparticles to drilling fluids is to achieve optimized drilling fluid with the aim of reducing costs through less but more effective drilling fluid consumption [32]. Table 12.2 shows the positive effect of various nanomaterials on drilling mud properties.

5.1 Enhancement oil and gas recovery

During the life of an oil reservoir, oil can be extracted from the tank during three periods. In the first stage, oil extraction forms by using the natural pressure of the reservoir, and the pressure of the reservoir will cause the oil to come out. Over time, a decrease occurs in oil extraction because of the decrease in natural pressure inside the reservoir. In this stage, which is called the second stage of extraction, by injecting conventional fluids such as water or gas into the reservoir, the pressure inside the reservoir can be restored and the production and extraction of oil can be returned to the normal state. Again over time and after the extraction of more oil from the reservoir, the pressure of the reservoir decreases again and as a result the oil production decreases again. In this stage, which is called the third stage of oil extraction, there is a need to apply new methods for oil extraction that is named by EOR.

The importance of developing and optimizing different methods of EOR becomes more important by knowing this fact that only one-thirds of the available oil in oil reservoir can be extracted from the reservoir by using conventional methods [30].

In addition to the older methods of EOR such as thermal or chemical methods or newer methods such as microbial methods, the use of nanomaterials with the aim of optimizing the process of EOR has received special attention in the past years. The above-mentioned EOR methods, besides the relative improvement of oil extraction, have some disadvantages. For example, chemical methods, due to the deposition of injected materials, cause the pores of the reservoir rock to close, and as a result, it becomes impossible to extract oil from the reservoir over time. But this problem does not exist in nanomaterials, and their injection into the reservoir does not cause the pores to close because nanomaterials have much smaller dimensions than normal materials. Also, the use of nanomaterials and nanoparticles along with other methods such as combining the use of thermal methods and nanomaterials with the aim of EOR will cause faster and better heat transfer in thermal methods.

The strategy of different companies to use nanotechnology in oil and gas industries and specifically in EOR is different from each other. One of the most used cases is the use of nanofiltration in various stages of the oil extraction process, among which it is possible to use nanofiltration to purify water and control its salinity before the start of the water injection and flooding process inside the reservoir. Applying nanofiltration made it possible to use unusable water such as sea salt water or sewage for the purpose of flooding the oil reservoirs. Another application of nanofiltration is the use of this technology in order to separate and recover the polymers injected into the oil reservoir after the polymer flooding process, which makes it possible to use and reinject these polymers for reflooding.

Next to the above-mentioned application of nanotechnology in EOR, using nanoparticles and nanofluids in EOR has also had successful experiences. In recent years, with the progress made in nanotechnology and the emergence of potential capabilities, many researchers have turned to using nanoparticles to improve the oil extraction [34]. Nanoparticles have special and unique properties and can cause easier and more effective separation of oil (Fig. 12.5). Adding these particles to the base fluid improves the properties of the flooded fluid in properties like density, viscosity, and thermal properties [35].

Many of Iran's reservoirs, especially Iran's oil reservoirs, are fractured carbonate type. The fractured carbonate type can cause limitations in the oil extraction process from the reservoirs. The presence of many cracks in the reservoir can cause the injected water to be wasted in the process of EOR, or in other words, the injected water will increase, so the water along

Application of nanofluids and future directions 305

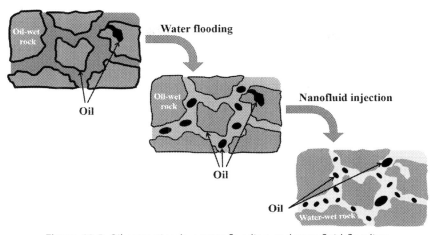

Figure 12.5 Oil extraction by water flooding and nanofluid flooding.

with the oil will increase, in which case the process of EOR will be disrupted and the separation of water from oil on the surface will be very difficult. To avoid this problem, various companies applied polymer gels based on nanoparticles. The use of polymer gels in the flooding process causes the gaps and cracks in the reservoirs to be closed by these gels and so prevents excessive water injection into the oil reservoir. Swelling of the gels in contact with water closes the cracks to prevent excess water injection. The use of nanoparticles in the structure of polymer gels will control their gelation speed and finally their better performance in oil reservoirs.

Also, the use of smart nanofluids in the flooding process will control and direct the fluid movement in the gaps and cracks in a reservoir. This work causes more surface of the reservoir to be in contact with the injected fluid in the process of EOR, and finally, this increases the sweeping efficiency of the oil extraction process. In fact, the presence of large and small cracks in the oil reservoir can cause the flooding fluid to deviate from the main paths and reduce the efficiency of flooding. Therefore, the task of nanofluids and nanoparticles used in the flooding process is to control the flow of injected fluid in the reservoir and in fact they will be used to homogenize the behavior of the flooded reservoir. In fact, these particles will reduce the size of the opening of large cracks and also control the opening of small cracks, and in result flooding will be done with the least amount of excess injected fluid into the oil reservoir.

Injecting nanofluids into the reservoir causes the wettability of the reservoir rock to change from oil-wet to water-wet, and in this way, they cause the oil to move more easily through the cracks and voids of the reservoir rock toward the production well. Also, nanoparticles are able to significantly reduce the interfacial tension between oil and injection fluid [36]. Therefore, the trapped oil in the void can be extracted with the help of various methods such as change in wettability, reduction of surface tension, etc. [37,38].

EOR with the help of nanoparticles uses mechanisms such as reducing the viscosity of oil, increasing the viscosity of injection fluid, reducing interfacial tension, and changing wettability to increase oil extraction. It should be noted that every nanoparticle does not lead to an increase in recovered oil, and the selection of the most appropriate nanoparticle should be analyzed considering various parameters such as: temperature, salinity, type and size of nanomaterial, and contact angle. Despite the wide variety of nanomaterials, the most used nanoparticles in researches are silica.

6. Future directions

The current research has focused on the application of nanofluids in various industries, especially industries related to the field of energy and energy systems, which include power plant industries, oil and gas industries, automotive industries in the fields of improving heat transfer and lubrication, as well as renewable energy systems and the present and potential applications of nanofluids in each of the mentioned industries were described in detail.

Nanotechnology and specifically nanofluids can have an undeniable effect on increasing the life and efficiency and reducing the costs of construction, implementation, operation, and maintenance of various energy systems. Finding optimal combinations of nanoparticles to synthesize new nanofluids with the aim of achieving the highest level of heat exchange and solving the cooling problem in various power plant equipment such as cooling towers, transformers and generators, equipment requiring cooling in the oil and gas industries and also for new energy systems, in addition to paying attention to the economic aspect of synthesized nanofluids, can be considered as one of the important foresights in the development of nanofluids application in energy systems.

Some of the open challenges in the field of electricity and energy systems, which always have capacity for improvement and development, are: the low life of some equipment in the electricity and energy industries due to high temperature operating conditions, the low efficiency of some equipment in energy systems due to lack of the appropriate heat transfer mechanism and the weakness of applied working fluid in heat transfer ability, the high dimensions and weight of energy system equipment due to poor heat exchange, and as a result, the complex placement of the equipment in the energy systems. By applying suitable and optimized nanofluids, all of these limitations can be fixed.

Nomenclature

Abbreviation
EOR Enhancement oil recovery
INIC Iran Nanotechnology Innovation Council
WS2 Tungsten disulfide

References

[1] S. Choi, Enhancing thermal conductivity of fluids with nanoparticles, ASME FED 231 (1995) 99—102.
[2] S.K. Das, S. Choi, W. Yu, T. Pradeep, Nanofluids: Science and Technology, 21, John Wiley & Sons, Inc., 2008, pp. 5—8.
[3] V. Segal, A. Hjorstberg, A. Rabinovich, D. Nattrass, K. Raj, AC (60 Hz) and impulse breakdown strength of a colloidal fluid based on transformer oil and magnetite nanoparticles, IEEE Int. Symp. Electr. Insul. (1998) 619—622.
[4] P. Kopcansky, L. Tomco, K. Marton, M. Koneracka, I. Potocova, M. Timko, The experimental study of the DC dielectric breakdown strength in magnetic fluids, J. Magn. Magn. Mater. 276 (2004) 2377—2378.
[5] M. Chiesa, S.K. Das, Experimental investigation of the dielectric and cooling performance of colloidal suspensions in insulating media, Colloids Surf. A Phys. Eng. Aspects 335 (2009) 88—97.
[6] S.S. Botha, P. Ndungu, B.J. Bladergroen, Physicochemical properties of oil-based nanofluids containing hybrid structures of silver nanoparticles supported on silica, Ind. Eng. Chem. Res. 50 (2011) 3071—3077.
[7] A.H. Dhiaa, M.I. Abdulwahab, S.M. Thahab, Study the convection heat transfer of AL_2O_3/water nano fluid in transformers, Eng. Tech. Journal 33 (7) (2015) 1319—1329. Part (E).
[8] http://indnano.ir/kalanano/product/0140/.
[9] https://nanoproduct.ir/company/112/%D9%86%D8%A7%D9%86%D9%88%20%D9%BE%D9%88%D8%B4%D8%B4%20%D9%81%D9%84%D8%B2.
[10] F. He, G.X. Xie, J.B. Luo, Electrical bearing failures in electric vehicles, Friction 8 (1) (2020) 4—28.
[11] K. Holmberg, A. Erdemir, Influence of tribology on global energy consumption, costs and emissions, Friction 5 (3) (2017) 263—284.
[12] J.B. Luo, X. Zhou, Superlubricitive engineering—future industry nearly getting rid of wear and frictional energy consumption, Friction 8 (4) (2020) 643—665.
[13] V. An, Y. Irtegov, C.D. Izarra, Study of tribological properties of nanolamellar WS2 and MoS2 as additives to lubricants, J. Nanomater. 2014 (2014) 865839.
[14] R.A. Al-Samarai, Y. Al-Douri, A.K.R. Haftirman, Tribological properties of WS2 nanoparticles lubricants on aluminum-silicon alloy and carbon steels, Walailak J. Sci. Technol. 10 (3) (2013) 277—287.
[15] T. Hisakado, T. Tsukizoe, H. Yoshikawa, Lubrication mechanism of solid lubricants in oils, J. Lubricat. Technol. 105 (2) (2015) 245—252.

[16] J. Padgurskas, R. Rukuiza, I. Prosyčevas, R. Kreivaitis, Tribological properties of lubricant additives of Fe, Cu and Co nanoparticles, Tribol. Int. 60 (2019) 224–232.

[17] A. Papadaki, K.V. Fernandes, A. Chatzifragkou, E.C.G. Aguieiras, J.A.C. da Silva, R. Fernandez-Lafuente, S. Papanikolaou, A. Koutinas, D.M.G. Freire, Bioprocess development for biolubricant production using microbial oil derived via fermentation from confectionery industry wastes, Bioresour. Technol. 267 (2018) 311–318.

[18] H.J. Song, X.H. Jia, N. Li, X.F. Yang, H. Tang, Synthesis of α-Fe_2O_3 nanorod/graphene oxide composites and their tribological properties, J. Mater. Chem. 22 (3) (2018) 895–902.

[19] P.K. Nagarajan, J. Subramani, S. Suyambazhahan, R. Sathyamurthy, Nanofluids for solar collector applications: a review, The 6th International Conference on Applied Energy – ICAE 2014, Energy Proc. 61 (2014) 2416–2434.

[20] M.M.A. Khana, N.I. Ibrahim, I.M. Mahbubul, H. Muhammad, Evaluation of solar collector designs with integrated latent heat thermal energy storage: a review, Sol. Energy 166 (2018) 334–350.

[21] I.S. El-Mahallawi, A.A. Abdel-Rehim, N. Khattab, N.H. Rafat, Effect of nanographite dispersion on the thermal solar selective absorbance of polymeric-based coating material, in: Energy Technology, 2018, https://doi.org/10.1007/978-3-319-72362-4_49.

[22] K. Khanafer, K. Vafai, Applications of nanomaterials in solar energy and desalination sectors, Adv. Heat Tran. 45 (2013), https://doi.org/10.1016/B978-0-12-407819-2.00005-0.

[23] H.H. Al-Kayiem, S.C. Lin, Performance evaluation of a solar water heater integrated with a PCM nanocomposite TES at various inclinations, Sol. Energy 109 (2014) 82–92.

[24] V. Msomi, O. Nemraoui, Improvement of the performance of solar water heater based on nanotechnology, in: 6th International Conference on Renewable Energy Research and Application, 2017.

[25] E. Natarajan, R. Sathish, Role of nanofluids in solar water heater, Int. J. Adv. Manuf. Technol. (2009), https://doi.org/10.1007/s00170-008-1876-8.

[26] X.F. Gao, J. Jang, S. Nagase, Hydrazine and thermal reduction of graphene oxide: reaction mechanisms, product structures, and reaction design, J. Phys. Chem. C 114 (2) (2010) 832–842.

[27] J. Theron, J.A. Walker, T.E. Cloete, Nanotechnology and water treatment: applications and emerging opportunities, Crit. Rev. Microbiol. 34 (2008) 43–69.

[28] T. Hillie, M. Hlophe, Nanotechnology and the challenge of clean water, Nat. Nanotechnol. 2 (2007) 663–664.

[29] R.M. Amin, M.B. Mohamed, M.A. Ramadan, T. Verwanger, B. Krammer, Rapid and sensitive microplate assay for screening the effect of silver and gold nanoparticles on bacteria, Nanomedicine 4 (6) (2009) 637–643.

[30] R. Mofidian, M. Jahanshahi, S.M. Hosseini, H. Eini, The practical and economic fields of nano in the optimization of upstream oil industries, Oil, Gas & Energy Monthly Magazine (2015).

[31] X. Kong, M.M. Ohadi, Applications of micro and nano technologies in the oil and gas industry- an overview of the recent progress, in: Paper SPE 138241, International Petroleum Exhibition & Conference, Abu Dhabi, 1–4 November 2010, 2010.

[32] T. Azam, M. Askari, Transformation in Upstream Oil Industries Using Nanotechnology, 2013.

[33] Application of Nanotechnology in Cement and Drilling Fluid, Mehrvision development and technology research group, Special Headquarters for the Development of Nanotechnology, Industrial Report No.81, 2015. (http://indnano.ir/)

[34] H.C. Lau, et al., Nanotechnology for oilfield applications: challenges and impact, J. Petrol. Sci. Eng 157 (2017) 1160–1169.

Application of nanofluids and future directions

[35] S. Al-Anssari, et al., Retention of silica nanoparticles in limestone porous media, in: SPE/IATMI Asia Pacific Oil & Gas Conference and Exhibition, Society of Petroleum Engineers, 2017.

[36] O.A. Alomair, K.M. Matar, Y.H. Alsaeed, Nanofluids application for heavy oil recovery, in: SPE Asia Pacific Oil & Gas Conference and Exhibition, Society of Petroleum Engineers, 2014.

[37] J. Saien, A.M.J.J.o.M.L. Gorji, Simultaneous adsorption of CTAB surfactant and magnetite nanoparticles on the interfacial tension of n-hexane–water, J. Mol. Liq. 242 (2017) 1027–1034.

[38] E.A. Taborda, et al., Experimental and theoretical study of viscosity reduction in heavy crude oils by addition of nanoparticles, Energy Fuels 31 (2) (2017) 1329–1338.

Index

'*Note:* Page numbers followed by "f" indicate figures and "t" indicate tables.'

A

Aggregation approach, thermophysical properties for, 17—19, 17t
Artificial intelligence—based Levenberg—Marquardt approach, 14—16
  mathematical formulation, 16—19
  results and discussions, 20—27
Artificial neural networks (ANNs), 20, 20f

B

Bejan number, 216
Biological preparation, 76
Brinkman number, 216
Brownian motion, 4—5, 5f, 103—104

C

Car radiator, geometry of, 173t
Carbon nanotube with multi-wall (MWCNTs), 136—138
Carbon nanotube with single-wall (SWCNTs), 136—138
Carbon nanotubes flow and heat transfer, 135—138, 137f, 140—146, 141f
  declaration of curiosity, 145—146
    Nusselt number, 145—146
    skin friction coefficients, 145
  nonuniform heat source, 143
  Rosseland approximation for radiation, 143—144
  similarity analysis, 144—145
Chemical preparation, 71—75
  microemulsion, 75, 76f
  precipitation, 71—73
  thermal decomposition, 73—75
Copper oxide, 169—184
  grid independent performance, 175, 179f
  heat transfer coefficient estimation, 174
  heat transfer performance characteristic, 176—178, 181t
  modelling and simulation, 172, 173f
  nanofluid

  base fluid and thermophysical properties of, 176, 179f
  physical properties of, 172—174
  simulation assumptions, 174—175
  validation of data, 179—184, 180t

D

Darcy Forchheimer
  magnetohydrodynamics flow, 135—140
  carbon nanotubes, 135—138, 137f, 140—146, 141f
    declaration of curiosity, 145—146
    nonuniform heat source, 143
    Rosseland approximation for radiation, 143—144
    similarity analysis, 144—145
  nonlinear thermal radiation, 138—139
  porous space, 139—140, 139f
  solution, numerical methods of, 146—149
    implementation of methods, 148—149
    Runge—Kutta—Fehlberg Method, 146—147
    shooting technique, 147—148
Dispersion light scattering (DLS), 200—201

E

Electric arc, deposition by, 68, 69f
Electrical generator, nanofluids in, 292—293, 293f
Electromagnetic force, 109
Electron beam evaporation, 65, 67f
Energy and electricity sector, 289—294, 290f
  electrical generator, nanofluids in, 292—293, 293f
  industrial products in Iran, 293—294
  power plant cooling towers, nanofluids in, 289—291
  transformer, nanofluids in, 291—292
Engine oils and nano-sized additives, 297

Index

Enhance heat transfer, different methods of, 62f
Enhancement oil and gas recover, 303–306
Entropy generation, 216

F

Ferrofluid, 64–76, 65f
  biological preparation, 76
  chemical preparation, 71–75
    microemulsion, 75, 76f
    precipitation, 71–73
    thermal decomposition, 73–75
  physical preparation, 65–71
    flame spray pyrolysis (aerosol), 70–71, 72f
    gas-phase deposition, 65–68
    laser-induced pyrolysis, 68–70, 70f
    powder balls milling, 70, 71f
  preparation, 63–76
  stability of, 63–64
  synthesis of, 64–76
  thermophysical properties of, 77–83
Ferrohydrodynamics (FHD), 106–108, 122–128, 124f, 255–256
Flame spray pyrolysis (aerosol), 70–71, 72f
Flow modeling, 260–265, 260f
Four-fifth RKF technique, 146–147

G

Gas-phase deposition, 65–68
Grid independent performance, 175, 179f

H

Heat source/sink, effect of, 280–282, 282f
Heat transfer coefficient estimation, 174
Heat transfer enhancement using ferrofluids, 86–95
  ferrofluid preparation, 63–76
    stability of, 63–64, 64t
    synthesis of, 64–76, 65f
  mathematical formulation of FHD, 83–86
  thermophysical properties of ferrofluid, 77–83
Heat transfer performance characteristic, 176–178, 181t

High field frequency parameter, effect of, 272–274, 272f
Homogeneous methodology, 125
Hybrid nanofluids, 7–8, 258–259

I

Induced magnetic parameter, 217–219, 218t
Industrial products in Iran, 293–294
Interfacial layer, effect of, 243, 243f

L

Lambert–Beer law, 202–203
Laser-induced pyrolysis, 68–70, 70f
Lorentz force, 15

M

Magnetic nanofluid flow, 105–106, 209–213, 216–228
  entropy generation, 216
  induced magnetic parameter, 217–219, 218t
  literature review, 209–213
  magnetic Prandtl number, 221–225
  magnetophoresis parameter, 225–228
  mathematical formations, 213–215
  numerical solution methodology, 216
Magnetic Prandtl number, 221–225
Magnetohydrodynamics (MHD), 108–109, 140, 211
  elliptical porous blocks, 109–114
  mixed convection in open cavity, 116–122, 116f
Magnetophoresis parameter, 225–228, 226f
Mathematical formulation, 16–19
  of FHD, 83–86
  thermophysical properties for aggregation approach, 17–19
Melting parameter, effect of, 243–244, 244f
Microemulsion, 75, 76f
Miscellaneous applications, 54–55
Molecular beam epitaxy, 68, 69f
Mono nanofluid and hybrid nanofluid, 266t

Index

313

N

Nanoabsorbents, 53—54, 53f
Nanoadditives in lubricants, 295—297,
    296f
Nanofluids, 1, 9—10, 101—104, 196—200,
    203—204, 233—236, 241—250,
    289—294
  as secondary fluid, 38—45, 38f
  base fluid and thermophysical properties
    of, 176, 179f
  brief history of, 2
  classification of, 7—8
  energy and electricity sector, 289—294,
    290f
    electrical generator, nanofluids in,
      292—293, 293f
    industrial products in Iran, 293—294
    power plant cooling towers, nanofluids
      in, 289—291
    transformer, nanofluids in, 291—292
  interfacial layer, effect of, 243, 243f
  mathematical formulation, 236—240,
    236t, 237f
  melting parameter, effect of, 243—244,
    244f
  nanofluids simulation approaches,
    103—104
    homogeneous approach, 103
    nonhomogeneous approach, 103—104
  nanoparticle diameter, effect of, 241—242,
    241f
  nanoparticles, 197—199, 197f, 198t
  nanoparticles in lubrication
    improvement, 294—297
    engine oils and nano-sized additives,
      297
    nanoadditives used in lubricants,
      295—297, 296f
  numerical experiment, 240
    numerical methodology, 240
    testing of code, 240
  oil and gas well, 299—301, 302t,
    303—306
  parameter δ1, effect of, 245—247, 246f
  parameter δ2, effect of, 247—249, 248f
  parameter on engineering coefficient, 250,
    250f—251f

physical properties of, 172—174
preparation methods, 2—3
radiation parameter, 244, 245f
scanning electron microscopy (SEM),
    203—204
simulation approaches, 103—104
  homogeneous approach, 103
  nonhomogeneous approach, 103—104
solar water desalination, 299—301
solar water heaters, 297—299
stability, 3—7, 200—204
  by backscattered light, 201
  by UV spectrophotometry, 202—203
  image analysis, 203—204
thermophysical properties of, 8—9, 8f,
    143t
transmission electron microscopy (TEM),
    203
use of, 189f
zeta potential, stabilization of nanofluids,
    199—200
Nanofluids—magnetic field interaction,
    101
  applications, 109—128
    FHD mixed convection of a ferrofluid
      flow, 122—128, 124f
    MHD forced convection in a
      corrugated channel with elliptical
      porous blocks, 109—114
    MHD mixed convection in an open
      cavity, 116—122, 116f
  ferrohydrodynamic, 106—108
  magnetic nanofluid, 105—106
  magnetohydrodynamics, 108—109
  nanofluids, 101—104
Nanolubricants, 50f, 51—53
Nanomaterials, drilling mud properties,
    303t
Nanoparticle copper oxide (CuO), 174t
Nanoparticles
  concentration, 282—283, 283f
  concentration against thermal
    conductivity, 178f
  diameter, 241—242, 241f
  functionalization, modification of the
    surface of, 198—199
  in base fluid, 191—196, 193t, 196f

Nanoparticles (*Continued*)
  in lubrication improvement, 294—297
    engine oils and nano-sized additives, 297
    nanoadditives used in lubricants, 295—297, 296f
    surfactant, modification of surface of, 197—198, 197f, 198t
Nonhomogeneous approach, 103—104
Nonlinear thermal radiation, 138—139
Nonuniform heat source, 143
Numerical experiment, 240
  numerical methodology, 240
  testing of code, 240
Numerical method and code validation, 269—270
Numerical methodology, 240
Numerical solution methodology, 216
Nusselt number, 145—146, 215, 280t, 281f

O
Ohm's law, 108
Oil and gas, 299—301, 302t
Optimal thermal applications, 187—196
  nanofluid stability, 196—204
    by backscattered light, 201
    by UV spectrophotometry, 202—203
    image analysis, 203—204
    nanoparticles, 197—199, 197f
    zeta potential modification, 199—200
  nanofluids discussion, 187—196
    nanoparticles dispersed in base fluid, 194—196, 196f
    nanoparticles in base fluid, 191—194, 193t
    techniques for preparation of nanofluids, 187—191
Ordinary differential equations (ODEs), 269—270

P
Parameter $\delta 1$, effect of, 245—247, 246f
Parameter $\delta 2$, effect of, 247—249, 248f
Parameter on engineering coefficient, 250, 250f—251f
Phase-change material (PCM), 55

Photothermal conversion efficiency setup, 95f
Physical preparation, 65—71
  flame spray pyrolysis (aerosol), 70—71, 72f
  gas-phase deposition, 65—68
  laser-induced pyrolysis, 68—70, 70f
  powder balls milling, 70, 71f
Physical quantities for various parameters, 218t
Physical Vapor Deposition (PVD), 190—191
Porous blocks' permeability, 111t, 112f
Powder balls milling, 70, 71f
Power plant cooling towers, nanofluids in, 289—291
Pulsed laser deposition (PLD), 66—68, 68f

R
Radiation parameter, 244, 245f
Radiation, effect of, 280—282, 282f
Radiative Ag—MgO—water hybrid nanofluid, 255—259
  mathematical formulation, 260—269
    boundary conditions, 265
    flow modeling, 260—265, 260f
    similarity transformation, 265—269
    thermophysical properties, 265, 265t
  numerical method and code validation, 269—270
  results and discussion, 270—283
    heat source/sink, effect of, 280—282, 282f
    high field frequency parameter, effect of, 272—274, 272f
    nanoparticles concentration, effect of, 282—283, 283f
    radiation, effect of, 280—282, 282f
    slip parameter, effect of, 275—278, 276f
    temperature ratio, effect of, 280—282, 282f
    unsteady parameter, effect of, 278—280, 279f
Refrigeration and air-conditioning, 35—37
  challenges and future scope, 55—56

Index

miscellaneous applications, 54—55
nanoabsorbents, 53—54, 53f
nanofluids as secondary fluid, 38—45, 38f
nanolubricants, 51—53
RK-4 base shooting method, flow chart
of, 270f
Rosseland approximation for radiation,
143—144
Runge—Kutta—Fehlberg Method,
146—147

S

Scanning electron microscopy (SEM),
203—204
Sherwood number, 215
Shooting technique, 147—148
Skin friction, 215, 217t, 275f, 275t
coefficients, 145
Slip parameter, effect of, 275—278, 276f
Sol—gel process, 73, 74f
Solar water desalination, 299—301
Solar water heaters, 297—299
Sputtering, 65—66, 67f

T

Temperature ratio, effect of, 280—282,
282f
Tetramethylammonium hydroxide
(TMAH), 78—79
Thermal decomposition, 73—75
Transformer, nanofluids in, 291—292
Transmission electron microscopy (TEM),
203

U

Uniform magnetic field (UMF), 15
Unsteady parameter, effect of, 278—280,
279f

V

Vacuum evaporation process, 65, 66f

Z

Zeta potential modification, 199—200

Printed in the United States
by Baker & Taylor Publisher Services